'Just in case you were thinking that religion is a thing of the past, I highly recommend that you read the rich collection of case studies in this eye-opening book. They make it abundantly clear that world religions are not only thriving, but they are also making a positive impact on safeguarding nature in the age of humans.'

**Shonil A. Bhagwat,** *Professor of Environment and Development, Head of School, Social Sciences and Global Studies, The Open University, UK*

'This important volume of case studies from around the world provides an invaluable service by highlighting the key role that mainstream religions and Indigenous traditions have to play in protecting nature. Without the deep kind of support described in Religion and Nature Conservation: Global Case Studies, programs of environmental and cultural conservation will not succeed in being equitable, inclusive, and sustainable.'

**Edwin Bernbaum,** *Co-Chair, IUCN Specialist Group on the Cultural and Spiritual Value of Protected Areas (CSVPA) and author of Sacred Mountains of the World, 2nd edition*

'This book is an invaluable contribution that illustrates the critical role of religions in the conservation movement. It documents remarkable examples from around the world that will inspire further engaged religious environmentalism. It is indeed a unique and indispensable monograph for the growing field of religion and ecology.'

**Mary Evelyn Tucker**, *Yale University, Yale Forum on Religion and Ecology*

# RELIGION AND NATURE CONSERVATION

This book presents a broad array of global case studies exploring the interaction between religion and the conservation of nature, reflecting on both successes and failures from the viewpoints of the religious practitioners themselves.

With conservation and religion often being championed as allies in the quest for a sustainable world where humans and nature flourish in harmony, this book provides a much needed compendium of detailed examples where religion and conservation science have been brought together. Case studies cover a variety of religions, faiths and practices, including traditional, Indigenous, Buddhism, Christianity, Hinduism, Islam, Jainism, Judaism, Shinto and Zoroastrianism. Importantly, this volume gives voice to the religious practitioners and adherents themselves, where they discuss their personal motivation as conservationists and how religion energises their commitment to the conservation of other species and ecosystems. Beyond an exercise in anthropology, ethnobiology and comparative religion, the book is an applied work, seeking the answer to how in a world of more than seven billion people we might help our own species to prevent the extinction of life.

This book will be of great interest to students and scholars of nature conservation, environment and religion, cultural geography and ethnobiology, as well as practitioners and professionals working in conservation.

**Radhika Borde** is a Lecturer at the School of Geography, University of Leeds, UK. She has published on sacred natural sites and sustainability and is a member of international professional associations working on these issues. She also supports local communities in India with conservation and sustainability projects.

**Alison A. Ormsby** teaches Environmental Studies at the University of North Carolina Asheville, USA. She is a human ecologist with 30 years of experience working with people and protected areas, environmental education and sacred natural sites. She has conducted research on sacred forests in Sierra Leone, Ghana and India.

**Stephen M. Awoyemi** is an Independent Researcher and past President of the Religion and Conservation Biology Working Group, Society for Conservation Biology. He holds a PhD from the Central European University (CEU), Austria.

**Andrew G. Gosler** is Professor of Ethno-ornithology in the Department of Biology and the School of Anthropology at the University of Oxford, UK. He is the author/editor of multiple books, including *Ethno-Ornithology: Birds, Indigenous Peoples, Culture and Society* (Routledge, 2011).

## Routledge Studies in Conservation and the Environment

This series includes a wide range of inter-disciplinary approaches to conservation and the environment, integrating perspectives from both social and natural sciences. Topics include, but are not limited to, development, environmental policy and politics, ecosystem change, natural resources (including land, water, oceans and forests), security, wildlife, protected areas, tourism, human-wildlife conflict, agriculture, economics, law and climate change.

**Threatened Freshwater Animals of Tropical East Asia**
Ecology and Conservation in a Rapidly Changing Environment
*David Dudgeon*

**Conservation Effectiveness and Concurrent Green Initiatives**
*Li An, Conghe Song, Qi Zhang, and Eve Bohnett*

**Religion and Nature Conservation**
Global Case Studies
*Edited by Radhika Borde, Alison A. Ormsby, Stephen M. Awoyemi and Andrew G. Gosler*

**Jackals, Golden Wolves, and Honey Badgers**
Cunning, Courage, and Conflict with Humans
*Keith Somerville*

**Case Studies of Wildlife Ecology and Conservation in India**
*Edited by Orus Ilyas and Afifullah Khan*

For more information about this series, please visit: www.routledge.com/Routledge-Studies-in-Conservation-and-the-Environment/book-series/RSICE

# RELIGION AND NATURE CONSERVATION

## Global Case Studies

*Edited by*
*Radhika Borde, Alison A. Ormsby,*
*Stephen M. Awoyemi and Andrew G. Gosler*

Routledge
Taylor & Francis Group
LONDON AND NEW YORK

earthscan
from Routledge

First published 2023
by Routledge
4 Park Square, Milton Park, Abingdon, Oxon OX14 4RN

and by Routledge
605 Third Avenue, New York, NY 10158

*Routledge is an imprint of the Taylor & Francis Group, an informa business*



*British Library Cataloguing-in-Publication Data*
A catalogue record for this book is available from the British Library

ISBN: 978-1-032-15900-3 (hbk)
ISBN: 978-1-032-15899-0 (pbk)
ISBN: 978-1-003-24619-0 (ebk)

DOI: 10.4324/9781003246190

Typeset in Bembo
by codeMantra

# CONTENTS

# FIGURES

# TABLES

# BIOGRAPHIES

**Stephen M. Awoyemi** earned a doctoral degree from the Department of Environmental Sciences and Policy, Central European University (CEU), Vienna, Austria. His doctoral research focused on how sociological theory can help explain and solve the conservation problem of trade in vulture parts for belief-based use in Nigeria. He holds a master's degree in Conservation Leadership from the University of Cambridge and currently serves as the chair of the University of Cambridge Conservation Leadership Alumni Network Council. His research interests broadly include conservation social science, conservation policy, religion, and conservation. Before starting this PhD program, Stephen worked with the Nigerian Conservation Foundation, the foremost conservation organisation in Nigeria as Conservation Policy and Campaign Officer/Head of Abuja Office and has been a long-time volunteer with the Society for Conservation Biology (SCB). He served as the President of two groups (Africa Section and Religion and Conservation Biology Working Group) concurrently, within the SCB, from 2015 to 2017. In September 2020, Stephen was awarded the CEU Presidential Scholar Award for academic excellence and leadership proficiency. E-mail: sawoyemi@gmail.com

**Edmund Barrow** grew up on one of the, then, few organic farms in Ireland, and went to Trinity College Dublin where he studied natural science. Edmund has worked for over 45 years, mainly in Africa, in 20+ countries focusing on community-based natural resource management, and has extensive field experience with sustainable development in different ecosystems, ecosystem restoration, participatory approaches to conservation, and governance. He pioneered village-level environmental and land use planning at local and landscape levels, and forest landscape restoration. He lives in Kenya with his wife and two sons. As a six-year-old, Edmund planted 300 chestnut trees on their organic farm in Ireland. All survive. Working in Turkana (northwest Kenya) 40 years ago, Edmund visited the Holy Land where there is similar vegetation. This sparked his interest in sacred trees and sacred groves. In the presence of the Fortingall Yew (Perthshire, Scotland), between 4,000 and 6,000 years old, Edmund tried to understand how this sentinel survives and what lessons it could share. Edmund believes that sacred trees and sacred groves are a positive force for conservation and our lives. We can hold a tree to be sacred. Sacred trees and sacred groves are portals to engage and reconnect. E-mail: edmund@barrow.co.ke

**Purnima Devi Barman** set out to pursue a PhD in ecology, but on seeing the state of Greater Adjutants in Assam, India, she put her PhD on the line for this species. She worked with the local community understanding their local knowledge, beliefs, and way of life to build a conservation plan from grassroots level. Purnima earned her PhD in 2019 and has saved the species from being extirpated. She drew inspiration from fledglings of Hargila, local women, and converted housewives into conservation champions for the Greater Adjutants.

**Stanley Baya** has worked for 19 years as the community conservation manager for A Rocha Kenya (ARK). His role at ARK has evolved over the years, starting as the coordinator of A Rocha Kenya's first community conservation project, the Arabuko-Sokoke Schools and Ecotourism Scheme (ASSETS, www.assets-kenya.org) that supports needy secondary school children with proceeds from ecotourism. His involvement in training local farmers in creation stewardship and Farming God's Way, also popularly known as Regenerative Agriculture, has exposed him to the striking duality in church teaching regarding matters of faith and people's livelihood activities, especially farming. Stanley is a Kenyan-trained schoolteacher with a deep passion for environmental conservation which arises mainly from his exposure to nature's complex interactions during his upbringing in the village. This led to his appointment as the patron of the environmental club in the school he taught. However, it wasn't until he met with Colin Jackson, the Director of A Rocha Kenya, that he made a connection between his Christian faith and his passion for environmental conservation. Stanley is married to Carol Muthoni and together they have two children, Jordan (11) and Eli (6).

**Shonil A. Bhagwat** (PhD) is a Professor of Environment and Development at the Open University, UK. Shonil became interested in sacred natural sites in the late 1990s when he travelled across India looking for traditional conservation practices in the remote rural countryside. Since then, he has shaped the field of research on sacred forests by combining the ecological understanding with the cultural practices associated with these forests. His pioneering work on sacred groves in Kodagu in South India has highlighted the important role these groves play in nature conservation outside formally protected areas. His work on long-term ecological history of these forests has identified the social, economic, cultural and political factors that might lead to revitalisation of faith-based practices of forest protection. Shonil's early work in India has informed his later research in a number of other places around the world where faith-based, traditional, cultural and spiritual practices have been vital for the protection of biodiversity whilst simultaneously providing a wide range of ecosystem service benefits to people safeguarding sacred natural sites. His work has contributed to the wider discussions of the role that faith groups can play in supporting environmental conservation and international development through compassion towards fellow human beings and towards other species.

**Kalzang Dorjee Bhutia** (PhD) is a Visiting Scholar in the Asian Studies Program at the University of California, Riverside in the United States. He is originally from west Sikkim, India, and completed his PhD in Buddhist Studies at the University of Delhi. He is currently completing a monograph on the environmental history of Sikkimese Buddhism. He was inspired to engage with Sikkimese Buddhist concepts of the environment by his upbringing in a Buddhist family lineage and his initial education at Pemayangtse Monastery in west Sikkim, where he learned to approach the land he lived in as a "sacred habitat", full of seen and unseen agents. His research is inspired by his concern with the rapid environmental changes underway in west Sikkim – including the impact of climate change, unplanned urban development, and hydroelectric projects – that have been worsened by a lack of political and public engagement with diverse indigenous and religious perspectives related to the land from different communities in Sikkim. Kalzang Dorjee hopes that his research can contribute to discussions across epistemologies about the importance of care for the sacred habitat of the Himalayas. E-mail: kalzangdbhutia@yahoo.com

**Revd. Dave Bookless** (PhD) is the Director of Theology for A Rocha International. His doctoral research at the University of Cambridge examined the value of wildlife and ecosystems through the interface of biblical theology and conservation biology. His academic interests include ecological eschatology, Indigenous Christian theologies, and cross-cultural missiology. Dave was born in India, lives in London, pastors a local Anglican church, and has lectured in over 40 countries across six continents. His writings on faith and environment have appeared in over 20 books, including *Planetwise*, translated into six languages.

**Radhika Borde** (PhD) is a Lecturer at the School of Geography, University of Leeds in the UK. She has been a Lecturer and Researcher at the Department of Social Geography and Regional Development at the Charles University in Prague. She has published journal articles and book chapters on indigenous social movements against mining, indigenous culture and religiosity, sacred natural sites in India, rural sanitation and waste management, activist media and indigenous women's movements. She has a PhD from Wageningen University in the Netherlands and is a steering committee member of the International Union for Conservation of Nature (IUCN) specialist group on the cultural and spiritual values of nature (CSVPA). She is also a published poet and author of short fiction, as well as the founder of a social enterprise in India. E-mail: r.borde@leeds.ac.uk

**Apela Colorado** (PhD) of Oneida-Gaul ancestry has dedicated her life's work to bridge Western thought and indigenous worldviews. As a Ford Fellow, Dr. Colorado studied for her doctorate at both Harvard and Brandeis Universities and received her PhD from Brandeis in Social Policy in 1982. She founded the Worldwide Indigenous Science Network (WISN) in 1989 to foster the revitalization, growth, and worldwide exchange of traditional knowledge and to safeguard the lives and work of the world's endangered traditional culture practitioners. In 1997, Dr. Colorado was one of twelve women chosen from 52 countries by the State of the World Forum (http://www.worldforum.org) to be honoured for her role as a woman leader. In addition to WISN's many projects—which span the globe and range from research on migration stories of Indigenous peoples in Central Asia to big cat conservation—Dr. Colorado founded the Indigenous Science and Peace Studies, UN University of Peace, the first fully accredited advanced degree program taught from an indigenous perspective that consciously integrates western knowledge. She has also authored numerous articles, including several in peer-reviewed journals, and is the author of *Woman Between the Worlds (2021)*. E-mail: apela@wisn.org

**Mark Coreth** is an Animalier Sculptor represented by the Sladmore Gallery in London, where he has exhibited for more than 20 years (www.sladmore.com). Born in London in 1958, he grew up on the family farm in Kenya. He was educated at Ampleforth College before going to the UK's military academy at Sandhurst. He served with the Blues and Royals of the Household Cavalry. While still a soldier and drawing heavily on experiences gained during his early years in Kenya, he taught himself to sculpt. Mark works directly with his subject in the field, which affords him a direct understanding of his subject. He has undertaken numerous high-profile commissions. E-mail: mark@markcoreth.com

**Anant Deshwal** is an Assistant Professor of Biology at Bradley University and has dedicated his life to conservation. He gave up a career as a civil engineer to follow his heart. In pursuit of effective and sustainable conservation, he worked with several Indigenous communities in India to understand their beliefs and way of life. He has always been fascinated by the spiritual connection between humans and non-humans. This understanding led him to successfully conserve snakes from being persecuted in south-eastern India. He has also worked on the conservation of tigers, crocodiles, antelopes, and birds. He strives to achieve conservation by bridging the divide between modern science and traditional ecological knowledge. E-mail: adeshwal@fsmail.bradley.edu

**Padma Dolker** is a Naropa Fellow and is from the Samad Rokchen region of Ladakh, India, and hails from a Changpa nomadic family. Presently working as a birding guide, she is also assisting an independent consultant to carry out conservation research in Ladakh. At the same time, a budding entrepreneur, she is in the process of setting up her start-up around pashmina wool with nomads in Changthang, Ladakh. A keen wildlife enthusiast, and growing up in a biodiverse region, living with nature and exploring has been a way of life for Padma. Stories of the lake creation, birds and mountains have been a collective knowledge that she has inherited from her grandparents. This interest has guided her present work and future plans of working with nomads to bring out not only their stories but also give a glimpse of their way of life to the outside world. Threats and lurking ecological changes in the landscape and to black-necked cranes that have been old friends of the Changpa nomads, have only further instilled the need of conservation and sustainability in her. She wants to ensure that the mosaic landscape offers its diversity to the future generations as well.

**Beth Duncan** is the former Executive Director for the Worldwide Indigenous Science Network (WISN), where she helped facilitate the interchange of indigenous and Western science for conservation of apex predators, sacred sites, and traditional knowledge systems. After receiving an MBA from Thunderbird School of Global Management, she served as a diplomat in China and on special assignments in South America, and she has also held a variety of higher ed marketing communications positions in the education and private sectors. She walked away from corporate America in 2012 to participate in a three-year leadership program affiliated with the Global White Lion Protection Trust in South Africa through which she met Dr. Colorado. Her work with WISN afforded her the opportunity to apply her direct experience to the areas about which she is most passionate – conservation of animals, sacred sites, and ways of life, intercultural bridge building, storytelling, empowerment, and connecting people. She recently completed a children's novel with a focus on apex predators, ancestral wisdom, and interspecies communication.

**Brittany Ederer** fights plastic pollution and equips everyday Christians to conserve nature in their own backyard in her role as a Conservation Ecologist for A Rocha USA; she also serves as the Project Coordinator for Climate Stewards USA. She received her BS in Wildlife Ecology and MS in Environmental Conservation from the University of Wisconsin-Madison. Brittany has been curious about the science versus faith divide since her college years when she experienced a reckoning that they belong together. E-mail: brittany.ederer@arocha.org

**Alexander Elizondo** is a Maleku ethnobotanical specialist and *Tócu* believer and named *Chayu* (great-grandfather) in *Tójifa Carráco* (Tonjibe), Maleku Territory, Guatuso, Costa Rica. Since childhood, Alexander has accompanied his mother collecting plants in their field near the Tonjibe *palenque*, where he taught him about the healing power of many plants. Growing up with elders and other young people, he learned respect for natural sacred sites when they would go around the river headwaters to look for the suita palm leaves needed for the roofs of the traditional houses. Since then, Alexander has specialised in health care with medicinal plants, builds traditional houses, and promotes ecotourism in the community, which is why he defends the conservation of sacred sites in the face of restrictions by government authorities and opposes the dispossession of their lands by non-Indigenous people.

**Leonel Elizondo** (RIP) was traditionally named *Afí Siúru* (Afí's father-in-law). He was a Maleku elder, traditional fisher, *Tócu* believer. *Carónco* (El Sol), Maleku Territory, Guatuso, Costa Rica. The journey to contribute to the struggle for the conservation of Maleku sacred sites starts with his first memories of fishing. When he was four years old in his traditional house located next to Sol River, during a very intense rainy season, the river had a high flow and his family had almost no food. Then, with a rope and hook with bait, he approached the river's shore and *Tócu* blessed the family with a large guapote fish (*Parachromis dovii*) and they ate for three days. Since then, he fished in rivers and lakes left by *Tócu* for Maleku sustenance, but the limitations of the government authorities and the refusal by the farmers to give access forced the Maleku people to fight to preserve the sacred sites where they must fish.

**Alemayehu Wassie Eshete** (PhD) was born in South Gondar, Ethiopia. He attended Alemaya University of Agriculture, Ethiopia where he received a BSc degree in forestry in 1992. He then received an MSc in Farm Forestry from Swedish University of Agricultural Science for his research on the roles of Ethiopian Orthodox churches in the conservation of forest biodiversity. He continued his research on church forests and earned his PhD in 2007 from Wageningen University, the Netherlands. He worked in different capacities both in government and in NGOs from 1993 to 2000. He then moved to the Organization for Rehabilitation and Development in Amhara (ORDA) and served in various positions: since January 2019 as an executive director of the organisation. Along with his development career, Dr. Alemayehu serves as an adjunct professor at Bahir Dar University and has mentored dozens of Ethiopian graduate and undergraduate students in addition to his administrative responsibilities. Dr. Alemayehu has been working with many international universities to further study and raise international awareness of church forests since 2005. Beyond his academic research, Dr. Alemayehu Wassie is also a writer in the world of literature. He has published mystic novels named 'Emegua', 'Ziguara', 'Merbebit' and 'Sebez'.

**Érika Fernandes-Pinto** is an Independent Researcher with a degree in Natural Sciences and a PhD in Social Sciences. She has been working for over 20 years in the management of protected areas, seeking the convergence of nature conservation policies with the rights of Indigenous and traditional peoples. Her doctoral thesis on Sacred Natural Sites was a pioneer in highlighting this theme in Brazil's context and received three academic merit awards. She is the creator of an Initiative that since 2013 maps and promotes the recognition and protection of these areas in Brazil and Latin America. She is a member of the Group of Experts on Cultural and Spiritual Values of Protected Areas, of the International Union for Conservation of Nature (CSVPA/IUCN). She works with the federal agency for the management of protected areas in Brazil, where she coordinates a program to integrate cultural values of nature into conservation strategies. She also publishes stories about sacred places on a blog (snsbrasil.blogspot.com), combining academic knowledge and experience in public management with spiritual wisdom. With her writing, chants and charms, she seeks to inspire people to open their minds and hearts to experience re-enchantment in the connection between culture and nature. E-mail: erika.icmbio@gmail.com

**Felipe Gomez** is a Mayan Healer and Spiritual Guide from the Quiche district. He is the director of Oxlajuj Ajpop, the National Association of Spiritual Leaders. He has been involved with the organisation since 1991. He served as the advisor and coordinator of the National Commission to Define Sacred Sites established after the Peace Agreements. Felipe also coordinated the development of a Law on Sacred Sites at which we take a closer look in this chapter. Felipe also managed a network of NGOs, universities and communities under the umbrella of COMPAS Central America. The network focused on supporting endogenous development, in education, resource management and healthcare. In Guatemala, this led to the development of a socio-environmental charter. Felipe currently serves as a Central American focal point to the Indigenous and Community Conserved Areas Consortium and is involved with supporting local and Indigenous communities with governing their territories.

**Revd. Prof. Andrew G. Gosler** holds a joint position as a Professor of Ethno-ornithology between the Department of Biology's Edward Grey Institute of Field Ornithology and the Institute of Human Sciences in the School of Anthropology and Museum Ethnography at the University of Oxford, UK. He is also the Director of Studies and Fellow in Human Sciences at Mansfield College, Oxford, and the Research Director for the Ethno-ornithology World Atlas project which he co-founded with colleagues BirdLife International. He has held a deep interest in birds from childhood, which led him not only into a research career as a bird ecologist, but to an interest in the human fascination with birds more generally and with a spiritual connection described by John Stott as *Ornitheology*. Studying and teaching about birds led him to become concerned about the growing disconnection between people and nature worldwide, and their lack of knowledge of the plants and animals around them. Resonant with a declining knowledge and understanding of religion, his research focus shifted to study the relationships between people and birds (ethno-ornithology), including the diverse spiritual contexts of those relationships. He is a Third Order Franciscan and was ordained as a minister in the Church of England in 2018. E-mail: andrew.gosler@zoo.ox.ac.uk

**Sara Hamann** is currently the Scientific Director at Gulf Offshore Research Institute in Mandeville, Louisiana. Sara has an interdisciplinary background in environmental science, conservation governance, research, and education. She has 5 years of experience working in not-for-profit organizations in Mexico, England, and the USA. She holds an MSc in Biodiversity, Conservation and Management from the University of Oxford. For her thesis work in the Biodiversity Institute at the University of Oxford, she researched Eastern Orthodox church forests in Ethiopia to measure their ecological value. She also studied the effects of habitat fragmentation and buffer zones on endangered species, conservation heritage, and regenerative agriculture. Sara is particularly interested in sustainable development, human-wildlife conflict, illegal wildlife trade, responsible sourcing, and climate change adaptation.

**Darla Hillard** has worked closely since 2000 with Indigenous partners throughout South and Central Asia as a Land of the Snow Leopard (LOSL) Facilitator with the Snow Leopard Conservancy (SLC) to bring conservation education to rural communities where functioning schools were sometimes non-existent. Where there are schools, lessons are mostly taught by rote; experiential learning (which SLC promoted through nature clubs and wildlife festivals) was a revelation for both teachers and students. The effectiveness of this approach has been proven. In 2010, SLC was introduced to a Central Asian community with strong connections with ancient indigenous practices, a community for whom the Snow Leopard is a sacred totem animal – protector and unifier of humanity. Out of that meeting, guided by many revered elders and keepers of traditional wisdom, has come the LOSL Network, whose members are inherently conservationists and who are deeply committed to saving both the spiritual and the terrestrial snow leopards.

**Nisa Khatoon** is working as a Senior Lecturer with the Education Department of the Government of Ladakh, India. She previously worked with WWF for nearly 10 years on different conservation research projects in Ladakh, including work on black-necked cranes and snow leopards. She has co-authored field guides on the floral diversity of Ladakh and published research papers on biodiversity from the region. She completed her Masters in Environmental Sciences from Jammu University and worked as a Research Scholar for the Wildlife Department and Aligarh Muslim University. Being from Ladakh, and having worked on black-necked cranes, she has interacted with local communities and is aware of the spiritual status of certain species in the region. While science has been the core of her work, its religious and cultural nuances have always fascinated her. She also believes that in the Himalayan region, where nature and culture often walk together, understanding both worlds is extremely important for conservation across the region.

**Ken Kitatani** is the Executive Director of Forum 21 Institute, a multidisciplinary research association for catalysing positive, integrative solutions and actions for human and environmental sustainability. Forum 21 works on all levels of society with a specific interest in three areas: (1) promoting sustainable development and uniting non-governmental organisations (NGOs) to support the adoption of the United Nations' Sustainable Development Goals and their implementation; (2) promoting eco-spirituality, eco-ministry, and eco-justice and supporting eco-ministry as an authentic and necessary form of service within the faith, interfaith, and interspiritual communities; and (3) sponsoring education and training programs on the local, regional, state, national, and international levels that deepen and broaden constituencies to foster sustainable practices and leverage sustainability policies at all levels. Ken is also the Principal Representative of the Forum 21 Institute United Nations NGO in special consultative status with the UN Economic and Social Council. Ken graduated from Columbia University with a BA in East Asian Studies.

**Radhika Kothari** is an Independent Consultant working in the Himalayan Landscape with local communities to support nature-culture conservation. Radhika has previously worked in the Himalayan regions with various organisations leading research, planning and management for social and community conservation and climate change adaptation. Radhika holds an MPhil in Conservation Leadership from the University of Cambridge and is a Dalai Lama Fellow. Her work and travels in nooks and corners of the Himalayas have taught her about a deeply seated reverence and awe of nature among remote mountain communities often reflected in their traditional ecological knowledge systems, cultural and spiritual or religious values. A worldview that has helped them survive the long passages of time and overcome hardships of climate, geography, and nature. Inherently knowing of interdependencies and a sense of harmony and balance with nature has profoundly impacted her, both personally and professionally. Her interest and goal are to combine nature-culture approaches in conservation. E-mail: radhikarvk@gmail.com

**Smitha Krishnan** (PhD) is a scientist at The Alliance of Bioversity International and Center for Tropical Agriculture (CIAT), Bengaluru, India. She is an expert in agrobiodiversity and ecosystem services. Smitha's research interests broadly include ecosystem services, pollination biology, plant-animal interactions, restoration, soil-plant relationships, and sustainable livelihoods. Although familiar with the concept of sacred forests as a child, she had her first experience as an ecologist while working on pollination services provided by wild bees that nest in sacred forest fragments of Kodagu located within the Western Ghats biodiversity hotspot region. While working on pollinators, the role that they play in food production and the indirect contribution that they make towards the functioning of various other ecosystem services piqued her interest. Smitha is particularly interested in the role that remnant forests play in providing ecosystem services to the surrounding landscapes.

**Yossi Leshem** is a Professor (Emeritus) at Tel-Aviv University and was the CEO of The Society for the Protection of Nature in Israel (SPNI), the leading conservation NGO in Israel. He has been studying bird-migration and the ecology of raptors for 50 years. As the Head of The International Center for the Study of Bird Migration, he initiated the use of Barn Owls as biological pest control agents in agriculture and carried the vision of connecting the region's countries through the project. So far, seven countries have joined.

**Jonathan Liljeblad** is an Associate Professor at the Australian National University College of Law. He holds a PhD and JD, both from the University of Southern California, and a BS from the California Institute of Technology. He was born in Myanmar and grew up in Sweden and the United States. He is a member of the Pa'Oh Indigenous peoples of Shan State, Myanmar. His research focuses on the promotion of international norms in developing countries, with a particular focus on human rights, indigenous rights, and environmental conservation. E-mail: jonathanliljeblad@gmail.com

**Jeremy A. Lindsell** (PhD) helps A Rocha – a faith-motivated conservation organisation – to develop and implement research and conservation actions to protect and restore threatened species and habitats. He draws on personal experience with conservation work in tropical Africa, the Middle East and Southeast Asia. Jeremy is an ornithologist by training and has published papers on tropical forest conservation, changes in forest biomass and wildlife populations and the conservation of a number of globally threatened mammals and birds.

**Meg D. Lowman** (PhD) Called the 'real life Lorax' by National Geographic and the 'Einstein of the treetops' by the Wall Street Journal, Dr. Meg Lowman is an author, explorer, scientist, arbornaut (translation: treetop explorer!), mom, Fulbright scholar, and change-agent for conservation. She has devoted over three decades to exploration and research on treetop secrets, as one of the first pioneers to launch canopy science and as a museum leader. Lowman has published 10 books and over 150 peer-reviewed publications, and holds a BA in Biology, MSc in Ecology, PhD in Botany, and Executive Management certificate from Tuck School of Business. She has received myriad prizes, including the Margaret Douglas Medal by the Garden Club of America, Roy Chapman Andrews Distinguished Explorer Award, Kilby Laureate, Odum Award for Excellence in Education by the Ecological Society of America, and Lowell Thomas Medal by the Explorers Club. Her recent projects included creating a UNESCO world heritage forest site in Malaysia and partnering with Ethiopia's Coptic priests to save their last 5% remaining church forests.

**Tweedy Malagian** has a Bachelor's Degree in Business Management and a Diploma in Community Development. He has been the Papua New Guinea

(PNG) country representative for the Canadian volunteer organisation CUSO-PNG. He also worked for Oxfam New Zealand as the Tari Coordinator based in Tari, Hela Province, PNG, working mainly on peace building. Tweedy is a Christian Environmentalist, with a passion for sacred natural sites and sacred landscapes. He has facilitated Environmental Stewardship and Land Tenure workshops where local land custodians were given the opportunity to rethink their purpose and the care they give to the vast land they have inherited from their ancestors. Tweedy has completed a two-month speaking tour with Restoring Eden (USA) – meeting and discussing environmental issues of climate change, greenhouse gas effects; he also spoke at several Christian Universities and Colleges on Christian Environmental Stewardship and Care and also on climate change and its effects on PNG with students and church leaders both in the US and in PNG. Tweedy assisted with the Sacred Land Film Project on issues affecting sacred landscapes in Papua New Guinea. E-mail: tmalagian@gmail.com

**Rachel Mander** started being interested in conservation when studying the political philosophy of climate change at the University of Cambridge, before working in the UK Parliament. She now holds a number of roles working to mobilise faith groups to take action on climate change and sees engaging with questions of values as integral to our existing environmental challenges and their solutions. She is a Theology and Churches Officer at A Rocha International.

**Ashley Massey Marks** (PhD) is a Science Educator for girls in New York, USA. She has conducted field research in Malaysian Borneo, the Gambia, and South Africa, and assessed data from Ethiopia and Japan using GIS and remote sensing. Ashley learned about sacred forests when she was charged by a warthog on her bicycle while serving as a US Peace Corps volunteer in the Gambia. When her neighbours heard where she had been cycling in the bush, they warned her about *ninkananko*, which they translated to 'dragons', living in local sacred forests. Ashley grew curious about how people conserve forests beyond government-demarcated parks and reserves in diverse cultural contexts around the world. Her work was most recently published in *The Routledge Handbook of Indigenous Environmental Knowledge* (2021), which received the Geoscience Information Society's Best Research Resource Award. Ashley has served on the boards of the Religion and Conservation Biology Working Group of the Society for Conservation Biology and the IUCN Commission on Environmental, Economic and Social Policy. She holds a DPhil in Geography and the Environment and an MSc in Biodiversity, Conservation and Management from the University of Oxford, and an AB in Environmental Studies from Dartmouth College. E-mail: ashley.massey@gmail.com

**Arvind Mishra** was the first to discover a breeding population of Greater Adjutant in 2006 and Lesser Adjutant in 2004 in Bihar, India. He founded the Mandar Nature

Club and was instrumental in setting up a Rescue and Rehabilitation Center in Bihar for Greater Adjutants. His understanding of traditionally gained ecological knowledge, an MSc in Zoology, and religious knowledge has been instrumental in him successfully channelling them for conservation efforts in Bihar, India.

**Hang Ryeol Na** is the Founding Professor of All Seeds Academy in Canada. He is a teacher and researcher of environmental health, forest governance and policy. Particularly, he is interested in the issues of international environmental or forest policy and governance, especially in Asia. Cyberinfrastructure for e-science is one of his areas too in the context of science, technology and society (STS). He is also a Christian missionary from Korea. Prior to the All Seeds Academy, Professor Na taught at several universities in the US including Rochester Institute of Technology, State University of New York at Geneseo, etc. E-mail: 1chr258@gmail.com

**Sandra Nogué** (PhD) is a Biogeographer at the University of Southampton, UK. Her research is focused on the use of fossils and modern datasets to determine rate of ecosystem change and the distribution of plants across a wide range of spatial and temporal scales (>100 years). In addition, she is interested in understanding biodiversity baselines and the relationship between biodiversity and human impacts on island ecosystems. Sandra started working on sacred groves with Shonil A. Bhagwat and Kathy Willis during her postdoctoral research at the University of Oxford. Sandra is the Vice President of Communications for the International Biogeography Society.

**Alison A. Ormsby** (PhD) teaches Environmental Studies at the University of North Carolina Asheville. She is a Human Ecologist with 30 years of experience working with people and protected areas, environmental education, and sacred natural sites. She is a member of the IUCN's Specialist Group for Cultural and Spiritual Values of Protected Areas, and has numerous publications about her work, including in the following books: *The Routledge Handbook of Indigenous Environmental Knowledge* (2021); *Asian Sacred Natural Sites, Philosophy and Practice in Protected Areas and Conservation* (2016); *Sacred Species and Sites: Advances in Biocultural Conservation* (2012); and *Sacred Natural Sites: Conserving Nature and Culture* (2010). Alison first became interested in the topic of sacred forests when she learned about sacred forests in the community where she lived in Sierra Leone, West Africa. Since then, she has conducted research on groves in Ghana, India, and in a return research experience in Sierra Leone. Alison served with co-editor Stephen M. Awoyemi on the Society for Conservation Biology's "Religion and Conservation Biology Working Group Board" from 2011 to 2013. She participated with co-editor Radhika Borde in 2017 in the Expert Workshop on the recognition and promotion of the cultural and spiritual significance of nature in the management and governance of protected areas, specifically World Heritage Sites, in Vilm, Germany. E-mail: aormsby@unca.edu

**Idit Pintel-Ginsberg** (PhD) is an Israeli Researcher in Jewish Folklore, focusing on Folk Literature, Jewish Cultural Symbolism, Rituals and Customs. She retired in 2015 from her former position as the academic coordinator of IFA – the "Israel Folktale Archives named in honor of Dov Noy" at the University of Haifa – a position she held for 11 years. She co-edited the following books: *The Power of a Tale, Stories from the Israel Folktale Archives* (together with Haya Bar-Itzhak) (Hebrew: Pardes 2008; English: WSUP 2019), *Masoret Haya, 33 Essays in Folklore in Honor of Professor Haya Bar-Itzhak* (together with Tsafi Zeba-Elran and Haya Milo) (Hebrew: Pardes 2020). She is the author of *The Angel and the Cholent: Food Representation from the Israel Folktale Archives* (Hebrew: Pardes 2016; English: WSUP 2021). Idit's interest in Tu Be'Shvat was inspired by modern rituals in Israel. Her research on the subject was published in the article: "Narrating the Past – 'New Year of the Trees' Celebrations in Modern Israel", *Israel Studies* in Spring 2006, and in the entry "Shvat, Fifteenth of (Tu Be'Shvat)" in the *Encyclopedia of Jewish Folklore and Traditions* in 2013 (Haya Bar-Yitzhak, editor). She is a member of Kibbutz Sasa in the northern Galilee. E-mail: iditpg@gmail.com

**Travis W. Reynolds** (PhD) is an Agricultural and Applied Economist with a background in public policy analysis, rural development, environmental studies, and organic farming. An Associate Professor in the Department of Community Development and Applied Economics (CDAE) at the University of Vermont (UVM), Dr. Reynolds holds a PhD in Public Policy and Management from the University of Washington, an MS in Applied Economics from the University of Vermont, and a BA in International Relations and French Civilization from Brown University. Dr. Reynolds is also a faculty affiliate of the UVM Food Systems Program, a Fellow at the Gund Institute for Environment, and currently serves as the Interim Director of the UVM Master of Public Administration (MPA) program. For the past decade, Dr. Reynolds has studied relationships between farm management, economic development, and ecosystem services – with an emphasis on poverty alleviation, sustainability, and resilience in low-income small-scale farming communities in sub-Saharan Africa. With grant funding from the National Science Foundation, USDA, and private and philanthropic organisations, Dr. Reynolds has led teams of international and interdisciplinary scholars, graduate students, and undergraduate students from diverse backgrounds in applied research at the intersection of agriculture, food security, and the environment.

**Alexandre Roulin** is a Full Professor at Lausanne University, Switzerland. His research on Barn Owls over 30 years has made him an internationally respected expert on the subject. In 2010, he joined the Barn Owl Project in The Middle East. As a citizen of Switzerland, his collaboration symbolically includes a politically neutral country in the project. Alex has contributed significantly to promote the vision of "Owls for Peace" between the different nations in the region, alongside scientific assistance for the project.

**Peter Rowe** is a PhD student in the Institute of Geography at the University of Edinburgh. His research explores the linkages between conservation, development, and Christianity, and questions how Kenya's diverse Christianities shape, and are shaped by, conservation practice. Specifically, he is interested in how faith shapes conservation relationships and the expectations that come to be attached to conservation and development initiatives. Relatedly, Peter is curious as to how shared faith between conservation programme participants and NGO staff shapes decision-making regarding participation in conservation and development schemes such as Farming God's Way. Peter's interest in faith-based conservation stems from the convergence of his Christian faith and formal training in Environmental Studies, beginning at Temple University in Philadelphia, USA. Building on this, while completing an MSc degree in Environment and Development at the University of Edinburgh, Peter undertook research into A Rocha Kenya's Farming God's Way project in the Dakatcha Woodland, Kenya to further explore the fruitful intersection between Christianity and conservation. Peter currently resides with his wife, Marygrace, in Edinburgh, Scotland. E-mail: peter.rowe@ed.ac.uk

**Jonathan Schorsch** is a Professor of Jewish Religious and Intellectual History at the Universität Potsdam, Founder and Director of the Green Sabbath Project (www.greensabbathproject.net) and Founder and Director of the Jewish Activism Summer School (www.jassberlin.org). He is trained as a historian and has been ecologically oriented since his teenage years. In recent decades, his concern about the world's worsening environmental crises has deepened. When Jonathan began teaching in a department of Jewish theology at the Universität Potsdam in 2015, he took advantage of the disciplinary opening to turn his scholarship also to ecological and ecotheological matters. His relevant publications include "Looking for an Ecological God," *Ehyeh Asher Ehyeh*, ed. Martin S. Cohen, Saul Berman and David Birnbaum (2019), "Olive Oil, Anointing and Ecology," *Ritual Dynamics in Judaism and Christianity*, ed. Claudia Bergmann and Benedikt Kranemann (2019) and *The Food Movement, Culture and Religion: A Tale of Pigs, Christians, Jews and Politics* (2018). As he taught and wrote on the relationship between Judaism and ecology—in community education settings, journalistic writings, and academia—the idea and practice of *shabbat*, sabbath, a weekly day of rest, a weekly earth day, kept returning to his mind. E-mail: jschorsch@uni-potsdam.de

**Nita Shah** (PhD) is the Deputy Director of the Bombay Natural History Society, presently heading its Bird Monitoring program using the Central Asian Flyway in the State of Bihar. She is a Nature Conservationist who has devoted herself to protecting the wild ass species and its habitat in the hot and cold arid landscapes of India (Ladakh, Sikkim, the Rann of Kutch) and in China (Tibet and Xinjiang). Her efforts to prevent the Rann of Kutch from being denotified led to the conservation of several characteristic species such as the Indian Wild Ass (*Equus* *hemionus khur*), flamingo breeding grounds, and the habitats of several endemic lower taxa. Her advocacy has been crucial to saving the Gyps vulture, the endemic Kashmiri Stag (*Cervus hanglu hanglu*, a subspecies of the Central Asian Red Deer), and the Tibetan Antelope (Chiru) (*Pantholops hodgsonii*) from the brink of extinction. She has effectively contributed to the conservation of *E. kiang* and marmots in the high-altitude cold desert of Ladakh. During her expedition to the unexplored Trans Himalayan areas of North Sikkim, she reported the occurrence of the Tibetan gazelle and had also established that the wild ass (*E. kiang polyodon*) was not locally extinct. She uses her field work to influence decision makers. She effectively acts as a catalyst for evolving policies to safeguard species and their habitats at the local and global scales. E-mail: n.shah@bnhs.org

**Zahed Shakeri** is a Vegetation Ecologist working with the research group "Social-Ecological Interactions in Agricultural Systems" at the Universities of Kassel and Goettingen. He earned his PhD in Forest Ecology in 2012 from the University of Tehran, Iran under the co-supervision of Daniel Simberloff and Reza Marvie-Mohadjer. He worked as an Assistant Professor at the University of Kurdistan from 2012 to 2018. His work focuses on anthropogenic and natural disturbances in forest ecosystems, vegetation dynamics, and biodiversity conservation. Growing up in rural Kurdistan, he learned a lot about traditional myths and taboos surrounding sacred natural sites. Nevertheless, in 2005, it was his friend and mentor, Hedayat Ghazanfari, who introduced him to the importance of Kurdish sacred groves for conserving biodiversity. Since then, he has been researching the ecological characteristics, vegetation composition, forest structure, as well as the conservation status of sacred groves in Kurdistan, Iran. Zahed's passion for Mother Earth is immense and he believes that nature is the best lab for scientists. Therefore, he spends a lot of time in the field, observing how humans affect the natural ecosystems and what we can do to keep our ecosystems healthy. E-mail: shakeri.zahed@gmail.com

**Manisha Sheth** is an Architect by training with a degree in Ecological Studies. She is a practising Hindu and spent six years living in a monastery during which time she studied the links between Hindu philosophy and nature. In 2006, she founded eCoexist after six years of training as an environmentalist with the Kalpavriksh Environment Action Group. Her work on the interface between faith and conservation led to her focus on several Indian festivals and a study of consumer behaviour around products associated with these festivals. As a designer, she has worked on sustainable materials and products, and as an entrepreneur, created markets for these. In 2009, she was invited to speak at the United Nations on the topic of Business and Biodiversity as her work addressed the economic aspects of conservation. More recently, Manisha has been exploring the idea of Consciousness in Nature and studying cultural paradigms that regard Nature as an intelligent and conscient energy. She is looking at art, ritual and healing traditions that are rooted in this understanding. E-mail: ecoexist@gmail.com

**Robert D. Sluka** had a love of the ocean from an early age, despite growing up in the midwestern USA. Watching Jacques Cousteau, SCUBA lessons, and trips to Florida and the Caribbean to dive convinced Robert that studying marine biology at the University of Miami was the path forward. Despite what he now knows as a wealth of material from the Christian tradition which links conservation to that faith, there was no connection between his beliefs and studies. His PhD focused on marine conservation and he worked on establishing the benefits of marine-protected areas for fish populations. Ten years in south Asia living in fishing villages and working with fishing communities helped Robert to link his conservation work to what he would now call justice; yet, he still saw the conservation work as a means to a spiritual end, rather than more integral to the outworking of his own faith. Eleven years in the UK helped Robert to interact with scientists with a more integral faith, and particularly being involved as a volunteer and now staff with A Rocha significantly altered his faith paradigm. Now back in the USA attempting to develop holistic projects, Robert continues to learn and grow each day discovering how amazing God's world is and how comprehensive the vision for reconciliation between God, people, and nature is. E-mail: bob.sluka@arocha.org

**C. Ryan Smith** was born in Lithia Springs, Georgia in the United States and has resided since 1998 in the Borgou region of Benin in West Africa. He is the Coordinator of the Antisua Forest Regional Council (CRFA), a locally headquartered Beninese non-governmental and non-profit organisation, which synchronises community-based management efforts for the Antisua Forest Nature Reserve. Ryan speaks the Baatɔnum language fluently and is a promoter of Baatɔnu culture and worldview, often sitting for interviews on local radio programs to discuss the Antisua Forest and Borgou current events. He is a parent and a member of a multi-religious extended Baatɔnu family. Ryan's academic background is in history and ecology at Emory University in Atlanta, Georgia, and natural resource sociology at the University of Parakou in the Beninese Borgou. He considers his near-quarter century of immersion in the socio-ecological, linguistic and spiritual realities of his adoptive country as an equally important component of his education. E-mail: perere_gourou@yahoo.fr

**David Solis-Aguilar** is an Independent Researcher, has a BSc in Political Science (Universidad de Costa Rica) and MSc in Human Geography (El Colegio de Michoacán, Mexico). His interest in the conservation of nature through faith began with an action-research project during undergraduate studies to facilitate dialogue between Bribri and Cabécar Indigenous communities in the Caribbean region of Costa Rica, to demand from the national government access to justice and environmental protection with respect to their culture. Subsequently, David's familiarisation with the Maleku community in northern Costa Rica was in support of community cultural organisation that allowed him to become acquainted with the beliefs about the *Tócu* caretakers of their people and the difficulties related to their conservation. This motivated David's Master's research in human geography on the territorialities of the Maleku people, which discusses the conflict over the Maleku's relationship with their ancestral territory articulated by sacred sites. E-mail: david.solis@colmich.edu.mx

**Rianne C. ten Veen** became interested in environmental issues from a young age – no doubt from growing up as the daughter of an oil man, living in various locations. She became a Muslim in her late 20s, attracted to what Islam stands for in regard to environmental justice. Since then, Rianne has at times spent less or more time increasing her knowledge on this subject, such as, in 2009, when she self-published a book on Islam and the environment and in 2021 when she was selected as a GreenFaith Fellow. Rianne has an LLM, an MA, an MSc, a PGCert, and a PGDip. For many years, she worked in disaster management, where the emphasis is often still more on response than on disaster risk reduction. She now mainly works in facilitating and training in intercultural exchange. E-mail: riannetenveen@gmail.com

**Will Tuladhar-Douglas** studies, practices, and teaches Buddhism, anthropology, and ecology. He does so as a Mahāyāna Buddhist who has studied with teachers in several traditions—especially Zen and Newar and Tibetan Vajrayāna. That informs Will's scholarship and activism. He sees no place, intellectually or morally, for human exceptionalism: all living beings are our kin and our collaborators in compassion. As Will explains in his contribution to this book, in Tibetan Buddhism, the vulture is a powerful symbol for the most extraordinary teacher. The two wings are said to represent scholarship and meditation, which combined in one great being can loft them over the greatest obstacles. Will has been fortunate to see many kinds of vulture, including some of the last wild California Condors and *Gyps* vultures in the Himalayas. E-mail: will@tending.to

**Bas Verschuuren** collaborates with Indigenous people, local communities, conservationists and researchers worldwide and published over 50 book chapters and scientific articles as well as five edited volumes on human-nature relationships, three with a focus on sacred natural sites. Since his youth, during which he explored the (pre-)Christian spiritual landmarks of the Dutch countryside, Bas continued a search for spirituality in human-nature relationships. During his studies in nature conservation, Bas encountered many different cultural notions of the sacred in various different ecosystems, mountains, wetland, and forests. This fascination with culture and conservation brought him to chair IUCN's specialist group on Cultural and Spiritual Values of Protected Areas (www.csvpa.org) and found the Sacred Natural Sites Initiative (www.sacrednaturalsites.org). Through these programmes, Bas gained experience in applied anthropological and multidisciplinary research on the cultural dimensions of nature. As an assistant professor with the Forest and Nature Conservation Policy group at Wageningen University, Bas researches the politics of multiple values, beliefs and worldviews in conservation governance and practice. Bas' chapter highlights some of the lessons learned in working with spiritual leaders in Guatemala on the need to be able to build common grounds that offer a space to different worldviews. E-mail: basverschuuren@gmail.com

**Ati Gunnawi Viviam Villafaña** is a young member of the Arhuaco Indigenous community of the Sierra Nevada de Santa Marta, Colombia. A graduate in Political Science from Javeriana University in Bogota, a TEDxTalk presenter, she combined her academic studies with working as a research assistant at Colombia's leading biodiversity science research center, the Humboldt Institute. Ati Viviam brings a young and eloquent indigenous voice to environmental and climate activism, and is a member of the coordinating team of Unite for Climate Action (https://www.uniteforclimateaction.com).

**Kathy J. Willis** (PhD) is a Professor of Biodiversity at the University of Oxford and the Head of the Long-term Ecology Laboratory (OxLEL). Previous roles include Director of Science at the Royal Botanic Gardens, Kew, Director of the Oxford University Biodiversity Institute, and a member of the UK Government's Natural Capital Committee. Kathy's research focuses on the development of biological datasets, models and innovative technologies to determine the diversity, distribution and abundance of plants and animals across global landscapes, in space and time. This evidence-base is then used to understand biodiversity baselines, the resilience of biological communities to external shocks, the role of local Indigenous communities and their practices in protecting and enhancing biodiversity (including Sacred forests), and the relationship between biodiversity and human health (good and bad). Most recently, Kathy has been involved in developing tools and models to determine the distribution of natural capital assets across global landscapes that are important for human well-being. Kathy has also been closely involved in the scrutiny of proposed 'nature-based solutions' to achieve global reductions in greenhouse gas emissions associated with International, UK and local government policies. Her work aims to ensure that these 'solutions' are also delivering benefits for biodiversity and residents of these landscapes.

**Fiona Wilton,** born in Cornwall, UK, has lived most of her life in South America, immersed in community-led programmes and policy to regenerate biocultural diversity, sacred natural sites and healthy oceans. Since the 1980s she has worked with the Gaia Foundation (www.gaiafoundation.org), global partners, local and Indigenous communities, on initiatives to protect and restore the Earth's vitality and diversity. While living in Colombia and working alongside Arhuaco leader Danilo Villafañe, she was introduced to the remarkable spirit and cosmovision of the guardians of the 'Heart of the World'. Fiona's contribution to this book draws from those years, along with recent interviews and collaboration with Ati Gunnawi Viviam Villafaña. E-mail: fiona@gaianet.org

**Chua Ying Xuan** is a recent graduate from Nanyang Technological University, where she majored in Environmental Earth Systems Sciences. She entered university as a fiery environmentalist and was involved in environmental advocacy initiatives both on and off campus. When she encountered Christ halfway into her university, she was at a loss how to reconcile her passion for the environment and her new-found faith. At the same time, she also stumbled upon a growing Creation Care Movement amongst Protestant Churches in Singapore. When a research opportunity sprang up, Ying Xuan was excited to further her understanding of Creation Care activity in Singapore. She was thankful for the Creation Care initiators who candidly shared their experiences of championing Creation Care in their respective Churches. She is also inspired by their faithfulness in stewarding God's Creations and hopes to care for God's Creations—whether in the big or small ways. E-mail: CHUA0811@e.ntu.edu.sg

**Konchok Zangpo** is an ordained monk from Phyang Monastery in Ladakh, India. At a very young age, Zangpo walked into the monastery, knowing fully about his path in life. Having gained philosophical and spiritual knowledge from various Masters in Ladakh, Zangpo now holds the reins of managing monastic affairs for the Phyang Monastery. With several years of meditation and contemplation, Zangpo also teaches philosophy. Zangpo's interest in ecological issues is as organic as his intuitive initiation into Buddhism. Having seen changes in the local environment and inherently knowing of the value of all beings, conservation values are well-aligned with the Buddhist religious outlook. Compassion and kindness that are cornerstones of Buddhism are also the need of the hour for the well-being of all life on Earth.

# PART I
# Overview

# 1

# INTRODUCTION

*Radhika Borde, Alison A. Ormsby, Stephen M. Awoyemi and Andrew G. Gosler*

That no person today needs to be told that the natural environment on Earth is threatened is not an overstatement. From extreme weather events to water taps running dry in various parts of the world, to high-level scientific reports that outline the grim outcomes that human-induced climate change is causing, the evidence is convincing, and many are convinced – though perhaps not enough to act. Part of the reason for this is that the scale of the action required can make the conservation efforts of individuals or small communities seem meaningless. Yet, it is as individuals and communities that humans exist on the planet, affect it, and elect or tolerate the leaders who have the power to make decisions relating to its well-being. For these constituencies to be mobilized, what is needed may quite simply be faith – faith that it is possible to make a difference by acting, faith that others would follow one's example of action, and faith that one is morally impelled to act. Religious faith is an example of this kind of faith and can be argued to be a deep-seated driver of values and a regulator of people's actions – if oriented towards the environment, it may indeed supply the energy for the action that the planet's crisis is calling for. This book presents stories of how religious faiths have supported actions aimed at the conservation of nature. This volume discusses the environmental values and initiatives of a range of religious faiths across diverse geographies – covering African Traditional Religions, Buddhism, Christianity, Hinduism, Indigenous Faiths, Islam, Judaism, Shintoism, Zoroastrianism, and also synergistic religiosities.

The environmental values of faith-based communities are attracting increasing research interest (Awoyemi et al., 2012; Bhagwat et al., 2011; Borde, 2018; Glaab and Fuchs, 2018; Gosler et al., 2013) and this is mirrored by their championing at the highest levels of environmental policymaking – as evinced by this statement by the former UN Secretary-General: "the world's faith communities occupy

DOI: 10.4324/9781003246190-2

a unique position in discussions on the fate of our planet and the accelerating impacts of climate change" (Ki-Moon, 2009).

In his book on faith, Fowler (1981: 321) "equates faith with individual meaning systems", arguing that faith "is often but not necessarily religious" and that faith is specific to the individual. Furthermore, Fowler (1986) argues that faith is relational and implies trust in another (whether person or entity), and also that it is imaginative, implying also the imagining of a better future, i.e. hope. It is this understanding of faith as a meaning-making, relational framework that runs through the diverse narratives presented in this volume – some of which discuss how the conflation of religious tradition and faith lends authority to actions aimed at environmental conservation, while others describe creative re-workings of ancient religious rituals and traditions to suit the current exigencies of the planet.

The 6th Assessment Report of the Intergovernmental Panel on Climate Change (IPCC, 2021) has been described as a "code red for humanity" by the current UN Secretary-General (Guterres, 2021), and the threat level can be described to be much higher for most of the other species that call Earth home, and which may not be able to mitigate the effects of the environmental disasters that the report predicts. However, consensus on how to address this threat is missing. The broad concept of conservation has been the subject of heated debate (Sandbrook, 2015), with discussions as to whether it implies a move towards ecosystemic stasis versus dynamism (Knapp, 2003). Other perspectives, such as those belonging to the OneHealth and EcoHealth approaches, have long been arguing that the health and well-being of all species on Earth are intimately entangled (Zinsstag, 2012) – these arguments have been borne out by the COVID-19 pandemic (Vanhove et al., 2020). The current environmental crisis may be an indication that the time has come for conservation in the context of the Earth to be re-interpreted simply as a matter of supporting the planet's capacities for health, with humans at the centre of this task for the reason that they bear greater responsibility for the damage done and wield greater agency, including moral agency, vis-à-vis the facilitation of processes of repair. If this is the task of the planet's peoples, without the resources of faith (religious or otherwise), which has been studied in relation to healing (Victor and Treschuk, 2020), it would be a heavier burden, with less chance of success.

It is also possible that the resources that religious faiths provide will be needed for more than the conservation of this planet. Postcolonial theorists have argued that human activity on Earth can be seen as a progression towards planetary colonization (Sidaway, Yuan Woon and Jacobs, 2014), and that in remedy, humans must re-imagine themselves as 'planetary subjects', i.e. subject to the ecosystemic principles that maintain equilibrium on Earth (Vinayaraj, 2016). But how does such a remedy apply in a situation where a hypothetical 'Earth subject' directs an imperial gaze towards other planets (not an unlikely scenario)? Indeed, currently, nature and its conservation are being imagined into a future where inter-planetary travel may be actualized for humans (see Peters, 2018). Central to such imaginings is the longstanding debate on what nature is (Ducarme and Couvet, 2020) – the diversity of concepts in this regard impacting the manner in which conservation is carried out (Ducarme, Flipo and Couvet, 2020). As per some strands of Christian theological participation in this debate, human culture mimics nature, insofar as both are seen as being creative acts (Leidenhag, 2020). Other worldviews are argued to see intimate entanglements between culture and nature and indeed no separation between them (Descola, 2013) – and indeed, the nature-culture dichotomies that modernity is understood to have erected, have been dismantled by social theorists (see Latour, 1993). One can only hope for synergies between culture and nature, i.e. for culture that supports nature, as has been argued religious texts exhort (Brown and Gosler, 2021) – this volume turns to the cultures that are embedded in the world's faiths for precisely this.

As an edited volume, this book, titled "Religion and Nature Conservation: Global Case Studies", can be read as a collection of narratives that present meaning-making, healing, and creative projects and ideas, from around the world – all of which are imagining/re-imagining nature, and involve relating and interacting with it in ways that question the construction of nature as a resource to be exploited simply for economic gain. These narratives have in common the message that we must tread more lightly upon what is not of our own making, i.e. nature, and that such an act, rather than just involving sacrifice, would be uplifting and empowering. Spanning each of the world's inhabited continents, the authors of this volume's chapters engage with the rich diversity of faith traditions that is the heritage of humanity as a whole, respond to the challenges of modernity, and express commitment to the cause of conserving nature and the rights of the peoples that ascribe it sacred value. Some of these narratives are chapter-length discussions. Others provide concise snapshots of faith-based conservation projects or concepts. Finally, it is important to state that our project is a postcolonial one. We have tried to ensure that the chapters in this book do not speak for 'others' amongst whom at least some of the authors cannot claim they belong – we have tried to ensure that most authors/co-authors themselves represent the regions and the faith traditions that are covered by their chapters. We acknowledge that even this is an exercise that is fraught with problems, but we would like to make at least a small attempt to refrain from rendering persons or communities subaltern, i.e. silenced by being 'spoken for' instead of 'speaking' (Kapoor, 2004; Alcoff, 1991). In the cases where the voices of others are 'translated' (see Maggio, 2007), we have tried to ensure that this is made explicit.

The chapters in this volume can be grouped according to several themes, and in some cases multiple themes cut across individual chapters. These themes are discussed subsequently with reference to the relevant chapters.

## Rights-Based Conservation Approaches

A report by the Intergovernmental Science-Policy Platform on Biodiversity and Ecosystem Services stated that the rate of environmental degradation on Indigenous lands is comparatively lower than on other lands (IPBES, 2019), and there are lively debates as to whether and how the religiosities of Indigenous peoples may support nature conservation. For an excellent collection of scholarship on Indigenous environmental ethics, please see the themed special issue of the *Journal for the Study of Religion, Nature and Culture* on Indigenous nature reverence and conservation (see Snodgrass and Tiedje, 2008). Importantly, it has been argued that in consideration of existing Indigenous environmental ethics, conferring rights to Indigenous peoples over the lands they have traditionally managed may indeed be an effective conservation strategy (Desmet, 2011; Howitt, 2018; Witter and Satterfield, 2018). This is reiterated by several of the chapters in this volume – see Chapter 7 by Barrow; Chapter 13 by Liljeblad; Chapter 6 by Solis-Aguilar et al.; Chapter 12 by Verschuuren and Gomez. It is important to add, however, that taken as a whole, these chapters strike a balance between emphasizing Indigenous rights aimed at the material security of Indigenous peoples, and conservation effects resulting from the recognition of these Indigenous rights. We are well aware of the criticism that has been directed against initiatives that impose environmental conservation on Indigenous peoples (Lindroth and Sinevaara-Niskanen, 2013; Radcliffe, 2014), and would like to contextualize the chapters that deal with rights-based approaches to conservation, within the lively debate that this topic has elicited.

## Philosophical and Theological Inspirations

In response to the environmental exigencies of the present, there is a growing trend of examinations within and across faith-based communities, for the resources that the world's religions may offer in this regard (see Harper, 2008; Jenkins, Berry and Kreider, 2018). Religious texts are being re-interpreted in search of the environmental ethics they may contain. To give an indication, there are several examples of this within the Christian tradition (Bauckham, 2011; Brown and Gosler, 2021; Horrell et al., 2008), Islam (Abd Rahman, Zabidi and Halim, 2020; Aboul-Enein, 2018), Buddhism (Capper, 2021; James, 2017), Hinduism (Framarin, 2011; Gairola, 2020), and Judaism (Tirosh-Samuelson, 2017; Yoreh, 2019). We would also like to mention the work of the past two decades, of the Yale Forum on Religion and Ecology, (FORE, 2022), and the book on religions and ecology by the Forum's founders Grim and Tucker (2014). The philosophical and theological inspirations for environmental conservation offered by the world's religious faiths are also explored by the authors of the chapters in this volume – see Chapter 2 by Bhutia; Chapter 10 by Rowe and Baya; Chapter 18 by Scorsch; Chapter 16 by Ten Veen. In each case, these inspirations are discussed in the context of derivative conservation projects.

## Species-Specific Conservation

From the protection of tigers via an Islamic *fatwa* (Mangunjaya and Praharawati, 2019) to the conservation of bird species via totemistic beliefs in African Traditional Religions (Mandillah and Ekosse, 2018) to the overlap between the perceived sacrality of specific sites and the conservation of the plant and animal species they contain (Pungetti, Oviedo and Hooke, 2012; Verschuuren et al., 2010), there is a rich body of literature on how faith traditions have aided in the conservation of the planet's biodiversity. The chapters in this volume are a contribution to this body of literature – more specifically, Chapter 3 by Kothari et al. discusses how faith traditions have led to the conservation of the black-necked crane (*Grus nigricollis*) and Chapter 19 by Deshwal et al. discusses similar issues in relation to the Greater Adjutant stork (*Leptoptilos dubius*) – both of these avian species are endangered. Other chapters, while not focusing on the conservation of a flagship species, discuss, for example, how habitats for bees are being conserved by the church forests of Ethiopia (Chapter 9 by Marks et al.) and also by the sacred groves of India (Chapter 8 by Ormsby and Krishnan), how habitats for fish that an Indigenous Central American community is dependent upon are being conserved via the ascription of sacred values to these habitats (Chapter 6 by Solis-Aguilar et al.), or how natural sites and landscapes constituting habitats for various species are conserved around the world via explicitly Christian initiatives (Chapter 15 by Ederer et al.).

## Innovations at the Intersection between Modernity and Religious Tradition

The place of religion in modern society is the subject of debate, but with some consensus that innovations within religions and spiritual traditions are playing an increasingly important role (Lawson, 2006; Lee and Ackerman, 2018). In many respects, the ecological articulations of the various faiths that this book presents, are innovations. In a few chapters, these innovations are made explicit and, in these cases, the religious innovations under discussion are responding to specifically modern problems such as the generation of toxic and non-biodegradable waste or carbon emissions – Chapter 5 by Sheth shows how concerns pertaining to eco-friendly materialities were introduced into an Indian religious festival; similarly, Chapter 2 by Bhutia discusses how a Buddhist monastery took on the task of managing waste in its region and Chapter 18 by Schorsch re-imagines the ancient Jewish ritual of the Sabbath as a mitigator of fast-paced, carbon-intensive lifestyles.

## Community Relations, Human Well-Being, and Economic Empowerment

The case studies presented in this volume's chapters do not ignore the important issue of the emotional and economic sustainability of greener lifestyles and conservation, and discuss how green faiths have played a part in the upliftment and empowerment of people and communities. Studies have pointed to the central role that religion plays in invigorating communities and supporting human well-being (Cohen and Johnson, 2017; Newman and Graham, 2018). In fact, it has been stated that caring for nature is in itself a component of human well-being (Jax et al., 2018). The chapters in this volume discuss these important points – see Chapter 14 by Fernandes-Pinto on the dependencies between synergistic religiosities and Brazilian sacred natural sites; Chapter 11 by Marks et al. on Shintoism in Japan; Chapter 10 by Rowe and Baya; Chapter 4 by Smith; and Chapter 18 by Schorsch. Chapters also discuss issues pertaining to the economic and social empowerment of communities involved in nature conservation via sacred ascriptions – see Chapter 8 by Ormsby and Krishnan on India's sacred groves; and Chapter 9 by Marks et al. on Ethiopian church forests. And finally, Chapter 17 by Coreth et al. discusses how inter-community relations and peace-building in a sacred and troubled land are secured via birds and acts of creativity and adventure inspired by them.

We invite an exploration of all the important themes that the case studies in this volume reflect. As may be clear by now, each case study is about more than religion and nature conservation, though their arguments do centre on this potential synergy. Rather, as a whole, the contributions present a re-imagining of how people can live in a modern world, with other species, with inspiration from religious traditions, with greater well-being as communities and individuals and with less inequality and injustice.

## References

Abd Rahman, N., Zabidi, F. N. M., & Halim, L. (2020). Integration of Tauhidic elements for environmental education from the teachers' perspectives. *Religions, 11*, 394. https://doi.org/10.3390/rel11080394

Aboul-Enein, B. H. (2018). "The earth is your mosque": Narrative perspectives of environmental health and education in the holy Quran. *Journal of Environmental Studies and Sciences, 8*(1), 22–31.

Alcoff, L. (1991). The problem of speaking for others. *Cultural Critique, 20*, 5–32.

Awoyemi, S. M., Gambrill, A., Ormsby, A. A., & Vyas, D. (2012). Global efforts to bridge religion and conservation: Are they really working? In T. Povilitis (Ed.), *Topics in conservation biology* (pp. 97–110). Rijeka, Croatia: InTech.

Bauckham, R. (2011). *Living with other creatures: Green exegesis and theology.* Waco, TX: Baylor University Press.

Bhagwat, S. A., Ormsby, A. A., & Rutte, C. (2011). The role of religion in linking conservation and development: Challenges and opportunities. *Journal for the Study of Religion, Nature and Culture, 5*(1), 39–60.

Borde, R. (2018). The significance of Indigenous nature spirituality. In B. Verschuuren & S. Brown (Eds.), *Cultural and spiritual significance of nature in protected areas: Governance, management and policy.* London: Routledge.

Brown, J., & Gosler, A. (2021). Who is my neighbour? The human animal in creation. In C. Nellist (Ed.), *Climate crisis and sustainable creaturely care* (pp. 17–34). Newcastle upon Tyne: Cambridge Scholars Publishing.

Capper, D. (2021). Intrinsic value, American Buddhism, and potential life on Saturn's moon titan. In O. A. C. Torres, T. Peters, J. Seckbach, & R. Gordon (Eds.), *Astrobiology: Science, ethics, and public policy.* Beverly, MA: Scrivener Publishing.

Cohen, A. B., & Johnson, K. A. (2017). The relation between religion and well-being. *Applied Research in Quality of Life, 12*, 533–547.
Descola, P. (2013). *Beyond nature and culture*. Chicago: The University of Chicago Press.
Desmet, E. (2011). Indigenous rights entwined with nature conservation. *International Law, 8*. Retrieved from http://hdl.handle.net/1854/LU-3007415
Ducarme, F., & Couvet, D. (2020). What does 'nature' mean? *Palgrave Communications, 6*(14): https://doi.org/10.1057/s41599-020-0390-y
Ducarme, F., Flipo, F., & Couvet, D. (2020). How the diversity of human concepts of nature affects conservation of biodiversity. *Conservation Biology, 35*(3), 1019–1028. https://doi.org/10.1111/cobi.13639
Forum on Religion and Ecology (FORE) (2022). Yale forum on religion and ecology. Retrieved from https://fore.yale.edu/
Fowler, J. W. (1981). *Stages of faith: The psychology of human development and the quest for meaning*. San Francisco, CA: Harper & Row.
Fowler, J. W. (1986). Faith and the structuring of meaning. In: C. Dykstra & S. Parks (Eds.), *Faith development and Fowler* (pp. 15–42). Birmingham, AL: Religious Education Press.
Framarin, C. G. (2011). The value of nature in Indian (Hindu) traditions. *Religious Studies, 47*(3), 285–300.
Gairola, S. U. (2020). Review article on relation between Hinduism and environment – A vedic approach. *Asian Journal of Environment & Ecology, 13*(3), 19–25.
Glaab, K., & Fuchs, D. (2018). Green faith? The role of faith-based actors in global sustainable development discourse. *Environmental Values, 27*(3), 289–312.
Gosler, A. G., Bhagwat, S., Harrop, S., Bonta, M., & Tidemann S. (2013). Chapter 6: Leadership and listening: Inspiration for conservation mission and advocacy. In Macdonald, D. & Willis, K. J. (Eds.), *Key topics in conservation Biology 2* (pp. 92–109). Oxford: John Wiley & Sons, Ltd.
Grim, J., & Tucker, M. E. (2014). *Ecology and religion*. Washington, DC: Island Press.
Guterres, A. (2021). Guterres: The IPCC report is a code red for humanity. Retrieved from https://unric.org/en/guterres-the-ipcc-report-is-a-code-red-for-humanity/
Harper, C. L. (2008). Religion and environmentalism, religion and the environment. *Journal of Religion & Society, 3*, 5–26.
Horrell, D. G., Hunt, C., & Southgate, C. (2008). Appeals to the Bible in ecotheology and environmental ethics: A typology of hermeneutical stances. *Studies in Christian Ethics, 21*(2), 219–238.
Howitt, R. (2018). Indigenous rights vital to survival. *Nature Sustainability, 1*, 339–340.
IPBES. (2019). *Global assessment report on biodiversity and ecosystem services of the intergovernmental science-policy platform on biodiversity and ecosystem services*. Bonn: IPBES Secretariat. https://doi.org/10.5281/zenodo.3831673
IPCC. (2021). *Climate change 2021: The physical science basis. Contribution of working group I to the sixth assessment report of the intergovernmental panel on climate change*. Cambridge: Cambridge University Press. Retrieved from https://www.ipcc.ch/report/ar6/wg1/#FullReport
James, S. P. (2017). *Zen Buddhism and environmental ethics*. London: Routledge.
Jax, K., Calestani, M., Chan, K. M. A., Eser, U., Keune, H., Muraca, B., … Wittmer, H. (2018). Caring for nature matters: A relational approach for understanding nature's contributions to human well-being. *Sustainability, 35*, 22–29.
Jenkins, W., Berry, E., & Kreider, L. B. (2018). Religion and climate change. *Annual Review of Environment and Resources, 43*, 85–108.

Kapoor, I. (2004). Hyper-self-reflexive development? Spivak on representing the third world 'other'. *Third World Quarterly, 25*(4), 627–647.
Ki-Moon, B. (2009). Speech to summit of religious and secular leaders on climate change. Retrieved from https://www.un.org/sg/en/content/sg/speeches/2009-11-03/speech-summit-religious-and-secular-leaders-climate-change
Knapp, S. (2003). Conservation: Dynamic diversity. *Nature, 422*, 475.
Latour, B. (1993). *We have never been modern*. Cambridge: Harvard University Press.
Lawson, S. (2006). *Culture and context in world politics*. London: Palgrave Macmillan.
Lee, R. L. M, & Ackerman, S. E. (2018). *The challenge of religion after modernity: Beyond disenchantment*. Abingdon: Routledge.
Leidenhag, J. (2020). On overcoming the culture-nature divide: A panpsychist proposal. *Scottish Journal of Theology, 73*(1), 43–54.
Lindroth, M., & Sinevaara-Niskanen, H. (2013). At the crossroads of autonomy and essentialism: Indigenous peoples in international environmental politics. *International Political Sociology, 7*, 275–293. https://doi.org/10.1111/ips.12023.
Maggio, J. (2007). "Can the subaltern be heard?": Political theory, translation, representation and Gayatri Chakravorty Spivak. *Alternatives, 32*, 419–443.
Mandillah, K. L. L., & Ekosse, G.-I. (2018). African Totems: Cultural heritage for sustainable environmental conservation. *Conservation Science in Cultural Heritage, 18*(1), 201–218. https://doi.org/10.6092/issn.1973-9494/9235
Mangunjaya, F., & Praharawati, G. (2019). Fatwas on boosting environmental conservation in Indonesia. *Religions, 10*, 570. https://doi.org/10.3390/rel10100570
Newman, D. B., & Graham, J. (2018). Religion and well-being. In: E. Diener, S. Oishi., & L. Tay (Eds.), *Handbook of well-being* (pp. 18–29). Salt Lake City, UT: DEF Publishers.
Peters, T. (2018). Toward a galactic common good: Space exploration ethics. In: D. Boonin (Ed.), *The Palgrave handbook of philosophy and public policy* (pp. 827–843). Heidelberg: Springer Verlag. https://doi.org/10.1007/978-3-319-93907-0_61
Pungetti, G., Oviedo, G., & Hooke, D. (2012). *Sacred species and sites: Advances in biocultural conservation*. Cambridge: Cambridge University Press.
Radcliffe, S. A. (2014). Plural knowledges and modernity: Social difference and geographical explanations. In K. Okamoto & Y. Ishikawa (Eds.), *Traditional wisdom and modern knowledge for the earth's future: Lectures given at the plenary sessions of the international geographical Union Kyoto Regional Conference, 2013* (pp. 79–102). Tokyo: Springer.
Sandbrook, C. (2015). What is conservation? *Oryx, 49*(4), 565–566.
Sidaway, J. D., Yuan Woon, C., & Jacobs, J. M. (2014). Planetary postcolonialism. *Singapore Journal of Tropical Geography, 35*, 4–21.
Snodgrass, J. G., & Tiedje, K. (2008). Indigenous nature reverence and conservation: Seven ways of transcending an unnecessary dichotomy. *Journal for the Study of Religion, Nature and Culture, 2*(1), 6–29.
Tirosh-Samuelson, H. (2017). Jewish environmental ethics: The imperative of responsibility. In J. Hart (Ed.), *The Wiley Blackwell companion to religion and ecology*. Oxford: Wiley Blackwell.
Vanhove, M. P. M., Hugé, J., De Bisthoven, L. J., Keune, H., Laudisoit, A., Thys, S., … Antoine-Moussiaux, N. (2020). EcoHealth reframing of disease monitoring. *Science, 370*(6518), 773.
Verschuuren, B., Wild, R., Mcneely, J., & Oviedo, G. (Eds.). (2010). *Sacred natural sites: Conserving nature and culture*. London: Earthscan.

Victor, C. G. P., & Treschuk, J. V. (2020). Critical literature review on the definition clarity of the concept of faith, religion, and spirituality. *Journal of Holistic Nursing, 38*(1), 107–113.

Vinayaraj, Y. T. (2016). *Dalit theology after continental philosophy.* London: Palgrave Macmillan.

Witter, R., & Satterfield, T. (2018). The Ebb and flow of Indigenous rights recognitions in conservation policy. *Development and Change, 50*(4), 1083–1108.

Yoreh, T. S. (2019). *Waste not: A Jewish environmental ethic.* Albany: State University of New York Press.

Zinsstag, J. (2012). Convergence of ecohealth and one health. *EcoHealth, 9*, 371–373.

# PART II

# Examples of Faith-Based Conservation from around the World

# 2

# CARING FOR THE LAND, CARING FOR THE DHARMA

## The Environmental History of Buddhism at Pemayangtse Monastery, Sikkim, as a Resource for Contemporary Conservation Initiatives

*Kalzang Dorjee Bhutia*

### Introduction

In 2017, the large Buddhist monastery Pemayangtse in the northeast Indian Himalayan state of Sikkim announced that patrons of the monastery could no longer offer plastic-wrapped food and goods for ritual donations to the monastery. This decision, communicated on large billboards and by lamas and monks[1] to local patrons through WhatsApp and Facebook posts and messages, was undertaken following the visit of the Swachh Bharat Mission (SBM) to the monastery. This central Indian government-supported programme was intended to propagate information about public sanitation and hygiene by eliminating open defecation and by implementing systems for more efficient solid waste management (Raman, 2020). At Pemayangtse, the ban on plastic offerings was significant, as it represented a vivid local example of what researchers have found elsewhere: that Buddhist institutions can be deeply effective collaborators with conservationists in the promotion of environmental awareness and education (Li et al., 2013; Karmapa, 2011; Knapp, 2012) (Figure 2.1).

Part of the reason for this success has been due to central Buddhist concepts that connect well with conservation goals. Scholars have noted a resonance between conservation and sustainability goals and the emphasis in Buddhism on the interdependence of and compassion for all sentient beings (Kaza and Kraft, 2000). However, this ecocentric interpretation of Buddhism has also been critiqued as ahistorical (Harris, 1991). Historian Johan Elverskog has demonstrated how Buddhism can be interpreted as a prosperity theology that used forest resources to build its institutions and encouraged Buddhists to pursue wealth and material resources as part of their spiritual goals (Elverskog, 2020). In the case of Tibetan and Himalayan forms of Buddhism, scholars have demonstrated how, historically, Buddhist individuals and institutions participated in hunting, resource

DOI: 10.4324/9781003246190-4

**FIGURE 2.1** Members of the Pemayangtse Monastic Shedra (Monastic School) wearing Swachh Bharat promotional material. Photo credits: Khenpo Wangyal, Pemayangtse Shedra.

extraction, and other activities that benefited human communities at the expense of local ecologies (Huber, 1997; Yeh, 2014; Cantwell, 2001).

Despite this, in contemporary times, many Tibetan and Himalayan Buddhist communities (and especially institutions such as monasteries) have become important participants of conservation and environmental awareness and implementation movements. Elverskog argued that this change is due to the adaptability of Buddhism to new cultural, economic, and political contexts, and is representative of the Buddha's teaching of the inevitability of impermanence (Elverskog, 2020, p. 120). He interpreted such a change as an encouraging example for the world to see how individuals and institutions have the agency to transform even deeply held worldviews in response to global environmental challenges. However, other scholars do not see this change as entirely radical, but instead based on older traditions and ideas. Anthropologist Karine Gagné has demonstrated that in Ladakh, traditions of care for the environment have been inspired by an interrelatedness between humans, nonhuman animals, and the land that developed from multispecies inhabitation of, and dependence on, a shared environment. These traditions are not based only on Buddhist philosophy, but also folk rituals, beliefs, and practical engagement with the environment (Gagné, 2018, p. xv). Such care functions as a foundation for facilitating concern for the environment, and thereby for conservation initiatives.

Understanding how this care connects to broader Buddhist concepts, such as interdependence and compassion, is important, because in Tibetan and Himalayan Buddhism, the environment is not just an environment, but can be sacred space, or a sacred habitat (Classical Tibetan: *gnas*[2]) that is home to many interdimensional residents, including deities and spirits that are not always seen, but who are often felt, and communicated with through ritual. This acknowledgement is important as it challenges divisions between nature and culture in what anthropologist Gillian Tan has called "an ecology of religiosities" (Tan, 2014).

Drawing on Gagné's insights, in this chapter, I will discuss historical practices related to care (Classical Tibetan: *gces*) for the sacred habitat of Sikkim in the region around Pemayangtse Monastery. I will analyse classical Tibetan-language ritual and historical texts and draw on oral histories gathered from community members and my experience as a ritual practitioner at Pemayangtse since my childhood as sources for understanding Pemayangtse's environmental history. Sikkim is a multiethnic, multicultural state in India where environmental protection has been co-opted by ethnic politics (Chettri, 2017; Acharya and Ormsby, 2017). Drawing on Buddhist heritage and ideals of the environment allows for an inclusive notion of Sikkimese identity that can be used in support of conservation aims (Wagh, 2017; Ramakrishnan, 2008). This chapter will provide examples of practices and concepts that may be generative for conservation goals and environmental protection in the eastern Himalayan region and beyond.

## Background: Sikkim as a Hidden Land and Sacred Habitat

Conservation and the environment more generally loom large in public discourse related to Sikkim. Lauded as the world's "first organic state" (Meek and Anderson, 2020), the small northeast Indian state has been consciously branded by the state government as "green" and environmentally aware. Much of this publicity dates from the long reign of Sikkim's fourth Chief Minister, Pawan Chamling (b. 1949), from 1994 to 2019. Chamling received awards for spearheading conservation initiatives, but his government also supported the planning and installation of large-scale hydroelectric projects that have been contested (Arora, 2009a; Gergan, 2017, 2020), and presided over the spread of unplanned urbanisation in the state which has led to unprecedented scales of concrete growth that has resulted in land erosion and landslides (McDuie-Ra and Chettri, 2020).

Despite this, Sikkim has continued to hold a reputation for environmentally conscious policy-making and popular support, and even though Buddhism is not demographically dominant, tourist literature continues to represent the state as a green Buddhist paradise (Arora, 2009b). This paradisiacal imagery has historical precedent. The Indigenous Lepcha ethnic community understood Sikkim to be the Mayel Liang, a paradise on earth. Tibetan Buddhists believed Sikkim to be a "Hidden Land" (Classical Tibetan: *sbas yul*), set aside by Guru Rinpoche in the 8th century to act as a safe haven for Buddhist practice in times of need. During conflict on the plateau, waves of Tibetans migrated to Sikkim, where they interpreted the mountainous forest landscape as a vast sacred habitat (Classical Tibetan: *gnas*), animated by deities, spirits, and supernatural forces, including lha (Classical Tibetan: *lha*), tsan (Classical Tibetan: *gtsan*), dud (Classical

Tibetan: *bdud*), and lu (Classical Tibetan: *klu*). Oral and written sources state that Guru Rinpoche converted these autochthonous beings into protector deities (Classical Tibetan: *chos skyong yul lha gzhi bdag*) of Sikkim and, by extension, Sikkimese Buddhism (Lhatsun, 2000). At the pinnacle of these interdimensional inhabitants was Kanchendzonga (Classical Tibetan: *Gangs chen mdzod lnga*, or, as it is often rendered, Kanchenjunga), the mountain deity that presided over the state (Balikci-Denjongpa, 2001). The descendants of these Tibetans, known as Lhopo (or, in contemporary Sikkim, Bhutia), collaborated with Indigenous Lepcha and other resident ethnic communities and promulgated Treasures, based on Guru Rinpoche's teachings that he had hid in the landscape, to develop distinctive Buddhist institutions and practices based around the distinctive ecology of Sikkim (Figure 2.2).[3]

In the 17th century, Lhatsun Namkhai Jikme (1597–1650/4), a Tibetan yogi, travelled to Sikkim after having a vision of the landscape. Once there, he took part in enthroning the first king of the Bhutia Namgyal dynasty, and promulgated a lineage of Nyingmapa Buddhism[4] based on visions inspired by the Hidden Land. He founded Pemayangtse Monastery in west Sikkim in the 1640s as the seat for his lineage. Pemayangtse is situated at 6,000 ft. above sea level, in a thick forest of medium-altitude alpine trees that include pine and juniper. This location had soteriological, political, and economic benefits. The soteriological benefit came from the remote location. When it was first built, the monastery's

**FIGURE 2.2** Kanchendzonga, the mountain protector deity of Sikkim. Photo credits: Chopel Dorjee Bhutia.

location reflected the aspirations of the institution to train monastics and ritual specialists in meditation, as it was located over an hour away by foot from the nearest human settlement and distractions of everyday life. Politically, the monastery was an "arrow's flight" from Rabdentse, the second capital of the royal Namgyal dynasty. The monastery's proximity to the royal capital was therefore significant for its legitimacy (Mullard, 2011).

The favoured position of Pemayangtse lamas in Sikkimese politics as advisors to the kings led them to be given a large estate that they were responsible for maintaining, along with their responsibilities to maintain good relationships with the interdimensional inhabitants of the sacred habitat in west Sikkim. The community at Pemayangtse functioned as lamas and advisors to the kings and their ministers. Senior members of the community were recruited from the Bhutia aristocratic clans, which made Pemayangtse distinct from other institutions that were made up of a variety of ethnic groups. The monastery is administered by the *Dor-u-cho-sum*, an institution made up of the three senior most qualified lamas: the Dorje Lopen (Classical Tibetan: *Rdo rje slop dpon*), or Vajra Master, a Tantric master with expertise in meditation and ritual; the Umzed (Classical Tibetan: *Dbu mdzad*), or Chant Master, who presides over rituals; and the Chotrimpa (Classical Tibetan: *Chos khrims pa*), or Discipline Master, who is responsible for the day-to-day maintenance of the monastery.

## Environmental Histories as Care for the Sacred Habitat at Pemayangtse

The responsibilities of maintaining multispecies relationships held by Pemayangtse can be divided into several areas. Here, I will discuss two: the responsibilities to maintain good relationships through ritual maintenance; and the responsibility of the Pemayangtse community to care for the forest and water in their vicinity, which includes the sacred habitat of west Sikkim, the region in the lap of Kanchendzonga. These historical examples demonstrate the complexity of multispecies relations in this area, and can be extended to other parts of Sikkim, since Pemayangtse was a powerful monastery with many satellite branches in other parts of Sikkim (and now, in northern West Bengal areas that were historically part of Sikkim), and was also located in the sacred habitat close to Kanchendzonga. Buddhist studies scholar Cathy Cantwell has written of how historical attitudes towards the environment in Tibetan Buddhist communities must be contextualised with care, since even though rituals often acknowledge the presence of other species in the sacred habitat they invoked, they often did not manifest any material benefits for the environment (Cantwell, 2001). This is also the case at Pemayangtse, where multispecies relationships often favoured human endeavours. However, these practices have alternative interpretations, which have recently inspired participants in conservation and environmental protection movements.

## Maintaining Relationships with the Sacred Habitat through Ritual: The *Nesol*

From the initial foundation of Pemayangtse, the community of practitioners there maintained an active ritual calendar throughout the year in order for them to fulfil their mandates as effective intermediaries between the different residents of Sikkim's animated sacred habitat. Several of these rituals are significant for their representation of the sacred habitat, and concern to outline appropriate acknowledgement, and even care, of other beings present in it.

The most significant of these rituals is the *Nesol* (English: *Earth Propitiation Rite*; Classical Tibetan: *gnas gsol*). Developed from a Treasure revealed from the landscape of Sikkim by Lhatsun Namkhai Jikme, this ritual describes the sacred habitat and lists all the beings within it. These beings are divided into three sections: the beings of the mountains, including the mighty mountain deity Kanchendzonga; the beings of the mid-level hills, including all of the choskyong yullha zhidak (Classical Tibetan: *chos skyong yul lha gzhi bdag*), or dharma protectors, such as tsan spirits of the trees; and the beings of the valleys and rivers, including lu, serpent spirits that bring wealth and auspiciousness to the region (Lhatsun, 2000). After outlining the residents of each section, the *Nesol* outlines appropriate care for each area. Some of these outlines work as specific and detailed guidance and warnings about what happens if the deities are not cared for appropriately. For example, the *Nesol* warns that,

> People should not burn impure substances, such as meat, or commit acts of violence and pollute the environment by cutting down trees. They should not disturb or pollute the lakes or the oceans, or tunnel into or break the mountains and rocks. If we disturb you in these ways, we apologise and admit to doing this, and ask for protection for all sentient beings.
>
> **If humans do not keep this promise, you may take their lives to punish them**
>
> *(Lhatsun, 2000, p. 31; translation and emphasis my own).*[5]

This final comment demonstrates that historical forms of Buddhist interdimensional relationships were not always based on peaceful interactions. In this and many other instances, threats of supernatural retribution functioned as a reminder for humans to behave appropriately or incur the wrath of the deities. *Nesol* rituals are performed throughout the year on the eighth, tenth, and fifteenth days of the lunar calendar in Pemayangtse and at other Pemayangtse satellite institutions, and during the large annual festival of *Pang Lhabsol* (Classical Tibetan: *Dpang lha gsol*), which is celebrated throughout the state. During this festival, lamas and monks from Pemayangtse are dispatched to the four sacred caves of Sikkim to give offerings to the landscape. In contemporary Sikkim, enactments of *Nesol* have been interpreted by scholars as signifiers of local identity and nationalism (Vandenhelsken, 2011). However, historically *Nesol* united the Buddhist community across different ethnic groups in the propitiation of the sacred habitat of Sikkim. This sacred habitat, presided over by Kanchendzonga, remained sacred due to its many interdimensional inhabitants.

A powerful example of the use of the *Nesol* to promote environmental awareness has been its invocation in anti-hydroelectricity movements. In 1994, the Sikkim state government announced plans to develop a large dam on the Rathong River. This river falls in the heart of the sacred habitat in west Sikkim. Pemayangtse lamas were important synergists in the anti-dam movement, and spoke at rallies, attended government meetings, and gave teachings to local communities where they brought together Buddhist and conservation discourses about the damage the dam would bring to the sacred habitat. They invoked the section of the *Nesol* above, stating that the guardian deities would be agitated if the dam went ahead, which would lead to disasters, illness, and poverty for the state of Sikkim. Some activists said that they would burn their *Nesol* texts if the dam went ahead, since the sacred habitat that the *Nesol* was intended to protect would be compromised (Balikci, 2008, p. 239). Eventually, in 1997, the state government cancelled the project. However, anti-dam movements have continued as the state has continued to announce other alternative projects. Lepcha communities in North Sikkim have undertaken hunger strikes, mass protests, and a coordinated online campaign to secure public support to have dams cancelled. Indigenous Lepcha traditions have been important resources for educating the government and public about the impact dams will have on multispecies inhabitants in Sikkim (Arora, 2009a; Gergan, 2017; Lepcha, 2020). The *Nesol* is also an important resource, as the use of Buddhist concepts in it complements indigenous cosmologies.

The *Nesol* also inspires other members of the Pemayangtse community to participate in conservation initiatives. A vivid example is the efforts by members of the community to coordinate tree and flower planting drives and the anti-plastic initiative mentioned above (Figure 2.3).

### *Asking Permission and Apologising for Environmental Interference: Sadaglunyen*

Another important ritual that is undertaken by Pemayangtse lamas and monks is *Sadaglunyen* (Classical Tibetan: *Sa bdag klu gnyan;* English: *The Propitiation of the Lords of the Land and the Nagas*), a ritual performed out in the community at buildings sites. *Sadaglunyen* is a prayer that invokes the local deities of the place (the *sadag*, or Lords of the Land), particularly the Nagas, a powerful class of supernatural beings, and supplicates them to grant blessings and favours to the human community there. Since there are Lords of the Land and Nagas throughout Sikkim's landscape, the *Sadaglunyen* is undertaken frequently throughout the year in response to a variety of events, including undertaking new building projects as a preliminary consecration (Classical Tibetan: *rab gnas*), or as a response

**FIGURE 2.3** A food offering with no plastic packaging at Pemayangtse Monastery. Photo credits: Khenpo Wangyal, Pemayangtse Shedra.

to illness in the community. It functions as a type of Environmental Impact Assessment, as whenever there will be any interaction with the land – be it through construction, dwelling, or even just passing through – local deities can be angered and can create obstacles. The *Sadaglunyen* involves invoking the local deities and offering drinks (Classical Tibetan: *gser skyems*) and incense (Classical Tibetan: *bsang*) either to acknowledge their presence, or after something has gone wrong, to apologise (Bhutia, 2021). Common obstacles that demonstrate supernatural dissatisfaction with human behaviour include weather storms, landslides, excess rain or drought, loss of wealth, illness, and social discord (Figure 2.4).

## *Caring for the Forest*

Beyond ritual forms, Pemayangtse lamas and monks also participated in other activities that were intended to protect and conserve the environment. Along with tax collection for the kings of the Namgyal dynasty, one of the primary revenue generators for the Pemayangtse community was the forest around the monastery. The forest was part of the *chikor* (Classical Tibetan: *phyi skor*), the outer pilgrimage circuit around the monastery. As well as demarcating sacred space, this forest was significant as it was full of interdimensional protectors of the monastery. Some of these protectors took the form of trees and were known as tsan and lu and received prayers and offerings throughout the ritual calendar year. In this way, some parts of the forest resembled Devithans, Hindu sacred

**FIGURE 2.4** An incense (sang) offering made from branches of local trees. Photo credits: Chopel Dorjee Bhutia.

groves, found elsewhere in Sikkim, that were also considered to be animated and therefore could not be interfered with (Acharya and Ormsby, 2017).

However, other trees were not part of this animated landscape, and could be removed to be used for construction projects at the monastery, and as firewood for the monastery's kitchens and ritual altars. Since the monastery had an active ritual calendar that meant wood was needed frequently, monastic officials were careful to supervise any trees removed from the forest in order to prevent this important resource from being depleted. Before trees were removed, the *Dor-u-cho-sum* had to be informed and offered guidance where the wood could be taken from. The kitchen caretaker (Classical Tibetan: *Spyi gnyer*) would work with kitchen staff (Classical Tibetan: *ma byan*) to use trees that had died or fallen over due to the weather, instead of chopping down healthy trees. When removing cuttings for making incense, both monastics and laypeople took care not to take too much, and their cuttings often helped to rejuvenate growth on older trees. This acted as a form of local forest conservation (similar to forms found in Bhutan by Roder et al., 2003). Local villagers were expected to make formal requests to remove wood for their own use to monastic authorities, and could be fined if they were caught chopping down trees without permission. These traditions of interaction with the forest demonstrated how this system of management can also be understood as a system of care, as the forest literally and symbolically provided the fuel for Pemayangtse's ritual activity.

## *Waste Management Systems in the Sacred Habitat*

Since Pemayangtse was high up on a spur and a challenging walk from the nearest settlement, access to fresh water to be used by the community for drinking, bathing, and cooking was carefully managed and there was concern not to waste water. The maintenance of hygiene and disposal of human waste was also cautiously monitored, since local protector deities could be angered by the inappropriate disposal of human waste that could contribute to the accumulation of ritual pollution (Classical Tibetan: *sgrib*) that, in turn, led to obstacles, including illness, accidents, and conflict. The issue of ritual pollution has been taken up by state-led and conservation-based initiatives throughout the Himalayas in connection to waste management (Allison, 2014, 2019).

As there were no running streams within the monastic compound, until about the 1940s water had to be carried from Dothum Pond, located in the forest. Dothum was home to a naga and was especially reserved for the monastic community to use as their primary source of freshwater. Animals consumed water from outflows of the lake, and laypeople took their water from streams closer to their villages. Dothum was thereby kept clean. The water was carried to the monastery in bamboo containers and boiled for consumption. In the 1940s, a water line was connected to the monastery from Tsangey, from the mountains of west Sikkim. To dispose of human waste, there were two areas that could be used as general toilets with deep pits on the edge of the forest near the monastery (senior monastic community members had private pits beside their residences that they would keep clean using ash). When the general pits were filled, they would be covered with ash and mud, and new areas would be selected, with great care shown to ensure that no unseen residents were present there and also to maintain a distance from areas of human habitation and food preparation. With the opening of west Sikkim for tourism in the 1980s, new toilet facilities were installed at the monastery. These are still situated in places that will not attract ritual pollution and maintain hygiene.

Historically, there was not a lot of non-biodegradable waste, since food packaging was made from bamboo and would be re-used or recycled. It was only with the entrance of more imported goods after Sikkim became part of India in 1975 that plastic became more common. The SBM recommended new forms of waste sorting for other nonhuman waste. Plastic has been banned on monastic premises, wet waste is used for animal fodder in nearby villages, and dry waste is collected by the Municipal Garbage Collections trucks from the nearby town of Gezing. This integration of Pemayangtse into state-led initiatives has been important for promoting conservation on a local level, but the monastery has historically also practised its own local ethics, underpinned by a concern for other seen and unseen inhabitants of the area, which also compels the community at Pemayangtse to remain vigilant in their care for the sacred habitat.

## Conclusion

Sikkimese Buddhist communities are currently compelled to consider the impact of their lifeways on the sacred habitat they live in because of the rapid changes they see around them: new dams being installed along Sikkim's sacred rivers; the lack of coordinated waste management in Sikkim's growing urban centres; the impact of climate change on Sikkim's crops and on human-animal relations. The deleterious impacts of these events – landslides, cloudbursts, and crop blight – are often interpreted through the lens of interdimensional imbalance. The guardian deities have been angered or driven away (Gergan, 2017; Bhutia, 2021). Even at Pemayangtse itself, in 2020, there was a heavy rain that resulted in a huge landslide that damaged the foundations of the monastic kitchen. These events act as important catalysts to encourage people to organise and support conservation efforts, and to be more aware of the environment in general as another alternate way to understand the sacred habitat.

This organisation has needed creativity and flexibility (Gergan, 2017). Historically, the community at Pemayangtse relied heavily on resources from the world around them to support their religious lives. Currently, a reorientation is underway where historical examples of caring for the environment – including traditions related to forest, waste, and water management – can be reoriented to reflect ecocentric rather than anthropocentric concerns. Certain rituals such as *Sadaglunyen* have not yet been explicitly invoked in conservation initiatives, but can provide important inspiration to communities. The participation of Pemayangtse lamas in government and local conservation initiatives encourages other local community members to support these initiatives, and demonstrates how historical concepts of caring for the sacred habitat can be extended into the present. There is currently an active discussion about how conservation can work together with development, since tourism is an important source of revenue for human communities in the Pemayangtse region. Whether or not Sikkimese Buddhists can retain their environmental awareness in the face of economic pressures is a question that is still open for debate. It appears that the coordinated conservation efforts drawing on Buddhist concepts are just beginning in Sikkim, and it will be generative to monitor these efforts to see how they respond to broader environmental changes in coming years.

## Acknowledgements

My thanks to the seen and unseen beings of west Sikkim, the sangha of Pemayangtse, and the editors for invaluable feedback. This research was generously supported by an ACLS/Robert H. N. Ho Research Fellowship in Buddhist Studies.

## Notes

1 The terms "monastery" (Classical Tibetan: *dgon pa*) and "monk" (Classical Tibetan: *tra pa*) here need some explanation. Historically, Pemayangtse housed both celibate, ordained monastics (Classical Tibetan: *dge slong*) and married Tantric practitioners (Classical Tibetan: *sngags pa*). In contemporary Sikkim, most of the teachers, ritual specialists, and meditation practitioners at Pemayangtse are not celibate. However, the English terms "monastery" and "monk" are used in Sikkim to refer to the institution and the people who care for it. For more on Pemayangtse in contemporary times, see Vandenhelsken (2003).

2 This chapter discusses Sikkimese Buddhist traditions. Sikkim is a multiethnic, multicultural, and multilingual state. Many different ethnic groups participate in Buddhist practices. A commonality between different Buddhist institutions in Sikkim and elsewhere in the Himalayas is the use of Classical Tibetan as the canonical language. For that reason, in this chapter, I will provide the Wylie transliteration for all Classical Tibetan terms.

3 A comprehensive anthropological overview of Sikkimese Buddhism is found in Balikci (2008), which is based on research from North Sikkim.

4 There are four well-known traditions of Tibetan and Himalayan Buddhism: Nyingma (Classical Tibetan: *Rnying ma*), Kagyu (Classical Tibetan: *Bka' rgyud*), Sakya (Classical Tibetan: *Sa skya*), and Geluk (*Dge lugs*). These traditions have been established by different teachers and promote distinctive historical and ritual lineages of Vajrayana, or Tantric, Buddhism. They have been promulgated throughout the Himalayas and Inner Asia, and are distinct from, though connected to, Vajrayana Buddhist traditions practised by the Newar community in Nepal.

5 An alternative translation to this text is found in Balikci (2008, p. 93).

## References

Acharya, A., & Ormsby, A. (2017). The cultural politics of sacred groves: A case study of *Devithans* in Sikkim, India. *Conservation and Society, 1*(2), 232–242. https://doi.org/10.4103/cs.cs_14_29

Allison, E. (2014). Waste and worldviews: Garbage and pollution challenges in Bhutan. *Journal for the Study of Religion, Nature and Culture, 8*(4), 405–428. https://doi.org/10.1558/jsrnc.v8i4.25050

Allison, E. (2019). The reincarnation of waste: A case study of spiritual ecology activism for household solid waste management: The Samdrup Jongkhar initiative of rural Bhutan. *Religions, 10*(9), 1–19. https://doi.org/10.3390/rel10090514

Arora, V. (2009a). 'They are all set to dam(n) our future': Contested development through hydel power in democratic Sikkim. *Sociological Bulletin, 58*(1), 94–114. https://doi.org/10.1177/0038022920090106

Arora, V. (2009b). Framing the image of Sikkim. *Visual Studies, 24*(1), 54–64. https://doi.org/10.1080/14725860902732710

Balikci, A. (2008). *Lamas, Shamans and ancestors: Village religion in Sikkim.* Leiden: Brill.

Balikci-Denjongpa, A. (2001). Kanchendzonga: Secular and Buddhist perceptions of the Mountain Deity of Sikkim among the Lhopos. *Bulletin of Tibetology, 38*(2), 5–37. https://www.repository.cam.ac.uk/bitstream/handle/1810/242513/bot_2002_02_01.pdf?sequence=1.

Bhutia, K. D. (2021). Purifying multispecies relations in the valley of abundance: The *Riwo Sangcho* ritual as environmental history and ethics in Sikkim. *MAVCOR, 5*(2), 1–25.

Cantwell, C. (2001). Reflections on ecological ethics and the Tibetan Earth ritual. *The Eastern Buddhist, 33*(1), 113–119.

Chettri, M. (2017). Ethnic environmentalism in the Eastern Himalaya. *Economic and Political Weekly, 52*(46), 34–40.

Elverskog, J. (2020). *The footsteps of the Buddha*. Philadelphia: University of Pennsylvania Press.

Gagné, K. (2018). *Caring for glaciers: Land, animals, and humanity in the Himalaya*. Seattle: University of Washington Press.

Gergan, M. (2017). Living with earthquakes and angry deities at the Himalayan Borderlands. *Annals of the American Association of Geographers, 107*(2), 490–498. https://doi.org/10.1080/24694452.2016.1209103

Gergan, M. (2020). Disastrous hydropower, uneven regional development, and decolonization in India's Eastern Himalayan Borderlands. *Political Geography, 80*, 102175. https://doi.org/10.1016/j.polgeo.2020.102175

Harris, I. (1991). How environmentalist is Buddhism? *Religion, 22*(2), 101–114. https://doi.org/10.1016/0048-721X(91)90058-X

Huber, T. (1997). Green Tibetans: A brief social history. In F. Korom (Ed.), *Tibetan culture in the diaspora* (pp. 103–119). Vienna: Verlag der Österreichischen Akademie der Wissenschaften.

Karmapa, O.T. D. (2011). Walking the path of environmental Buddhism through compassion and emptiness. *Conservation Biology, 25*(6), 1094–1097. https://doi.org/10.1111/j.1523-1739.2011.01765.x

Kaza, S., & Kraft, K. (Eds.). (2000). *Dharma rain*. Boulder, CO: Shambhala.

Knapp, K. R. (2012). Buddhist sacred natural sites conservation: A meeting ground between international and local. In K. Ura & D. Chophel (Eds.), *Buddhism without borders* (pp. 122–135). Thimphu: Centre for Bhutan Studies.

Lepcha, K. (2020). The Teesta hydro power projects: A historical analysis of the protect movement in North Sikkim. Ph.D. dissertation, Sikkim University.

Lha btsun nam mkha' 'jigs med. (2000). *Rig 'dzin srog sgrub*. Delhi: Chos spyod dpar skrun khang.

Li, J., et al. (2013). The role of Tibetan Buddhist monasteries in Snow Leopard conservation. *Conservation Biology, 28*(1), 87–94. https://doi.org/10.1111/cobi.12135.

McDuie-Ra, D., & Chettri, M. (2020). Concreting the frontier: Modernity and its entanglements in Sikkim, India. *Political Geography, 76*, 102089. https://doi.org/10.1016/j.polgeo.2019.102089.

Meek, D., & Anderson, C. R. (2020). Scale and the politics of the organic transition in Sikkim, India. *Agroecology and Sustainable Food Systems, 44*(5), 653–672. https//doi.org/10.1080/21683565.2019.1701171

Mullard, S. (2011). *Opening the hidden land*. Leiden: Brill.

Ramakrishnan, P. S. (2008). Demojong: A sacred landscape within the Sikkimese Himalayan landscape, India. In J. Mallarach (Ed.), *Protected landscapes and cultural and spiritual values* (pp. 159–169, Vol. 2). Heidelberg: IUCN, GTZ and Obra Social de Caixa Catalunya.

Raman, P. (2020). The politics of visibility in urban sanitation: Bureaucratic coordination and the Swachh Bharat Mission in Tamil Nadu, India. Ph.D. dissertation, Massachusetts Institute of Technology.

Roder, W., Dorji, K., & Gratzer, G. (2003). Nutrient flow from the forest – Source of life for traditional Bhutanese agriculture. *Austrian Journal of Forest Science, 1*, S65–S72.

Tan, G. (2014). An ecology of religiosity: Re-emphasizing relationships between humans and nonhumans. *Journal for the Study of Religion, Nature and Culture, 8*(3), 307–328. https://doi.org/10.1558/jsrnc.v8i3.307

Vandenhelsken, M. (2003). Secularism and the Buddhist monastery of Pemayangste. *Bulletin of Tibetology, 39*(1), 55–73. http://himalaya.socanth.cam.ac.uk/collections/journals/bot/pdf/bot_2003_01_05.pdf

Vandenhelsken, M. (2011). The enactment of Tribal Unity at the Periphery of India: The political role of a new form of the Panglhabsol Buddhist Ritual in Sikkim. *European Bulletin of Himalayan Research, 38*, 81–118. http://himalaya.socanth.cam.ac.uk/collections/journals/ebhr/pdf/EBHR_38_04.pdf

Wagh, S. (2017). The Kanchendzonga sacred landscape: Nature-Culture linkages and local associations with place. *Journal of World Heritage Studies*, Special Issue, 56–61. DOI: 10.15068/00148450

Yeh, E. (2014). The rise and fall of the Green Tibetan: Contingent collaborations and the Vicissitudes of harmony. In E. Yeh & C. Coggins (Eds.), *Mapping Shangrila: Contested landscapes in the Sino-Tibetan Borderlands* (pp. 255–278). Seattle: University of Washington Press.

# 3

# OF THE WHEEL OF LIFE AND GUARDIAN DEITIES

## How Buddhism Shapes the Conservation Discourse in the Indian Trans-Himalayas

*Radhika Kothari, Nisa Khatoon, Padma Dolker and Konchok Zangpo*

### Introduction

The wheel of life is at the heart of Buddhist philosophy. It emphasises the concept of dependent co-rising in conjunction with existence's relative truth, the nature of all phenomena, the web of interdependencies and how each affects the other in a non-linear and casual way. The wheel of life is also represented by both human and non-human beings that include animals, birds, and spirits and deities. From Buddha's perspective, the path to awakening takes several lifetimes. Therefore, it is fundamental to follow the '*dharma*' of cultivating kindness and compassion towards all other sentient beings. The wheel of life presented by the Buddha himself is a common sight at Buddhist monasteries across the Himalayas. It is often a symbolic representation of a moralist compass to uphold meritorious deeds and optimise the treasured human birth.

This ability of Buddhism to view the world of a non-linear causality, where phenomena affect each other in a reciprocal interaction, is similar to the messages now echoed in deep ecology and conservation movements (Macy, 1991). With the growing understanding of visible and invisible impacts of human actions on the natural systems and the long-term consequences, the lens of interdependencies, interconnections and the principles of cause and effect is often at play. This common hermeneutics of Buddhism and the working of natural systems are now considered a crucial bridge to engage Buddhist communities in conservation in the Himalayas and South Asia. It resonates with moralistic responsibilities and intertwines with the day-to-day behavioural practices and interactions with the natural environment (Chakraborty and Chakraborti, 2011).

Similarly, in the northernmost part of India, in the Ladakh Trans-Himalayas, we look at the role indigenous ontologies coupled with Buddhism play in shaping the conservation message and discourse. We specifically examine the case

DOI: 10.4324/9781003246190-5

of the wetlands of Changthang and the flagship species – black-necked cranes, the efforts of the local community, the local monastic group and WWF-India to support conservation. We also examine the fluidity of religio-cultural elements in a radically changing world. We conclude that while people and their religious rationale may change with more interactions with the modern world, there are lessons in Buddhism, especially the 'middle way', that can be enrolled in conservation planning and strategy.

## Setting the Context

### *The Land*

Ladakh forms the north-western end of the Indian Trans-Himalayas, between the Tibetan Plateau and the Karakoram. The region has an altitudinal gradient ranging from 3,000 to over 7,000 m above sea level (asl) and is characterised by rocky terrain, rugged mountain valleys and cold desert conditions, regardless of which, Ladakh is a biogeographic zone of importance. It holds vital glacial and wetland ecosystems that support several endangered mammals like the rare snow leopard, Tibetan wolf and Tibetan gazelle. It also hosts several migratory birds like the black-necked crane, ruddy shelduck, bar-headed goose and numerous species of plants, altogether creating a unique assemblage of high-altitude flora and fauna (Namgail et al., 2010).

### *The People*

Despite the limited resources, the altitude and tundra weather, people have inhabited the region for 4,000 years (Rabgias, 1977). In the past, Leh city of Ladakh had been a vibrant business hub on the silk route, thus showcasing both Tibetan and South Asian influences. While Buddhism has moulded human temperaments and social behaviours, Islam has also coexisted with a similar cultural binding of resilience to nature's severities. People have developed an exclusive ecological interface with a land-based economy model, seasonal subsistence farming, a set of values, customs, rituals and practices to survive some of the harshest conditions on Earth (Norberg-Hodge, 1993; Bhasin, 2012).

### *The Sacred Complex*

The religio-cultural elements or the sacred complex of Ladakh are incomplete without references to *Bonchos* (Bon religion). *Bonchos* is generally believed to be animistic religion based on nature worship and constitutes sorcery, mysticism and non-human realms (Handa, 2001). People of Ladakh believe in spirits, guardian deities and protective entities that live in the mountains, streams, valleys and high passes to govern daily life and actions. Any negligence or pollution of the natural world can anguish these spirits and bring wrath to the household or the whole community (Bhasin, 2005). Wildlife, too, has a meaningful place in the life of Ladakhis. For example, the Asiatic Ibex *Capra ibex sibrica* is still considered a symbol of good luck, prosperity and often seen in monastic dances. Throughout Ladakh, there are horns of ungulates piled and displayed near houses as an offering to various spirits and deities to maintain crops, grazing conditions and prevent disasters (see Figure 3.1) (Fox et al., 1994).

And, as in different indigenous traditions, Ladakhi oral and folk literatures, including songs and music, are filled with human interactions with the natural world and instil a sense of oneness to withstand the intense climatic conditions.

The advent of Buddhism and its philosophy of interconnectedness and cyclic existences have further fortified the narrative of deep-seated respect for different forms of life and nature. It is common to see *tarchok* (prayer

**FIGURE 3.1** A pile of ungulate horns with prayer flags in a Ladakhi village. Photo credits: Radhika Kothari.

flags) represented by the elemental colours: Red (fire), Yellow (Earth), White (wind), Blue (sky) and Green (water) scripted with sacred Buddhist *mantra* (words) near meadows, mountain passes, rivers, lakes and other natural elements. It is a symbol of sanctity, therefore important not to pollute. Both *bon-chos* and Buddhism have influenced each other, thereby completing the holy complex and the Ladakhi worldview that are practised even today (Golomb, 1995). There is a thorough acknowledgement of the more practical needs of daily life with the background of a cosmological matrix of the natural and supernatural worlds to maintain steady relations with guardian deities *(lha)*, spirits *(lha srog)*, the Buddha *(sangyas)* and other *Boddhisattva* (see Figure 3.2) (Dinnerstein, 2013).

## The Sacred Bird

The grounding of religio-cultural elements has also found its way in the case of the black-necked crane (*Grus nigricollis*) in Ladakh. Locally known as '*Cha thung thung*', the black-necked crane is believed to have carried the re-incarnations of Dalai Lama on its back, therefore highly venerated across the region. The cranes are also considered a sign of good omen, good luck and often seen accompanying deities (see Figure 3.3). They are also found in *thangka* (textile paintings) and wall paintings across monasteries in Ladakh. There are special mentions of the black-necked cranes even in the indigenous folklore, including the *chartses* – a dance

**FIGURE 3.2** An old chorten (stupa) overlooking a frozen Tso Moriri lake. Photo credits: Radhika Kothari.

form, where dancers perform the movements imitating the birds and the popular poem for children by Gyan Jamyang Gyalson (2007):

*Cha daa thung thung karmo*
*yid tuu yong vaa cha jik*
*Changthang chikpo man pa*
*najye khampa skyotdang*

Dear bird, black-necked crane,
remember this bird,
please visit our home as well
not only Changthang.

**FIGURE 3.3** A wall-painting showcasing mi-tsering interacting with black-necked cranes. Photo credits: Radhika Kothari.

Another folkloric song of a grandmother requesting the black-necked cranes to show the courtship dance is well-known locally.

*cha thung thung karmo tses shig tong;*
*aba leb pa khara tangin,*
*ama leb pa toyos tangin,*
*cha thung thung karmo tses shig tong.*

Oh, black-necked crane, please dance;
when my father comes, I'll give you candies,
when my mother comes, I'll give you roasted barley,
Oh, black-necked crane, please dance.

The Changpa nomads of Samad believe that the black-necked cranes and the ruddy shelducks *Tadorna ferruginea* are like *Lama* (monks). The birds circumambulate the sacred mountains of *chan-do-ar, ril-cha-pun sum, charim* and the Thukje Monastery at Tso Kar while arriving and departing on auspicious days like the 8th and 15th days of the month in the Tibetan Calendar.

## Religion and Conservation in Ladakh

Religion and indigenous ontologies have long shaped the human-nature relationship in Ladakh. It has also become a handy tool in the conservation practitioners' toolkit in the region. While religio-cultural values align with the conservation agenda, there are also caveats to the narrative like the disapproving attitude towards predators like the snow leopards and Tibetan wolves. This has been tackled by a local organisation – Snow Leopard Conservancy India Trust – which has made pioneering efforts to recourse the human-predator interactions in Ladakh using Buddhist ideals of tolerance and compassion (Jackson et al., 2003). Another example is of neutralising *shadongs* (wolf-traps) into a *Stupa* (Buddhist shrine) or *Lhato* (shrine of a local guardian deity) to promote harmony and coexistence. Religion and communal values to engage communities are considered valuable tools for achieving conservation targets in Ladakh (Bhatia et al., 2017).

### *The Wetlands of Changthang and the Conservation of Black-Necked Crane*

The eastern Ladakh landscape, also known as Changthang, is marked with wetlands that have long nourished the cold desert region. Rigul Tso and Startsapuk Tso form the Tso Kar wetland complex; Hanle, Chusul and Tso Moriri wetlands are essential for water, fodder, fuelwood, ground-water discharge and other critical ecosystem services. These wetlands are also home to many migratory birds, including the flagship species – black-necked cranes (see Figure 3.4). Black-necked cranes, categorised as vulnerable on the IUCN red list, are also a Schedule I species under the Wildlife (Protection) Act of 1972 in India. These birds hold the elusive

**FIGURE 3.4** A pair of black-necked cranes (Cha thung thung) near Tso Kar wetland. Photo credits: Radhika Kothari.

record of being one of the last cranes discovered by the scientific community, adding to its age-old charm. The very first sighting was at Tso Kar in 1919 by the naturalist F. Ludlow. Adding to the clandestine appeal is also the region's inaccessibility, making black-necked cranes ever more mysterious (Chandan et al., 2005). The black-necked cranes arrive in Ladakh in late March or early April and migrate by late October. The wetlands of Ladakh are considered to be among the most important breeding grounds for these birds (Pfister, 2001).

WWF-India initiated a conservation project called the 'Conservation of High Altitude Wetlands of Ladakh' in 1999. The primary goal was to develop a strategy and action plan for the conservation of high-altitude wetlands of Ladakh. These fragile wetland ecosystems were among the most crucial breeding grounds for a range of high-altitude avifauna, including the black-necked crane. The project was in response to the pressures that the wetlands had begun to face.

Changthang was opened to tourism in 1994, directly leading to a sudden surge of interest in the wetlands. The most isolated and remote corners of Changthang saw a gradual increase of tourists, trekkers, campers and pop-up joints, all of which were new to the ecosystem and the people. The lack of adequate infrastructure, haphazard development, road construction and increasing tourism prospects in the summer months (June–September) coincided with the season of the migratory birds, making it imperative to conserve (Mishra and Humbert-Droz, 1998).

The project pioneered many firsts in terms of black-necked crane studies in the region. The primary objectives were to establish the population status, identify specific habitats, breeding productivity, nesting and feeding sites of black-necked cranes in Changthang. An additional objective was to formulate strategies with the local stakeholders for long-term conservation and monitoring.

The initial scientific and socio-economic studies within Changthang highlighted the potential threats, making the stakeholders' involvement that ranged from local community members, religious leaders, security forces, political, administrative and tourism stakeholders indispensable. The engagements initiated dialogue and coordinated actions that were the basis of the site-based interventions. The priority of including the understanding and knowledge of local communities, who held a wealth of customs, beliefs and practices associated with the migratory birds and the wetlands, turned out to be a catalytic collaboration (Gujja, 2005). The programme has had far-reaching impacts on community conservation initiatives for black-necked cranes and the wetlands of Changthang.

The intervention identified previously unknown sites creating a much-needed database on the black-necked cranes in Ladakh. It also recorded some of the highest ever sightings of the bird in the region. The extensive study also reported an increase in the population of black-necked cranes, attributed mainly to new breeding sites and the decline in breeding, attributed to increasing human activities, and secondary impacts, like direct killing by feral dogs and unplanned developmental activities in the region (Chandan et al., 2005, 2007).

The community engagement and collaboration increased awareness for further protection of the sacred species and wetlands. People were made aware of the lurking threats and mounting pressures on the habitats of the black-necked cranes and other migratory birds. In an immediate response to the matter, one of Changpa groups at Korzok and its monastery inhabiting the Tso Moriri wetlands was spearheading the cause of conservation with WWF-India. The forming of the Tso Moriri Conservation Trust was a prime outcome. The *Chagzot* (Head Lama) of the Korzok monastery and the local school Principal presided as the Chairperson and Vice President. Given *Chagzot's* status in the community, his support for conservation incredibly sustained the cause. The Trust adopted several conservation measures to protect the wetland complex and undertook various education and awareness programmes focusing on high-altitude wetlands with WWF-India, also partaking in the scientific studies. The Trust also encouraged livelihoods that were less impactful to the fragile ecosystems, like local handicrafts and carpet weaving.

Subsequently, the Tso Moriri Wetland Complex was successfully selected as a Ramsar Site under the tutelage of the monastic leaders, local community and concerned Government Departments in the administration and only recently, in January 2021, Tso Kar was also declared as a Ramsar Site. Tso Moriri, which holds significance in the Korzok community's cultural and spiritual realm, was pledged as a 'Sacred Gift' for a living planet by the *Chagzot*, the elected councillor, and WWF-India. This only reinforced the local conservation efforts for critical habitats and migratory birds (WWF-India, 2001).

Since the wetlands came under pressure due to increasing summer activity involving tourism, efforts were made to implement corrective measures at Tso Kar and Tso Moriri. Ecotourism standards were introduced, emphasising the need for low-impact tourism in such fragile ecosystems. Regulations, including designation of camping sites, vehicle parking, garbage collection and disposal, sanitation system and tourist fee collection, were developed and implemented by the communities. Scattered camping near the lake, bird habitats and central areas within the wetlands are disallowed even today.

The project led to greater awareness among various stakeholders, considerably enhancing the scientific understanding, improving the management of wetlands and underlining the long-term impacts of climate change in the region with a particular reference to the avifauna population. The project also led to reinvigorating the sacred status of black-necked cranes and wetlands of Changthang.

Despite the efforts of WWF-India and the monastic group, today, unplanned infrastructural development, growing military presence and burgeoning tourism industry continue to threaten the wetlands and avifauna in Changthang (Prins and Namgail, 2017).

## A Changing Landscape

In recent years, Ladakh has made rapid transitions with more contact with mainstream development. Urbanisation, unplanned expansions, mass tourism and inadequate policies fail to safeguard traditional ways of life and further threaten the very foundations. With socio-economic changes, the cultural fabric is also under dire stress. In the context of such metamorphosis, many elders believe that urban centres, commercialisation and desire for money anger the *Lha*. Such resentments echoed prominently during the 2010 flash floods that caused widespread destruction in Ladakh. Believed to be prophesied by *Guru Padmasambhava*, the floods were a sign of retribution from '*Lha Lu*', the guardian deity of water, for increasing pollution, declining moral values, disturbances of the elements and societal demerits (Butcher, 2013).

Furthermore, the ecological alterations due to climatic change are adding strain to the whole dynamics. The patterns of warming winters, changing temperatures and precipitation have led most local herders to believe that the population of herbivores like the Tibetan wild ass *Equus kiang* have increased and are drying grasslands, thus causing fierce competition with their livestock. Although stringent rules forbid people from killing wildlife, the vulnerability invokes a highly negative attitude towards wildlife that mar the mostly relaxed human-nature relationship (Interview, October 2020).

While wetlands and certain avifauna continue to be considered sacred, they may not necessarily reflect people's interaction with nature in changing times. Sabharwal (2016) calls out the conservationist's narrative of an 'imagined ecological balance' before the economic surge in Changthang as overly simplified. The author also emphasises the need for a multidimensional view of various dynamics at play in the region, also highlighting the fluidity of religio-cultural systems in the passages of time.

## The Middle Way Forward

There is undeniable evidence that we are confronting the 'converging extinction crisis' of degrading natural and cultural systems (Harmon, 1996). Still, the amalgam path to conserving ecosystems and protecting people's cultural diversity and well-being is long and steep. In regions like Ladakh – where humans and natural systems are so intimately tied, lines often blurred, worldviews far more composite, contextual and non-linear, make it even more mystifying.

Engaging local communities where ontologies are well-aligned to global priorities is fundamental to the conservation discourse. Yet, there is also danger in isolating individual cultural practices that align while ignoring other cultural, socio-economic complexities, multiple and sometimes mutually incompatible values of nature that vary over time and space (Schneider, 2018). Faith-based conservation most certainly adds a vital tool in the conservationist's toolkit. However, the degree of engagement has many caveats and is a continuous work-in-progress. Finding a common ground that serves both ecosystems and communities requires a thorough understanding of ecological and religio-cultural thresholds. Moreover, it requires keeping biases or rigid sectoral agendas aside to fully realise place-based definitions of well-being.

The inherent understanding within the schools of Buddhism of dependent co-rising, the past actions to present consequences, the relation of self to society and nature can constitute the philosophical and moralist grounding for community conservation efforts (Macy, 1991). It is the Buddhist principle of '*Madhyamaka*' or the middle way that offers a more practical tool for the conservation practitioner. The middle way is the creative potential found in between two opposites. In other words, a middle way is the way of reconciliation of differences and aversion of extremities that helps maintain a certain equilibrium, a path of conservation that needs nurturing in the 21st century. It helps illuminate the complexity of the human-nature relationship in the emerging world and offers possibilities for the entire spectrum of stakeholders involved. It also helps to withdraw either ends in favour of the well-being of both human and natural systems. As aptly articulated below:

> Everything exists; this is one extreme.
>
> Everything does not exist; this is the other extreme.
>
> Avoiding both extremes and following the middle way, is the path of the Tathagata (Buddha)
>
> *Samyutta Nikaya*
>
> *(Adapted from Kalupahana, 1979).*

## Conclusion

Buddhism, as many religions, offers significant lessons of coexistence, kindness and compassion. More importantly, the emphasis on cycles of death and rebirth, and dependent co-rising thereof, has particular orientation in the consciousness of its followers and believers. In Ladakh, Buddhism and indigenous ontologies have been partially captured, whereas the Islamic viewpoints are yet to be fully understood. Nonetheless, the religio-cultural complex has favoured the conservation dialogue with evident impactful results.

Indeed, conservation has a rather arduous way forward with growing complications of a homogenised economic worldview. But, for conservation to forge alliances with communities for restoration and protection of some of the last pockets of natural and cultural diversity, new strategies and an empathetic understanding of the philosophical, moralistic and religious worldview of people are pivotal and the need of the hour.

## References

Bhasin, V. (2005). Ecology and health: A study among tribals of Ladakh. *Studies of Tribes and Tribals, 3*(1), 1–13.

Bhasin, V. (2012). Life on an edge among the Changpas of Changthang, Ladakh. *Journal of Biodiversity, 3*(2), 85–129.

Bhatia, S., Redpath, S. M., Suryawanshi, K., & Mishra, C. (2017). The relationship between religion and attitudes toward large carnivores in northern India? *Human Dimensions of Wildlife, 22*(1), 30–42.

Butcher, A. (2013). Keeping the faith: Divine protection and flood prevention in modern Buddhist Ladakh. *Worldviews: Global Religions, Culture, and Ecology, 17*(2), 103–114.

Chakraborty, R., & Chakraborti, C. (2011). Relevance of deep ecology and Aphorism for Sikkim Himalayan eco-system. In A. Balikci-Denjongpa & A. McKay (Eds.), *Buddhist Himalaya: Studies in religion, history and culture* (pp. 123–133). Sikkim: Namgyal Institute of Tibetology.

Chandan, P., Chatterjee, A., Gautam, P., Seth, C. M., Takpa, J., Haq, S., ... Vidya, S. (2005). Black-necked crane -status, breeding productivity and conservation in Ladakh, India 2000–2004. WWF-India and Department of Wildlife Protection, Government of Jammu and Kashmir.

Chandan, P., Chatterjee, A., & Gautam, P. (2007). The management planning of Himalayan high altitude wetlands. A case study of Tso Moriri and Tso Kar Wetlands in Ladakh, India. *Proceedings of Taal*, 1446–1452.

Dinnerstein, N. (2013). Songs, cultural representation and hybridity in Ladakh. *HIMALAYA, the Journal of the Association for Nepal and Himalayan Studies, 32*(1), 16.

Fox, J. L., Nurbu, C., Bhatt, S., & Chandola, A. (1994). Wildlife conservation and land-use changes in the Trans-Himalayan region of Ladakh, India. *Mountain Research and Development, 14* (1), 39–60.

Golomb, E. (1995). Oracles in Ladakh: A personal experience. In: Adler, L.L.; Mukherjee, B.R (Eds), *Spirit versus scalpel: Traditional healing and modern psychotherapy* (pp. 59–75). Westport/CT: Bergin & Garve/Greenwood Publishing Group.

Gujja, B. (2005). WWF International's regional approach to conserving high-altitude wetlands and lakes in the Himalayas. *Mountain Research and Development*, 25 (1), 76–79.

Gyalson J. G. (2007). Bog-Rig Reader III. State Board of School Education. *State Board of School Education*. J&K Government. Srinagar (pp. 18–19).

Handa, O. C. (2001). *Buddhist Western Himalaya: Part I. A politico-religious history.*

Harmon, D. (1996). The converging extinction crisis: Defining terms and understanding trends in the loss of biological and cultural diversity. In *Colloquium losing species,*

*languages, and stories: Linking cultural and environmental change in the Binational Southwest.* Tucson: Arizona-Sonora Desert Museum.
Jackson, R., Wangchuk, R., & Dadul, J. (2003). Local people's attitudes toward wildlife conservation in the Hemis National Park with special reference to the conservation of large predators. *Field Series Document,* 7, 1–29.
Kalupahana, D. J. (1979). The early Buddhist notion of the middle path. *The Eastern Buddhist, 12*(1), 30–48.
Macy, J. (1991). *Mutual causality in Buddhism and general systems theory: The dharma of natural systems.* Suny Press. State University of New York Press, Albany, USA.
Mishra, C., & Humbert-Droz, B. (1998). Avifaunal survey of Tsomoriri Lake and adjoining Nuro Sumdo Wetland in Ladakh, Indian trans-Himalaya. *Young, 56*, 14.
Namgail, T., Van Wieren, S. E., & Prins, H. H. (2010, August). Pashmina production and socio-economic changes in the Indian Changthang: Implications for natural resource management. In *Natural Resources Forum* (Vol. 34, No. 3, pp. 222–230). Oxford: Blackwell Publishing Ltd.
Norberg-Hodge, H. (1993). Learning from Ladakh. *The Trumpeter, 10*(4), 1–5.
Pfister, O. T. T. O. (2001). Birds recorded during visits to Ladakh, India, from 1994 to 1997. *Forktail 17*, 81–90.
Prins, H. H., & Namgail, T. (Eds.). (2017). *Bird migration across the Himalayas: Wetland functioning amidst mountains and glaciers.* Cambridge: Cambridge University Press.
Rabgias, T. (1977). Salient features of the history of Ladakh. *Voice of the Himalaya, 1*, 1–4.
Sabharwal, A. (2016). Contested affluence: Cultural politics of Pashmina wealth and wildlife conservation in Ladakh. In Wood, D.C. (Ed.), *The economics of ecology, exchange, and adaptation: Anthropological explorations (36).* Emerald Group Publishing Limited, Bingley UK (pp. 77–113).
Schneider, H. (2018). What role for culture in conservation? *Oryx, 52*(2), 199–200.
WWF-India. (2001). Sacred gifts for a living planet. *WWF-India Newsletter,* January.

# 4

# THE BORGOU AND ITS SPIRITS

## Nature and the Traditional Faith Practice of the Baatɔnu People of West Africa

*C. Ryan Smith*

***Note on non-English vocabulary in the text:*** Throughout the chapter the reader will encounter names, titles and terms written in Baatɔnum, the language of the Baatɔnu people who are the main subject of this document. The Baatɔnum language is written in standard Latin script (the alphabet of English and Western European languages) with the addition of accents, vowels and consonant combinations drawn from the International Phonetic Alphabet as defined by the Alphabet of Beninese National Languages.[1]

Baatɔnum contains two accents: ` for downward vocalisation, and ˜ for nasalization; and two extra vowels: "ɔ" pronounced as the English open o, and "ɛ" the open e.

Note: "**o**" in Baatɔnum is the English closed o and "**e**" the closed a; "**u**" is the hard closed u, "**i**" the closed e, and "**a**" the English ah. There are also two explosive sounds in Baatɔnum, **kp** and **gb**, and any double letter (aa, ll, etc.) requires the vocal pronunciation of both.

Baatɔnum orthography is prioritised in this publication so as to present the language to an English-speaking audience, certainly for the first time for many readers, and to give credence to the indigenous compositions of local words which are often poorly transliterated into French and English—the languages of the colonial powers from 1897 to 1960—sometimes to the point of incomprehension. A Baatɔnum word which is not a proper noun is generally written in italics or underlined the first time it appears within the chapter, accompanied by its definition or translation.

Otherwise, place names within the study area are written with the official Beninese French spelling ("Borgou," for example) except when specifically chronicling the Nigerian experience ("Borgu," as explained in the text). Ethnic groups and other non-Baatɔnu cultural appellations are denoted by commonly

DOI: 10.4324/9781003246190-6

recognized international terms ("Fulani," "Yoruba," "Shango," etc.). Finally, the scientific names of listed flora and fauna are italicised in Latin, as is the practice.

## A tree without roots is a dead tree

### *Baatɔnu Proverb*

During the month of November in the Borgou there is a short interlude between the end of the rainy season and the beginning of the dry season when time stands still.

The rains, which fall from May through October and produce verdant scenery under cool overcast skies, fill the creeks and rivers to bursting by early September. Frogs, crabs and fish emerge from their slumber within the mudflats, having been dried out for months, and provide food for birds and other wildlife. Fishing begins again among local human residents. Cattle herders return to their distant homes, no longer required to congregate around scarce watering holes. Vegetation conceals everything.

The Borgou rainy season commences with breath-taking fury. Tempests excite merciless winds from horizon to horizon, reminders that the hurricanes which strike fear in the heart of the so-called New World have their meteorological genesis over the Borgou. The sky is close at hand with steely grey tidal waves of precipitation thunderously rolling in from the east before pouring down upon the trembling land below. The Borgou night is filled with a chorus of chirps and croaks emanating from newly teeming wetlands, and the Milky Way is magnificent in its splendour stretching across the nocturnal vault—along with appearances by both the Southern Cross and the North Star—following a cleansing wash of the atmosphere earlier in the day.

These six months of refreshment are sufficient to nourish the Borgou wilderness and rejuvenate the annual cycle that animals and plants follow in this part of West Africa. Farmers are busy in their fields during the rainy season constantly looking to the skies to ascertain if water from the heavens will be forthcoming on a particular day. Simple activities like washing clothes or travelling to a neighbouring village are dependent on the continuous threat of rain.

Then, like whiplash, the dry season appears out of nowhere and is fully installed by December. It is as parched as the rainy season is soaked, with crisp nights, dehydrated noontime breezes, blinding sunlight, all-encroaching dust, and, a little later, oppressive heat. The two seasons could not be more different.

November, by contrast, is when the Borgou pauses to take a deep breath, to reflect upon the year half completed and that which is to come. The grasses are at their highest, drying out and in flower, with tiny, fragile blossoms affixed to billowing heads. Some trees are also coming into bloom, announcing six dramatic months of arboreal reproduction among various species. Fruits and seeds will soon be ready. The sun is hot at midday but not unbearable, as it will be by April, and evenings are comfortable. Creation is free from the charred detritus that will become ubiquitous following the man-made bushfires of February and March.

The tenor of the moment is calm. Nature is resting during this brief in-between time. One wishes to linger forever.

A November trek through the Borgou back-country gives the pleasant impression of being alone among the luxuriant foliage, but no: Lurking nearby, usually unseen, are countless animals, insects, fellow human beings and, above all, *spirits*.

The spirits exist in several forms and convey the ageless link between ecosystems, people and the divine. Human culture in the Borgou turns upon this very link. It is the triangulation between nature, society and faith that explains the collective tendency to sustainably conserve the landscape in this setting.

Conservation in the Borgou is not a singular act but the ongoing process of heeding precedence, including the use and upkeep of sacred natural portals spanning generations. That process provides legitimacy for customary leaders who are intermediaries among spirits, natural resources and fellow citizens, and who embody centuries of living history. It also ensures continuity of common identity for all residents by virtue of shared norms, values and heritage. The natural environment is a vector of human culture in this case but also a beneficiary of that same culture—a kinship with spirits at its heart.

The aim of this chapter is to familiarise the reader with the Borgou and its people and then to introduce the traditional religion of the region. The first half of the chapter presents the Borgou's geography and ecology followed by a brief historiography of its citizens. It subsequently explores the faith practice of the Baatɔnu people and the symbiotic relationship of that faith with nature. Finally, the chapter examines the societal architecture constructed around Baatɔnu faith and heritage which facilitates consensus building, solidarity, collective action and conservation.

## The Borgou

The territory called Borgou is at once an administrative region in the Republic of Benin, a game reserve in Nigeria (where the spelling is *Borgu*), and a socio-political entity dating from antiquity spanning the frontier between both countries. The latter stretches from the banks of the lower Niger River in the east to the southern edge of the Atacora Range in the west, from the Tchabé Kingdom on the southern edge to Dendi Country in the north. It is further bordered by the Yoruba homelands to the southeast, the Hausa emirates to the northeast, and a constellation of ethnic groups to the northwest and southwest including the Waama, Ditamari, Gourmantché, Lokpa, Pilapila, Anii and others (Figure 4.1)

The Baatɔnu people are the majority human inhabitants of the Borgou. They share the space with minority groups including the Fulani and allied Gando, who are semi-nomadic herders of the Borgou race of cattle (*Bos taurus brachyceros*) (Dehoux & Hounsou-Ve, 1993), and often reside in hamlets on the outskirts of Baatɔnu villages; and with the Boko, Mokolé and other hunting and farming

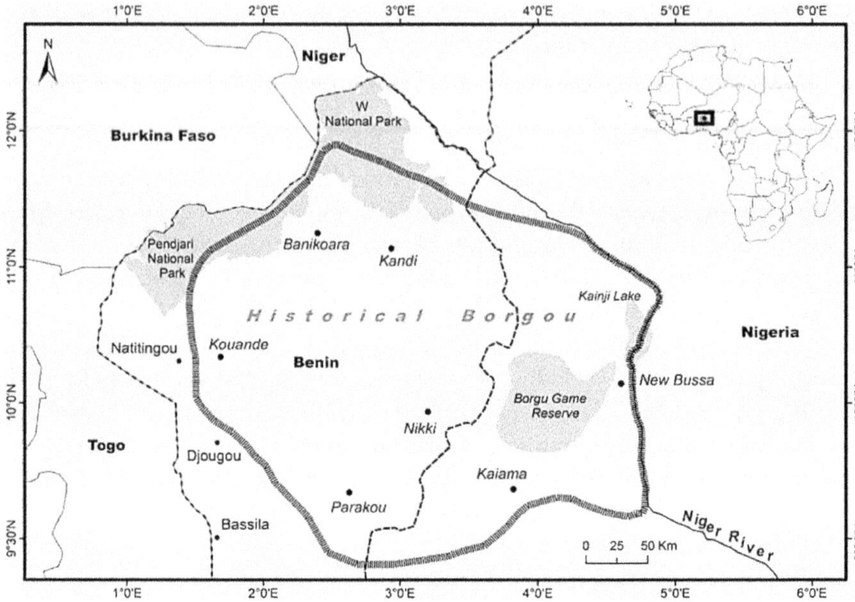


FIGURE 4.1 Map of the historical Borgou with contemporary nation states, selected cities and reserves. Credits: Cartography: K. Saddam Sahagui; Concept: C. Ryan Smith; Data source: IGN Bénin and literature review.

peoples whose ancestors migrated from elsewhere. The linguistic Borgou is heterogeneous but Baatɔnum, the language of the Baatɔmbu (plural of Baatɔnu), nevertheless dominates. The name Borgou is probably a deformation of the Baatɔnum *Baru Wuu*, meaning the realm of the Baatɔnu people.

The physical Borgou does not lend itself well to the grand vistas and dreamy, far-flung horizons sought by foreign tourists in Africa. It spreads out across a leafy undulating terrain, and though there are some hills and rocky outcroppings affording the human eye an occasional peek at the sprawling green topography, mostly the Borgou is bush country: dense savanna woodlands, marshy flood-plains bisected by palmate islands and enclosed by gallery forests, and stretches of grasses as tall as a house.

What the Borgou might lack in postcard views is more than made up for in habitats and hiding places for plants and animals. This region is part of the Guineao-Sudanian ecological zone and offers sanctuary to floral and faunal species otherwise found further south as well as further north, in more humid and arid environments respectively (Arbonnier, 2000; Sinsin *et al.*, 1997). The Borgou is blessed with diverse ecosystems, and is, quite likely, at the heart of a wildlife migratory corridor connecting today's west-central Nigeria to central and northern Benin.

## The Baatɔnu People

The Baatɔnu people are historically hunters. Their society's mythology, visual and oral legacy, and internal ethos are fashioned on images of the hunt, including the region's game. They have developed a complex hierarchy of customary leadership which draws its legitimacy from sacred and other environmental elements in the surrounding countryside.

The Baatɔnu milieu is a composite cultural expanse founded upon the social design of the Indigenous people of the Borgou and supplemented over many generations by foreigners who would eventually integrate the landscape of that group (Lombard, 1965). The ethnographic term *Bariba* is often used in English and French to refer to this amalgamated population (De Moraes Farias, 1998), but Baatɔnu is the name used by the people themselves. Those who identify as Baatɔmbu can be divided into three constituents: indigenous Barubibu, assimilated foreigners, and Wasangari (Bio Bigou, 1995).

Barubibu are of mysterious origin but are thought to have been resident in the Borgou by at least the 7th century CE, though human presence of unknown circumstance has been signalled in the area dating to much earlier (Bagodo, 1993). These inhabitants were joined over time, beginning in antiquity, by several waves of migrants from the ancient Ghana-Mali-Songhaï nexus along the middle Niger River to the northwest (Brégand, 1998). The migrants introduced new cultural practices into indigenous Barubibu society.

Non-Islamic Mandé hunters, for example, brought with them their religion based on animal and tree deities (Stewart, 1993), a faith which was absorbed by the Barubibu. Borgou religious leaders would come to be deemed priests of the Earth (*wuugi* or *wuun yẽro* in Baatɔnum) and a political structure grew up around priests which merged temporal authority with spiritual roles. Even today the concept of traditional governance in the Borgou is buoyed by a legitimacy derived from relations with spirits in nature.

The Wasangari are also of mysterious origin. The word *wasangari* means "foreigner" in the Hausa language and can refer in Baatɔnum to a class of princes as well as a general reference for adventurers and ne'er-do-wells. Legend places the provenance of the Wasangari princes in the Arabia of the Prophet Mohamed. Their leader, Kisra, had refused to abandon his traditional religion for Islam and thereafter sought refuge in Africa (Débourou, 1993). Historians, otherwise, disagree on the precise timing of the arrival of Kisra and his followers in the Borgou, citing the 14th to 18th centuries CE (Bagodo, 1993, *idem.*; Lombard, 1965, *idem.*), but the mid-15th century is a common reference based on contemporary writings from the neighbouring Songhaï Empire.

The descendants of Kisra usurped the political authority of Earth priests and established new thrones of their own. They were successful in doing so and in maintaining their power indefinitely by virtue of marriage and power sharing (Kuba, 1998). The Wasangari took Barubibu women as their wives and their bi-ethnic children became the heirs to the thrones. Earth priests maintained their spiritual authority which they exercised over the Wasangari political elite. Barubibu religious leaders held the critical responsibility of ensuring customary adherence in the choosing and enthroning of Wasangari chiefs, as well as of royal burials, both of utmost consequence for establishing political validity for the foreigners.

In fact, Baatɔnu civilization became an intricate network of intertwined leadership bridging the political and spiritual and shared by the Wasangari and Barubibu.

## Traditional Baatɔnu Religion

While interactions between the citizens of the Borgou and its natural resources are, first and foremost, utilitarian, providing nutrition, fodder for livestock and materials for building construction, medicine, and much more, the spiritual rapport is likewise unmistakable.

*Gusunɔ*—the Celestial Emperor (God in Baatɔnum)—is the sole creator and lord of all, He who reigns from on high. Spirits, who are not God or even gods but who do have powers of perception and implementation that surpass those of regular human beings, live in the same world as people, in trees, water courses and other special places in nature, and can forge enduring relationships with individual Baatɔmbu. Those relationships may be formalised into faith practice and then instruct everything from governance by customary leaders, maternity decisions, ritualistic burial and the rendering of justice, to the celebration of the acquired synthesis between humans and spirits.

## Spirits in the Borgou

There are many categories of Borgou spirits. *Yakasun tɔmbu* ("bush people") are like regular folks though visible only to a few humans, and live inside trees. They are not always a happy lot and can bring about turmoil if not placated. *Wɛrɛkunu* are mischievous entities sometimes encountered on lone journeys down isolated trails. A *Dɔ̃wiru*, translated as a "firehead," is eternally burning—not causing anyone harm, just burning, quietly observing, and unintentionally frightening passers-by. *Sĩiniba*, or genies, are an Arab import (*djinn*). They can help (or harm) people, usually Muslims, and are mollified by the intervention of Islamic sages, personal prayer and recitations of the Holy Koran.

*Gɔribu*, "the dead," are Baatɔnu ancestors who must be periodically greeted and soothed to keep the peace in a family. They are the first to be fed during the annual *tamkpaasu* ("new yam") harvest and are remembered at *Dɔ̃ Kɔ̃ru*, the Baatɔnu fire festival and lunar New Year when households are cleansed of the aforementioned malefic spirits.

Years after their initial funerals the souls of some ancestors return at the close of a *gɔɔkperu* ceremony, which is a second, or final, funeral observed sometimes decades after the first. The returned soul is called *tɛɛra* and, while inhabiting the body of an unwitting bystander, will venture to favourite locations in the village just as the ancestor would have done while alive, much to the surprise and delight of onlookers. A gɔɔkperu might be attended by thousands of family members and well-wishers over several jovial days and nights.

These spirits and ancestors all reside in the Borgou. None of them, however, are the immediate focus of Baatɔnu religion, though they can be the subject of religious intercessions.

## The Bũnu

Conversely, the *bũnu* spirits (singular = *bũu*) are the bedrock of Baatɔnu religion. They enter into long-term relationships with individuals, unions which can be passed on from generation to generation. *Bũsãaru* or *bũsɔmburu* are terms for Baatɔnu animism/spirit conjuring. A devotee of this spiritual practice is called a *bũnugi* (plural *bũnugibu*).

There are three bands of bũnugibu among the Baatɔmbu: the Sambaani, said to be of Baatɔnu roots; the Kaau, in some regions called Bukakaau, ostensibly shared with the Hausa and Djerma peoples and possibly related to the Mandé contribution (and seen as suitable for Borgou Muslims); and Wuuru, sometimes referred to informally as Samputo, which is appropriated from the Yoruba cult of Ogun, the divinity of iron and blacksmiths. Kaau and Wuuru, despite foreign derivations, are thoroughly part of mainstream Baatɔnu culture while recognition of their faraway birthplace is not lost on practitioners.

A bũu possesses the body of a Baatɔnu person who, after experiencing the physical sensation of *bũnu warama* ("falling for the spirit"), then undergoes months of secluded training in the art of faith practice for her or his particular band of bũsãaru. The training concludes with the new devotee's official presentation to the community in a *dii yaru* ("coming out of the room") ceremony whose highlight is the discovery by the new bũnugi of a sacred object hidden in the bush. Finding the object while entranced under the influence of the spirit—and under the vigilant gaze of elder spirit practitioners—means successfully passing a test, proving that one really is possessed by that particular bũu and not just pretending.

The dii yaru ceremony is filled with furious drumming and dancing. Participants are clad in beautiful hand-woven *barubekuru* cloth sometimes dyed with indigo (*Lonchocarpus cyanescens*) harvested in surrounding woodlands. The cloth designs suggest abstract imagery from nature, especially the *gɔna* (guinea fowl, *Numida meleagris*), a symbol of the Baatɔnu people. Barubekuru cloth is worn on the most deferential of occasions, including weddings, the enthronement of chiefs (also called dii yaru) and other royal commemorations, and by Barubibu upon interment. Bũnugibu are called upon during these and other events to ensure spiritual concord.

A bũnugi tends to inherit the spirit from an ancestor, though once a devotee passes away, it cannot be predetermined to whom within the extended family the bũu will next appear. The bũu makes that decision alone and the transition from one devotee to the next could take many years, or only days. Bũnu are spirits, certainly, but they have recognizable human-like forms and personalities, too. The bũnugi refers to her or his bũu as *yĩini* ("boss" or "lord") and the relationship is reverential, but also friendly, a partnership (Figure 4.2).

Sometimes the bũu is given the name of its first human devotee. As such, the ancestor is assured to be remembered among the living. In addition to her or his own sacred places of faith practice, a present-day bũnugi who has inherited the spirit from an ancestor continues to visit the ancestor's *bũyeru* (bũu "shrine") making offerings and sacrifices upon its *yãkuyeru* ("altar") and performing dancing and entrancement rituals. These annual pilgrimages are occasions for fellowship with other bũnugibu who travel to the event from far and wide.

An altar might include a statuette of its bũu surrounded by rocks, shells, guinea fowl feathers and sundry animal bones—objects from nature—as well as calabashes and pots in which offerings can be placed. Some shrines are located outside in natural surroundings while others encompass a small round room, often in proximity to several such rooms each housing an altar for a bũu. The round

**FIGURE 4.2** Bũu statuette and altar inside a shrine at the home of Wuuru spiritual leader Baan Gɔɔbi in the Beninese Borgou. Attached to the base are guinea fowl feathers. Cowries are affixed to the statuette. A *musu baka* (leopard, *Panthera pardus*) prowls on the wall. Photo Credits: C. Ryan Smith.

buildings, called *bawɔ*, mirror the forms of Baatɔnu kitchens, royal reception quarters, *taka* yam mounds on Borgou farmsteads, and the *baatɔnsikuru* tomb styles uniquely reserved for Barubibu. They are frequently painted in bright colours with images of wildlife, trees, sacred objects, and past and present devotees sprawling across the interior and exterior walls.

The shrine, with the altar at its heart, serves as a sacred space reserved for consulting the bũu on existential questions and for sacrificing animals to ensure good outcomes.

## Sina Gura

The traditional faith practice of the Borgou draws inspiration from the hunt. Baatɔnu hunters rely on bũnu to lead them to game and to protect them from harm in the bush as well as throughout all walks of life. A hunter is, therefore, hyper conscious of his surroundings in the forest and continually on the lookout for a bũu who will serve as his guide and friend.

Bũnu spirits often reside in trees. When the hunter is alerted to the presence of a bũu, and if that spirit accepts to guide him, the host tree becomes forevermore a shrine, a sacred site for that Baatɔnu man. He and his descendants will either take care of that tree and its bũu in the forest or, if the bũu lives in a waawura tree (*Lophira lanceolata*), bring the tree home to form a shrine which will be used by members of his family and neighbours during bũsãaru spirit conjuring ceremonies. These ceremonies bring practitioners together throughout the year and form the energising nucleus of bũnugibu alliances.

Bũnu can reside in several different kinds of trees as well as in various exceptional locations throughout the natural landscape, but it is the bũu from a waawura tree which, upon the tree's being transformed into a shrine, becomes the renowned and formidable *Sina Gura* (Figure 4.3).

If a couple fails to conceive, Sina Gura can help. When the baby is born, she or he will be given a name that signifies the special relationship with this powerful spirit. Sina Gura is also the senior spirit among bũnu, who often live in cooperative settings. All three bũsãaru bands revere Sina Gura, while other individual bũnu are venerated only within the confines of the specific band to which a particular bũnugi is a member. There is much collaboration between bũnugibu even though each band has its own sacraments, conventions and corresponding bũnu spirits.

Sina Gura means the Prince of the Rains, or the Celestial Prince. It is quite possible that Sina Gura, while being a Baatɔnu divinity, is related—albeit indirectly—to Shango, the divinity of thunder adored by the Yoruba people. Shango's symbol is the double-headed thunder axe. When he gets a notion to do so, Shango, with great bravado, hurls his axe down from the sky to the ground below producing a lightning flash and thunder clap (Quiroga, 1985).

When lightning strikes in the Borgou, bũnugibu go to the burned spot to recover the *gurugbã*, a rock believed to be delivered to Earth by the lightning.

**FIGURE 4.3** Waawura tree in bloom, a potential home for Sina Gura. Antisua Forest, Beninese Borgou. Photo Credits: C. Ryan Smith.

Gurugbã translates as "rain axe" or "celestial axe" (in the same vein as thunder axe), which, linguistically speaking, exposes affiliations between the Baatɔmbu and Yoruba that traverse the ages. Meanwhile, Shango survived the trans-Atlantic slave trade and was brought by African peoples to the Americas and Caribbean. Shango is present today in Cuban Santería, Brazilian Condomblé, as well as diverse faith communities across the diaspora (Bazinet, 2015; Murphy, 2010).

## Faith Practice and Becoming Nature

During periods of trance, which are integral to bũsããru ceremonies, bũnugibu—both women and men—are controlled by their respective bũnu who, in turn, confer upon their devotees the souls of wild animals from Borgou ecology. These human beings literally become those animals (hyena, python, baboon, etc.) and then head out into the bush roaring, barking, squawking, hissing and crowing, carrying with them sacred objects often crafted from iron, as they harvest plants, tree leaves and roots, and other ingredients which will help in the preparation of potions intended to protect people against harm (Figure 4.4).

Bũnugibu are identified throughout their communities as the animals they become during trances. Each animal has its own set of taboos. One might, for

**FIGURE 4.4** Devotees of the Wuuru band of bũnugibu returning from the bush with vegetation intended for protective potions, as divulged to them by their bũnu. Under trance, they are noisily manifesting the wild animals whose souls are within them. Gãku drummers lead, and Yerima, an elder bũnugi of the Sambaani band, is directing the group as she continually rings a sacred bell. The devotees carry other sacred metallic objects including rifles, which have been ensconced with animal hides and cowries. Photo Credits: C. Ryan Smith.

example, remind a cook that a visiting grandmother cannot eat duck meat. "Why not duck?" "Because she is a duck." "Oh, right, of course."

Baatɔnu devotees of the Kaau band of bũnugibu can become both a wild animal and a Fulani person. During ceremonies those devotees dress up in exaggerated traditional Fulani garb. This provides humorous scenes for all, especially actual Fulani neighbours, but the experience is not meant as mockery. The Fulani, as herders who spend significant amounts of time in the bush with their livestock, are seen as being that much more connected to nature, and are, thus, subjects of spiritual veneration for these bũnugibu.

## Traditional Baatɔnu Authority

The customary leaders of the Baatɔnu people uphold communal identity. They are the living bonds with a glorious past and provide grounding in a changing modern world. They serve as heads of families and villages, mediators of conflict, and decision makers over the most crucial of society's issues. Honour and respect are automatically accorded to these women and men.

As explained earlier in the chapter, traditional Baatɔnu authority stems from a network of historical relations between groups and individuals. The Sina Boko of Nikki is the emperor of the Baatɔnu people and supreme customary leader of the Borgou in both Benin and Nigeria. He is of Wasangari lineage but cannot rule unilaterally. The emperor is surrounded by a phalanx of Barubibu advisors and officials without whom he could not assume or maintain the throne, much less reign (Kuba, 1998, *idem.*).

Most Borgou villages have a chief, called a *sunɔ* in Baatɔnum, who is either of Wasangari or Barubibu lineage. He is the main ruler of his community, usually the descendant of its founder, though the sunɔ is not the village's lone traditional authority figure. There are many customary titles in a given village, including the *damasunɔ* (hunting chief), *ararisunɔ* (chief of the butchers), *yayigbe* (the female chief of brides), *kurɔmɔrɔnsunɔ* (chief of married women), *gãkusunɔ* and *barasunɔ* (chief drummers), *bũsunɔ* (chief of the bũnugibu), *timgi* (healer), and various associates of these and many more personalities who have been bestowed with special knowledge and authority.

These leaders, as well as village elders generally, work together to solve common problems and make decisions based on consensus—which must be considered by the sunɔ. Baatɔnu traditional authority may not be democratic by modern definition but it is representative and participatory. Most Baatɔnu families can count one of their elders as a customary leader of some sort (Alber, 1997).

## Legitimacy Derived from Nature

The process of becoming a leader often follows a trajectory imbued with gifts from adjacent ecosystems. Take, for instance, village chiefs. To be selected as sunɔ, candidates for the throne must satisfy a series of sacrificial rites at bũu shrines and the burial places of certain ancestors. These rites incorporate components harvested from nature. The candidate who is selected by fellow elders is ceremoniously washed upon enthronement with water from springs which are made sacred by virtue of hosting bũnu. He is clothed in barubekuru cloth evoking the guinea fowl. His palace is adorned with images of wildlife and he receives visitors while sitting on the hide of a *kpasa* antelope (*Tragelaphus scriptus*) offered for that purpose by a local hunter (Figure 4.5).

Thereafter, the sunɔ collaborates with bũnugibu who, having been informed by their bũnu, will alert him if evil forces are present in the village or if imminent dangers are foreseen, against which he must take action using specific natural resources under the direction of those spirits. Annual sacrifices are also made following protocols established at the founding of the village which put the sunɔ into direct contact with spirits residing in sacred spaces. These royal shrines are typically located in forest groves and surrounded by remarkable trees.

The hunting chief decides where in the landscape communal hunting will be banned to allow for the replenishing of faunal populations. He also oversees the fair distribution of hunting spoils village-wide, while the sunɔ makes sure that

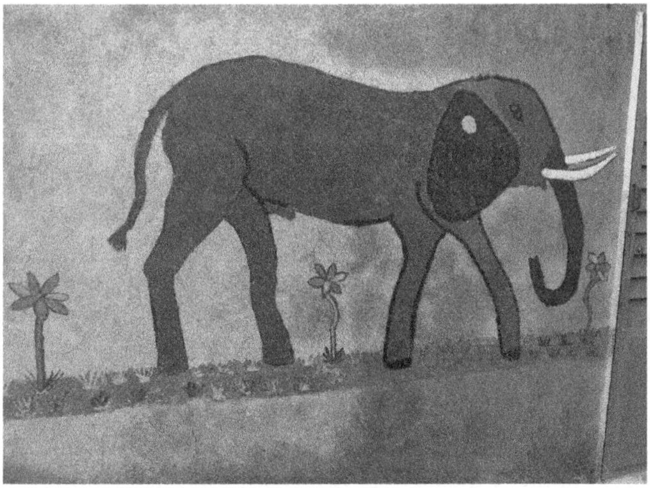

**FIGURE 4.5** Mural of a *sunnu* (African savanna elephant, *Loxodanta africana africana*). On the internal wall of the palace for Biɔ Buke Kpɛllɛn Sunɔ, Baatɔnu chief of Pèrèrè Centre in the Beninese Borgou.

his colleague's will is respected on both fronts. The two men also protect against the destruction of sacred sites, such as bũu shrines and burial grounds, which often anchor larger intact ecosystems.

When the sunɔ dies, his body is again washed with sacred water before being placed in a tomb together with cuttings from trees of ritualistic importance found in the groves supporting royal shrines and in woodlands surrounding his community. The leadership experience of the sunɔ has followed universally familiar custom nourished by nature and directed by spirits. The repetition of this customary sequence—to the letter—is legitimising for leaders in the eyes of the general population.

## Collective Action and Conservation in the Borgou

Collective action is a hallmark of Borgou village life, the fruit of Baatɔnu solidarity, and an instrument employed by customary leaders to accomplish tasks for the public good. Such tasks bring people together for purposes benefitting the whole, not the singular. They can be *mundane* (cleaning up roadsides, raising a house, attending consensus-building meetings, helping on someone's farm); *cultural* (supporting a chief during his dii yaru ceremony, flushing out malefic spirits on Dɔ̃ Kɔ̃ru, dancing at a marriage party, cooking food at a gɔɔkperu funerary event); or *dramatic* (cutting breaks against out-of-control bushfires, going to war to defend the country). Through collective action leaders galvanise the community for common purpose while often bolstering conservation directly or indirectly.

Hunting is a collective action which spawns conservation through spirituality in the Borgou. The *damaru* form of traditional Baatɔnu hunting necessitates the participation of several men from across the village, submitting to the direction of the damasunɔ hunting chief, operating during a particular moment in the dry season along pre-determined corridors within the landscape. The hunters crisscross the countryside every year and gain intimate knowledge of its ecosystems over the course of lifetimes. Some also encounter spirits, create shrines, reveal Sina Gura and initiate multi-generational religious and familial bonding opportunities.

Instead of an event, hunting is a process, an institution at the heart of Baatɔnu society. Damaru is a collective action managed by traditional authority which generates sharing of resources (Tatsua Kameda *et al.* 2003), ecological knowledge, and common identity among participants. Hunting is also the platform from which bũsãaru spirituality is born, another Baatɔnu collective action par excellence. While hunting is a proximate source of food in the Borgou, collective action in nature—first by hunters, then by spirit practitioners—is the ultimate source of Baatɔnu spirituality.

Engagement in the communal duties associated with the management of natural resources and wild landscapes to ensure the perpetuity of sacred spaces and a good hunt is likewise collective action. Baatɔnu people use plants and animals for their everyday survival but they also protect some habitats from excessive exploitation (those harbouring bũnu spirits, for example) while submitting to consensus-based decisions from traditional authority intended to promote sustainable use and social accord overall. Customary leaders organise collective action around many issues in a given community. Conservation of the sacred natural sites which grant them legitimacy is one such issue.

## Baatɔnu Heritage and Identity: Nature and Spirit

Baatɔnu people have long welcomed strangers into their midst and readily adopted useful cultural traits brought to the Borgou by others, including the divvying up of political power when it was seen as beneficial. The Baatɔmbu have refused, however, to cede moral and spiritual authority to anyone else, ever (Lombard, 1960). They have been steadfast in the protection of their traditions, which they call *deema*, and are exquisitely proud of their unique place in the world.

It is at the intersection of *nature* and human *culture* that Baatɔnu deema was created and then flourished for centuries in relative harmony with the environment. The shared norms generated by the nature-based social and spiritual heritage of the Baatɔnu people have created collective mechanisms at the family, community and national levels by which consensus could be established, and from consensus and mutual experience—solidarity. While deema is a noun, it is also a verb, meaning "to find" or "to come across." Tradition, as well as history and heritage for the Baatɔnu people, is that which one "finds" already established in the world upon one's birth.

Deema is the stuff of Baatɔnu identity and pride but also the key stimulus for conservation in their homeland. Indeed, as Baatɔnu people will readily tell you, deema is a gift entrusted by Gusunɔ, never to be dishonoured—just like the natural resources of the Borgou.

Adore the spirits, remember the ancestors, work well together, bequeath a bountiful land and a living culture to the next generation: To do so is deema.

## Acknowledgments

My personal relationship with the Borgou is intimate and profound, and also unexpected. I was not born in the Borgou but have spent my entire adulthood there—which is saying something, in the throes of middle age—embraced with open arms by a host community and family both of which I now claim as my own. I raised my adoptive Beninese children in the traditional manner of the Baatɔnu people with guidance from village elders and my kids' own extended biological family members. Baatɔnum became my language and through it I was able to experience the fullness of Borgou life. I have enjoyed being mentored by extraordinary leaders of Baatɔnu custom and indigenous spiritual practice, a privilege few people not originally from the Borgou have experienced.

This chapter seeks to relay some of the knowledge I have gained from my mentors over the decades. No cultural secrets have been divulged here. Everything enumerated in this text is of general familiarity in the rural Borgou but I offer my own analysis of the links between the spirituality of the place and nature, traditional authority, solidarity and collective action. I would like to thank the following confidants whose contributions to this chapter were central to its writing:

The Troukassa brothers, Nassirou and Gaston, men in their forties who are experts of Baatɔnu custom. Nassirou is a butcher and Gaston a drummer and both are enthusiastic educators of their cultural heritage; Yerima, their mother, who is a bũnugi of the Sambaani band; Baan Gɔɔbi, a renowned leader of Wuuru bũnugibu and trainer of new devotees, and her husband, Salomon Mɔra, who is a healer; Issa Sunɔ Bɔni, a 40-something ranger in a community-based nature reserve and son of a late village chief. Issa is sometimes affectionately called *Sina Wunɔ*, "chief's little brother," who assists royal personalities with daily errands and customary tasks. He is also a *wɔɔdɛɛ*, or member of the village volunteer militia, and an expert of all things relating to traditional Baatɔnu authority; and Masɔɔn Kurɔ, a Kaau bũnugi in her early 60s. Further, I thank Judy Blair McDonald in Georgia for her editing expertise.

Finally, I dedicate this chapter to the memory of three outstanding people whose teaching and encouragement, beginning in the 1990s when I moved to the Borgou, aided my integration into (and my understanding of) the cultural dynamic of the Baatɔnu milieu: Sina Yoru and Sunɔkpɛllɛgurugi, Baatɔnu village chiefs; and Yarun Kurɔ, a bũsunɔ of the Sambaani band.

## Note

1 Originally adopted in 1975 and updated several times since, including in 2008.

## References

Alber, E. (1997). Le Pouvoir Local Face aux Mutations au Niveau de l'Etat. Le Cas d'un Village Bariba. *Cahier d'Etudes Africaines 37*(1) (145), 137–155.

Arbonnier, M. (2000). *Arbres, Arbustes et Lianes Des Zones Sèches d'Afrique de l'Ouest.* CIRAD, MNHN, UICN.

Bagodo, O. (1993). Jalons et Perspectives pour une Approche des Problèmes dans l'Histoire du Baruwu (Borgou) Précolonial. *Afrika Zamani 1*, 125–148.

Bazinet, R. (2015). The sonic structure of Shango Feasts. *Ethnomusicology Review*, 20, 1–19.

Bio Bigou, L. B. (1995). *Les Origines du Peuple Baatonu.* Editions du Flamboyant, Cotonou, Bénin.

Brégand, D. (1998). Anthropologie Historique des Wangara du Borgou. In E. Bosen, C. Hardung, & R. Kuba (Eds.), *Regards sur le Borgou*, 245–263. L'Harmattan, Paris.

Débourou, D. (1993). Importance de la Valée du Niger dans l'Explication du Peuplement du Borgou. *Afrika Zamani 1*, 149–160.

Dehoux, J.-P., & Hounsou-Ve, G. (1993). Productivité de la race bovine Borgou selon les systèmes d'élevage traditionnels au nord-est du Bénin. *Projet de Développement Pastoral Intégré dans le Borgou, Phase II.* FAO.

De Moraes, P. F. (1998). For a non-culturist historiography of Béninois Borgu. In E. Bosen, C. Hardung, & R. Kuba (Eds.), *Regards sur le Borgou*, 39–69. L'Harmattan, Paris.

Kameda, T., Takezawa, M., & Hastie, R. (2003). The logic of social sharing: An evolutionary game analysis of adoptive norm development. *Personality and Social Psychology Review* 7(1), 2–19.

Kuba, R. (1998). Les Wasangari et les Chefs de Terre : Une Histoire d'Intégration Mutuelle. In E. Bosen, C. Hardung, & R. Kuba (Eds.), *Regards sur le Borgou*, 93–119. L'Harmattan, Paris.

Lombard, J. (1965). *Structures de Type "Féodal" en Afrique Noire : Etudes des Dynamismes Internes et des Relations Sociales Chez les Bariba du Dahomey* (Vol. 1). Mouton et Co., Berlin.

Lombard, J. (1960). La Vie Politique dans une Société de Type Féodal : Les Bariba du Dahomey. *Cahier d'Etudes Africaines 1*(3), 5–45.

Murphy, J. M. (2010). Yoruba religions in Diaspora. *Religion Compass 4*(7), 400–409.

Quiroga, M. A. (1985). Shamanic parallels in the thunder Deity Shango: From Yoruba to Cuba and on to Miami. *Athanor 4*, 69–75.

Sinsin, B., Daouda, I., Ahokpè, E., Téhou, A., Coubéou, P., Toko, I., & Fandohan, S. (1997). *Faune sauvage des forêts classées de Goungoun, Sota et des environs.* MDR/DFRN/ Projet UNSO/BEN93/G 31. Cotonou, Bénin.

Stewart, M. H. (1993). *Borgou and its kingdoms: A reconstruction of a Western Sudanese polity.* The Edwin Mellen Press, Lewiston, NY.

# 5

# REINTERPRETING BELIEFS AND TRANSFORMING RITUAL TRADITIONS AROUND THE GANESH FESTIVAL IN PUNE, INDIA

*Manisha Sheth*

## Introduction

This chapter provides an overview of the Ganesh Chaturthi festival in India. It analyses the environmental impact of this Hindu festival and the efforts made by several organisations in the city of Pune to mitigate the pollution caused by the festival by encouraging believers to reinterpret the tradition.

Lord Ganesh is a Hindu deity who is considered to preside over the element of Earth among the five elements of Nature. He has a human body with an elephant head, and he is a familiar figure in most Hindu households.

Since Independence from Britain in 1947, the city of Pune has become famous for its large-scale collective celebrations of this ten-day festival. The Ganesh Chaturthi festival is celebrated every year in late monsoon (August–September) to honour the coming of Ganesh to the earthly plane, in the form of an idol. In the ritual, as he is welcomed into the home in the form of a sculpted image, he becomes a beloved friend, who is decorated by the family with love, and to whom they surrender their challenges (see Figure 5.1). When the time comes for him to leave, they are reminded of the transient nature of all relationships and bid him farewell. Finally, there is a ritual of immersion of the idol into a water body, along with all the materials used in its worship.

In recent times, the original natural materials used to make the Ganesh idols have been replaced with synthetic and non-biodegradable substances, which, when immersed into water bodies, result in water pollution (Lokhande 2019; Bhattacharya et al. 2014).

For over a decade now, several environmental organisations have been engaged in campaigns to raise awareness about the impact of the festival on water bodies and propose alternatives to the manner in which the festival may be celebrated to make it more sensitive to Nature. In 2020, the Central Pollution Control Board

DOI: 10.4324/9781003246190-7

FIGURE 5.1 Worship of a Ganesh idol conducted by Sri Vivek Godbole. Photo credit: Manisha Sheth.

(CPCB, a national regulatory body) finally issued guidelines that banned the use of chemical substances for idols that are meant to be immersed (CPCB 2020).

## Ganesh: The Symbol

The son of Goddess Parvati, Ganesh carries several layers of meaning for his devotees, of which the most popular is that he is *Vighnaharta* – the remover of obstacles. According to popular mythology, Parvati formed Ganesh from her own body, independent of the involvement of her husband, Shiva. It was in a tussle between Shiva and Ganesh that her son was beheaded by Shiva in a fit of rage (Shastri 1970). Remorsefully, Shiva then gifted him with the head of an elephant and this is how Ganesh came to have an elephant's head (see Figure 5.2).

As a combination of both human and elephant forms, Ganesh embodies the virtues of both. This pot-bellied stout deity, who wears a serpent and rides a rat, is said to be easily appeased and is the first deity to be worshipped before the start of any new venture, to ensure its success.

While the symbol of Sri Ganesh can be tracked iconographically, dialogue with religious leaders is important for behaviour change and reinterpretation of symbolism. In a conversation with Sri Vivek Godbole, head of a Vedic school in Maharashtra, he suggested that a possible origin of the festival of Ganesh

**FIGURE 5.2** A Ganesh idol with an elephant head and human body. Photo credit: Umesh Kale.

Chaturthi was in the agrarian culture of Maharashtra. In a 2008 conversation with the author, he spoke of the fact that during the monsoon season, the banks of the rivers have sediments of rich fertile soil, which is of great significance to the farmer community (eCoexist 2015). He proposed the idea of bringing some of this soil home to worship, as a way of offering thanks to Mother Earth and praying for a good harvest. After the worship is over and the soil infused with prayers and blessings, it is returned to the source where it came from, the river. He pointed out that in Maharashtra, three festivals occur close to each other during the monsoon season – Bail Pola, Nag Panchami and Ganesh Chaturthi. In all three, soil is used to sculpt deities that are then worshipped.

Godbole suggests that the forms given to the soil are secondary, and that the prayers were primarily directed at the Earth itself. This explanation connects the worship directly to the health of the environment, the soil and the waters. In South India, devotees continue to follow this tradition of bringing a handful of soil home and sculpting the form of Ganesh with it themselves for the festival.

## Ganesh Chaturthi: Festival as a Social Connect

The name Ganesh, also known as *Ganapati*, is rooted in the Sanskrit word *gana* – which means people. The suffix *isha* means 'Lord', while the suffix *pati* means 'Protector'. One way of understanding Ganesh is as the leader, or Lord, of ganas – his devotees. In this form, his primary role is social, in bringing together community.

In the city of Pune in the state of Maharashtra, this festival is possibly one of the largest gatherings in the city, as it is home to Ashta Vinayaka – eight Ganesh temples of ancient significance that host various forms of Sri Ganesh. Devotees visit each other's homes to have '*darshan*', i.e. to receive blessings of the deity as he is embodied in each home. They exchange sweets and compare their creative efforts in decorations, spending time with each other's families. At the grand finale of the festival, they gather together in procession, to bid him an appropriate farewell with song and dance, and immerse him at various locations along the river (see Figure 5.3).

The festival of Ganesh Chaturthi emerged into a social movement because of Sri Balgangadhar Tilak, a freedom fighter (Gopinath 2019). Tilak invited the citizens of Pune, in 1893, to come out and collectively worship Lord Ganesh with

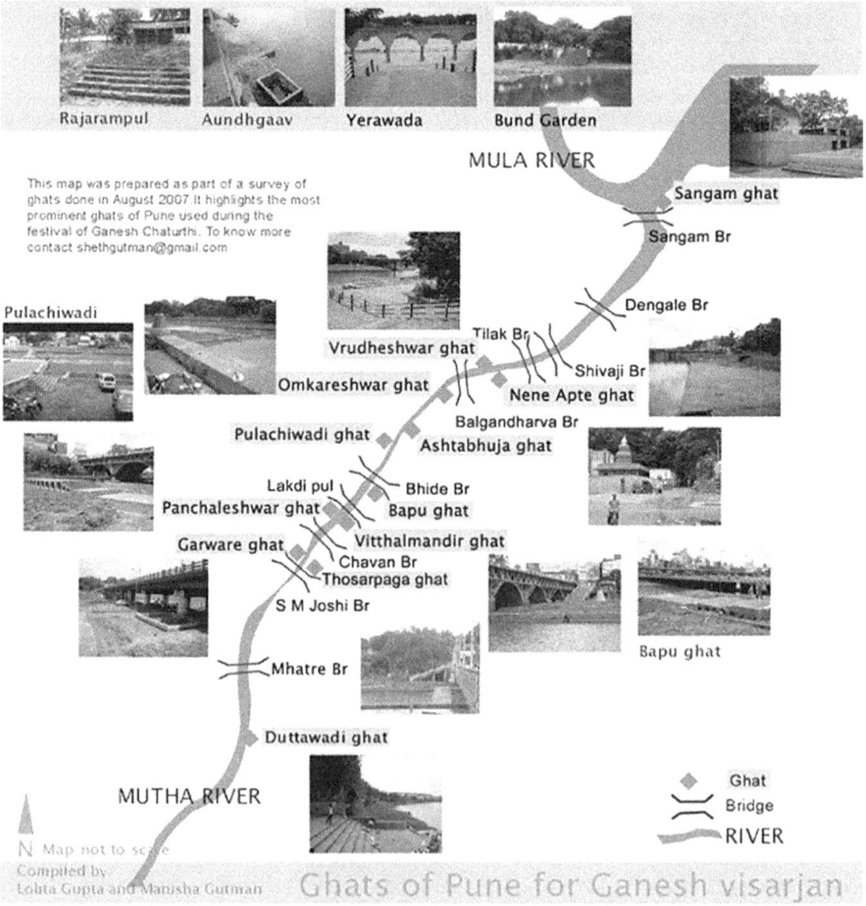


**FIGURE 5.3** Map of the immersion sites (ghats) along the Mula-Mutha Rivers, Pune. Credits: Compiled by Manisha Sheth and Lolita Gupta.

procession and fanfare, and openly express their pride in Indian culture to confront their British colonists. The ritual worship of Sri Ganesh, which had been an individual act so far, was thus transformed into a community event with this step.

The river Mula-Mutha in Pune is a confluence of two rivers that is considered an auspicious location. During the Ganesh festival, the river is where millions of devotees gather each year, to immerse their objects of worship, and yet it is a highly polluted water body which is only a trickle for most of the year. This map (see Figure 5.3) was prepared as part of a survey of immersion sites done in August 2007.

## Visarjan: The Act of Letting Go

In Hindu ritual, an idol meant for worship is imbued with a divine presence which is invoked in it through certain processes. This presence is believed to inhabit the physical idol for a period of time, and it is understood that while the idol is being worshipped, it is in fact the presence within the idol, that is sacred and the focus of such worship.

Once this worship is over, the presence may also leave the idol, beyond which it remains a mere object and is not considered sacred anymore. The Marathi term for the offerings made during worship, after the worship is over, is *Nirmalya*, which literally means 'that which is not soiled'.

At the banks of the river or sea, the Ganesh idol is ritually immersed into flowing waters. This act of *visarjan* primarily represents dissolution. The Marathi word comes from the original Sanskrit word *visṛj* which connotes sending forth, giving up or letting go. In physical action, *visarjan* involves the immersion of an idol into water.

It symbolises the transient nature of the material Universe and is a reminder to the devotee to be detached and surrender the form of Ganesh that he has come to adore so deeply, to where it originally came from, the Earth. However, there is a promise of a return again the following year, in true cyclical fashion; and prayers invite him to come back soon, the next year.

The act of letting go and allowing the idol to disintegrate is a key part of the immersion ritual. But what happens when the idol does not disintegrate?

## The Environmental and Social Dilemma

The main impact that the Ganesh festival has on ecology is due to the materials used in the idols and the decorations, compounded by the single use custom, and the immersion ritual. The Ganesh idols that were originally made using clayey soil were then replaced by the use of Plaster of Paris which were further adorned using bright chemical paints containing toxic heavy metals. This shift was a result of the emergence of a small-scale industry, as the demand for idols grew and the production of these was handed over to professional sculptors. Plaster of Paris is easier to mould, less fragile and cheaper to produce than clay. Similarly, decorations made of natural leaves and flowers were replaced by plastics and synthetics. As populations increased, the scale at which the festival was celebrated, and the number of idols reaching the rivers, also magnified. In 2015, the festival generated an estimated 20,000 crore rupees (approximately 2.8 billion USD) and was seeing an annual growth rate of 30% (Times of India, 15 September 2015).

The industry of Ganesh idols had its own challenges with the use of natural materials, that, while being biodegradable, were fragile and resulted in losses due to damage and weather during production and transport. Plaster of Paris offered a cheap and convenient solution; it was easy to mass produce, fast drying and more durable than natural clay. It required less skill in handling and was cheaper. The downside, though, is that Plaster of Paris does not biodegrade. The idols sink to the bottom of the water bodies and sit there, slowly

**FIGURE 5.4** Dead fish floating in a lake the day after the festival immersion ritual. Photo credit: Sunjoy Monga.

releasing the chemical paints into the water and the toxins kill aquatic life. The day after the immersion ritual, it is not uncommon to see dead fish floating on the surface (see Figure 5.4). Also, these plaster idols can be found on the river bed up to several years after they have been immersed (Jain et al. 2018). People who live downstream from the city where the celebration is conducted are also deeply impacted by the pollution of the river waters, as they sometimes bathe in and drink this water.

Apart from its ecological impact, the Ganesh festival is a major source of revenue and income. The tradition of hand sculpting Ganesh idols has provided a source of livelihood to artisanal communities for several decades now (see Figure 5.5). Traditionally, a bond is formed between the person who sculpts the idol and those who worship it. Often, families will insist on buying from the same sculptors only and come to rely on them for their annual requirement.

As explained by the artisans, only a few of the original master sculptors are around today, as the younger generation is choosing other careers, and the craft has become more of an industry with copies of the master design being 'printed' out by labourers. The paucity of labour poses a challenge to this small-scale industry, and mechanisation may soon set in. The materials chosen for the idols and decorations are also responding to the demographical changes in the artisanal community. While natural clay requires more skill to sculpt, Plaster of Paris can simply be poured into a mould by a labourer.

Any change to the festival or regulation of the production of items for the festival is bound to impact this community and for several years, this has deterred the government from imposing laws around the issue.

**FIGURE 5.5** A family of artisans in Pune whose main source of livelihood is the Ganesh sculptures. Photo credit: Manisha Sheth.

## Possible Alternatives and Solutions

To address the complexity involved in solving these issues, a multi-pronged approach emerged in the city of Pune, quite organically, as various organisations focused on different aspects of both ecological and livelihood issues.

Across all of these efforts, the underlying goal is to convince the Ganesh devotees to take ownership and responsibility for the impact that their religious choices were having on the environment. Using a combination of scientific facts, photographic evidence and with the support of religious leaders, the Ganesh campaign has evolved over ten years to finally result in a ban on the use of toxic chemicals in immersible idols (CPCB 2020).

### *Material and Product Design*

In 2007, eCoexist started to address the core of the problem by looking at the materials being used in the products sold for the festival (eCoexist 2021). eCoexist is a social enterprise that works simultaneously on environmental and social issues, looking for integrated solutions to both. Approaching the issue from a design perspective, eCoexist recognised that it was not the ritual itself that was problematic, but the fact that the materials had changed over the years that was the cause of the present-day pollution. The team looked for ways to revive traditional materials such as natural clay and natural pigments and re-introduce them to the market. Their challenge was twofold – first to convince the artisans to work with fragile and biodegradable materials, and second to ensure that there was a market for these biodegradable products.

eCoexist's approach was one of respect for the religion, its belief systems and its rituals. They invited Sri Vivek Godbole, as a religious leader, to speak to the citizens of Pune about the need to revere and protect Nature. Year after year, he spoke and sang about Ganesh, representing him as the deity that presides over the Earth, reminding devotees of this aspect. eCoexist also organised rituals of blessing by Sri Vivek, as he conducted a worship of the idols that were to be sold for the festival. These events became occasions of a revival not only of the religious symbolism itself, but also of a reconnection of this tradition with its origins in Nature.

With the blessing of a religious priest and Vedic scholar, citizens began to be convinced about the shift to natural material. Among the natural materials eCoexist explored were natural clay, paper mache and cow dung. Vegetable or mineral pigments drawn from turmeric and different types of earth replaced the toxic chemical paints. Plastic and styrofoam decorations were replaced with natural materials like rice husk, fabrics and paper. eCoexist proposed that devotees consider reducing the size of the idols they worshipped and also encouraged the use of a permanent idol that could be symbolically immersed in water at the end of the ritual, then saved to be reused again the next year. Most of all, eCoexist further advised that all immersions should take place at home only, in a clean pail of water, which could then be poured into potted plants or gardens.

## *Education and Marketing*

The campaign to change mindsets around the Ganesh idols also used scientific evidence to illustrate the urgency of the problem. Joining hands with eCoexist, Oikos, another green social enterprise (Oikos 2016), took on the responsibility of raising awareness around the issue in 2010. Posters were designed in English and Marathi, articles written and interviews given. The issue was introduced as a subject of study in schools so that children could discuss both sides of the problem. The paucity of statistical data about the pollution levels in the water was a major challenge. In collaboration with the Maharashtra Pollution Control Board (MPCB 2019), and with research conducted at the Indian Institute of Technology (IIT) (Asolekar 2007), the campaign brought the issue to the forefront before and during the festival.

This approach of using religious dialogue along with scientific data was effective in convincing the public to a degree. However, without ensuring access to natural alternatives, simply raising awareness would not be enough. The promotion and marketing of the eco-friendly Ganesh idols and decorations were taken on by these organisations as well. This involved buffering the risks faced by artisans, retraining the aesthetic expectations of the public and balancing the costs of the entire project to maintain affordability. In short, it involved redesigning and sensitising the entire supply chain of these products to the market.

## *Measuring Sustainability*

Measuring the ecological sustainability of a supply chain is not an easy task. It involves studying all the aspects of production, distribution and marketing and also needs to regulate waste management to design a truly circular economy (SEI 2019).

The Ecological Society (https://www.ecological-society.com/), an educational institution based in Pune that focuses on ecology and natural resource management, held a neutral space in 2019 to review the issue from various angles. They acted as a liaison between non-governmental organisations and companies to determine the pros and cons of the various alternatives available in the market, apart from being the guardians of a fair dialogue between all the stakeholders involved.

Educational institutions such as the Center for Environmental Education, the IIT and the Ecological Society provide a scientific counterpoint to the belief systems in place. In playing this role, they often question the religious belief systems, but with no other agenda except that of conserving Nature. This has the potential to come across as insensitive. Professor Asolekar (2007) of the IIT, whose focus area is water conservation, proposed possible solutions to the management of waste from Plaster-of-Paris idols. However, these met with a fair amount of resistance from religious groups at the time, as they involved crushing the idols after use. Twenty years later, as the importance of recycling has been integrated into public thought, there is much more openness to considering this possibility.

## *Waste Disposal*

The ultimate solution from an ecological perspective would be to use a permanent and reusable Ganesh idol for the immersion ritual. This is a radical change in tradition and could not be achieved without the blessing of religious leaders.

However, the most severe impact of the Ganesh Chaturthi festival is felt in the disposal of the 'waste' generated during it. The ritual of immersion, as explained earlier, marks the departure of the presence from the idol and after being immersed, the idol is no longer of any significance to the worshipper. However, the modern concept of waste is not applicable to religious waste, called *nirmalya*, which still demands a certain amount of respect in the way it is handled.

Ecologically, the scores of idols that gather at the bottom of the water bodies pose a large environmental threat. The disposal of such idols that do not dissolve in water and do not degrade for years has been a source of great stress, to the government who had no recourse except to collect them in the night and simply relocate them to the outskirts of the city and abandon them there, creating another ecological impact (Bengrut and Bidkar 2015).

SWaCH (https://swachcoop.com/) is a cooperative of women who collect garbage door to door (referred to as rag-pickers) in Pune City which brings together nearly 8,000 women who play the role of collecting and segregating garbage on a daily basis to complement the waste management systems set up by the municipal corporation.

In 2011, in collaboration with eCoexist, SWaCH began to operate on the banks of the rivers in Pune, inviting people to use man-made water enclosures (water tanks) to immerse their idols, rather than put them into the river. These women stood on the river banks (ghats), and collected all the materials that came along with the idols, and segregated them into categories, which were then appropriately recycled or composted as necessary.

Over the years, this effort by SWaCH was welcomed by the citizens of Pune who understood the need for appropriate waste management. Five years after the segregation activity on the ghats began, it was adopted by a leading industry based in the city and their employees started to volunteer for the work. The women rag-pickers visit many homes every day collecting garbage and pleading to citizens to make more eco-friendly choices for waste disposal. Nearly 400 tonnes of waste offerings, the *nirmalya*, were collected by the group in a 10-day period along the banks of the river (Times of India, 29 September 2015). In this way, they act as guardians of the river, ensuring that the waters are kept free of any man-made substances during the festival period.

## *The Sanctity of the Mula-Mutha River*

Jeevit Nadi is an organisation that came up in 2015 with the intention of reviving the dying river and bringing it back to life for Pune. This group took up the issue of the Ganesh immersions with the main intention of reducing pollution of water bodies due to different festivals and rituals which impact our natural water bodies. Their approach was scientific and ecological. Their work on the ground included reviving the idea of donating the used idol for the purpose of recycling it. In 2009, the same idea had been rejected by Hindu groups for being irreverent. Jeevit Nadi noticed a deep-rooted fear among believers which causes a resistance to change. They also observed that to many devotees, the festival was a social demonstration of their adherence to the faith, more than having a deep spiritual significance in their own lives.

With the help of *kirtankars* (minstrels) and *bhajan mandalis* (choirs), they attempted to bring the message of pollution to the devotees at the river banks. At the same time, Jeevit Nadi simplified scientific data to convey it to the masses and found that when the issue was explained in simple terms, even the lay person was willing to consider a shift in practice.

### *Ganesha as Gajanana: The Elephant*

A final aspect of the entire movement involves reminding devotees that Ganesha is half-animal and half-human. In this form, he demands a peaceful coexistence of these two forms of life, because the divinity is a combination of them both. eCoexist has recently taken up the Gajanana campaign, which aims to inspire Ganesh devotees to contribute to the conservation of wild elephant populations in India as a form of their worship of Ganesh. Along with the Assam Haathi Project, eCoexist aims to slowly direct the love for this symbolic deity to extend to a love for the actual elephants.

## Conclusion: Paths to Change

In the reinterpretation of ritual and tradition, resistance to change is expected as it threatens an established identity. The Ganesh festival is a multifaceted event that provides a sense of social cohesiveness to the Hindu community at various levels. The move to make it more eco-friendly required a change of material, of form, of process and of belief. In general, Hindu thought has space for questioning; however, this process of questioning needs to be self-driven and cannot be perceived to be a critique from 'outsiders'.

The Hindu community itself is a diverse heterogeneous group, whose beliefs vary from God as an abstract formless divine energy to a very personal relationship with a chosen deity. The fear was that any change to the ritual would render the worship futile.

The idea that an idol that was once broken had lost its divinity and could not be worshipped prevented devotees from shifting to more natural materials that were prone to damage. The belief that one always had to worship a larger idol than the previous years also created some resistance to putting a cap on the size of the idols used. Some of the religious groups argued that the immersion had to be done in flowing waters, and therefore the move to doing home immersions was not acceptable to them. The unwillingness to crush and recycle used idols resulted in a rejection of any constructive recycling proposals. Disagreements between environmentalists themselves as to the real impact of various materials also resulted in a level of confusion in the general public, and people simply wanted one clear answer as to how to proceed.

The politics of the festival involved the appeasement of the majority community by the politicians in power, as they portrayed the regulations on the festival as a form of religious prejudice. This presented one more source of resistance to change.

Gradually, as India's environmental problems grow more severe and awareness about the need to protect natural resources grows, change is evident. Religious groups have also started to integrate this message into their dialogue and become open to these eco-friendly messages.

Hindu philosophy speaks of the difference between belief (*shraddha*) and blind faith (*andha shraddha*) and invites practitioners to discern the difference between the two. Questioning and reinterpretation are welcomed in Hindu thought and the Ganesh campaign triggered this process of questioning within the community of Ganesh lovers. Often when a devotee felt panic over a damaged biodegradable/eco-friendly Ganesh idol, a discussion was initiated with them to help them go deeper into their belief systems and to address the fear it brought up. Debates took place between younger and older generations within families, as this questioning invoked a change of choice around how to celebrate the festival. And the expansion of the identity of Ganesh, not simply as the idol, but as the Lord of the Earth, who had taken temporary residence in the idol was helpful in turning believers towards their ecological responsibilities.

In recent years, the neighbourhood committees that organise the festival have started setting up entirely eco-friendly arrangements. Competitions and prizes in multiple cities were arranged by local newspapers to encourage this shift and provide incentive to the mandals, and conservation has become a collective endeavour (PTI 2016). The city of Pune, which once used the Ganesh festival to reassert its independence, is now slowly converting it into a vehicle for an ecological message. This has been possible because of the convergence of the scientific and the spiritual dialogue, towards a larger goal adopted by both, that of Nature conservation. In 2022, several organisations in Pune led by eCoexist, have come together to plan a massive city wide campaign to collect the clay sludge after the immersion of the natural clay idols and return it to the artisans that make the clay idols. In this way, they hope to minimise the amount of fresh clay being mined for this industry, as clay can easily be recycled and reused if collected in a systematic manner. This is the first time something like this will be attempted in the country and the willingness of citizens to come together and take this next step, indicates how much their mindsets have already changed.

To conclude, for faith to become a medium for a shift towards conservation, an **internal dialogue** has to emerge internally between members of the faith community. A process of **respectful questioning** has to be initiated, preferably by the leaders of the community themselves, supported by scientific data and visual evidence. Going **back to the origin** of the ritual or the tradition and finding a natural source also helps reorient the belief system to include Nature. Change takes time and there has to be a **persistent dialogue** over a period of time. **Alternative methods** of practice can be proposed to the community, which do not clash with their beliefs.

At all times, there has to be a respect and deference for the belief system with a genuine effort to understand its worldview and wisdom. Inspiring people to adopt a new way of practising their faith, and inviting them to step into a larger identity as citizens of planet Earth, works to dissolve resistance to change.

## References

Asolekar, S. R. (2007). Preventative environmental management plan for eco-friendly Ganpati festival. Retrieved 4 July 2021 from https://e-coexist.com/wp-content/uploads/2018/04/paper-by-asolekar.pdf

Bengrut, D., & Bidkar, M. (2015). Dumped in the dead of night. Retrieved 4 July 2021 from https://punemirror.indiatimes.com/pune/cover-story/dumped-in-the-dead-of-night/articleshow/49158932.cms

Bhattacharya, S., Bera, A., Dutta, A., & Ghosh, U. C. (2014). Effects of idol immersion on the water quality parameters of Indian water bodies: Environmental health perspectives. International Letters of Chemistry, Physics and Astronomy, 20(2), 234–263.

Central Pollution Control Board (CPCB), Ministry of Environment, Forests and Climate Change. (2020). Revised Guidelines for Idol Immersion. Retrieved 26 June 2021 from https://cpcb.nic.in/openpdffile.php?id=TGF0ZXN0RmlsZS8yOTVf-MTU4OTI3Mzk4Nl9tZWRpYXBob3RvMjQzNjUucGRm

eCoexist. (2015). The natural origins of Ganesh Chaturthi: Talking to a Vedic scholar. Retrieved 26 June 2021 from https://www.e-coexist.com/the-natural-origins-of-ganesh-chaturthi-talking-to-a-vedic-scholar/

eCoexist. (2021). Eco-Ganesh Campaign. Retrieved 4 July 2021 from https://www.e-coexist.com/ecoexist-foundation/eco-ganesh-campaign/

Gopinath, S. (2019). Heterotopic assemblages within religious structures: Ganesh Utsav and the streets of Mumbai. Open Cultural Studies, 3, 96–106.

Jain, A., Nayak, J., More, B., Mehta, M., Tamukuala, T., & Shah, G. (2018). Impact of idol immersion activities leading to deterioration of water quality. International Journal of Advanced Scientific Research and Management, 3(12), 71–77.

Lokhande. (2019). The effect of Ganesh Idol immersion on the water quality of Gorai Jetty, Mumbai – the environmental health perspective. International Journal of Trend in Scientific Research and Development, 3(3), 398–402.

Maharashtra Pollution Control Board (MPCB). (2019). Water quality monitoring Network in Maharashtra. Retrieved 4 July 20201 from https://mpcb.gov.in/water-quality

Oikos for Ecological Services. (2016). Welcome to Oikos. Retrieved 7 July 2021 from https://oikos.in/html/index.html

Press Trust of India (PTI). (2016, August 29). Vadodara Ganesh Mandals compete to keep festival eco-friendly. The Indian Express. Retrieved 18 July 2021 from https://indianexpress.com/article/india/india-news-india/vadodara-ganesh-mandals-compete-to-keep-festival-eco-friendly-3002244/

Shastri, J. L. (Ed.). (1970). The Śiva Purāṇa, Part II. Delhi: Motilal Banarsidass.

Stockholm Environment Institute (SEI). (2019). Transformational change through a circular economy. Retrieved 4 July 2021 from https://www.sei.org/publications/transformational-change-through-a-circular-economy/

Times of India. (2015, September 15). Ganesh festival to generate Rs 20,000 crore revenue across country. Retrieved 26 June 2021 from https://timesofindia.indiatimes.com/city/ahmedabad/ganesh-festival-to-generate-rs-20000-crore-revenue-across-country/articleshow/48964447.cms

Times of India. (2015, September 29). Over 500 tonne nirmalya collected in 10-day festival. Retrieved 4 July 2015 from https://timesofindia.indiatimes.com/city/pune/over-500-tonne-nirmalya-collected-in-10-day-festival/articleshow/49146388.cms

# 6

# THE CONSERVATION OF MALEKU PEOPLE'S SACRED NATURAL SITES IN COSTA RICA

*David Solis-Aguilar, Leonel Elizondo and Alexander Elizondo*

## On Being Maleku

The Frío River basin in northern Costa Rica is the ancestral territory of the Maleku, Indigenous people colonised by Nicaraguans towards the end of the 19th century (Edelman, 1998). The Maleku elders still believe in *Tócu* as the world creator, including spirits who inhabit rivers' headwaters and lakes. These waterbodies are inhabited by Maleku's deceased ancestors too, making them sacred natural sites (SNSs). Currently, 40% of the Maleku's sacred sites are under public administration in two natural protected areas (NPAs), while only 10% of the sacred sites are cared for directly by the Maleku within their territory demarcated by the government in 1976, and the remaining 50% of the sacred sites are privately owned by farmers. The Maleku currently inhabit a demarcated territory between La Muerte, Sol and Cucaracha Rivers, in San Rafael district part of Guatuso county, established in accordance with the 1977 Indigenous Act (IA), as Solis-Aguilar expose (2021) (see Figure 6.1).

According to the 2011 national census, Indigenous people in Costa Rica represent only 2.4% of the national population – 104,143 inhabitants – and only 34.5% of them live in one of the 24 demarcated territories. There are 701 self-identified Maleku across the country, 223 are *maleku lháica* speakers who live outside the demarcated territory and 478 live in three settlements or *palenques* named El Sol, Margarita and Tonjibe inside the demarcated territory (INEC, 2013).

The historical dispossession of lands and waterbodies in the Maleku ancestral territory changed their relationship with the SNSs, and motivated them to actively participate in the conservation of the *Toro lhámi* sacred site for the last three decades through their traditional fishing in Caño Negro Wildlife Refuge

DOI: 10.4324/9781003246190-8

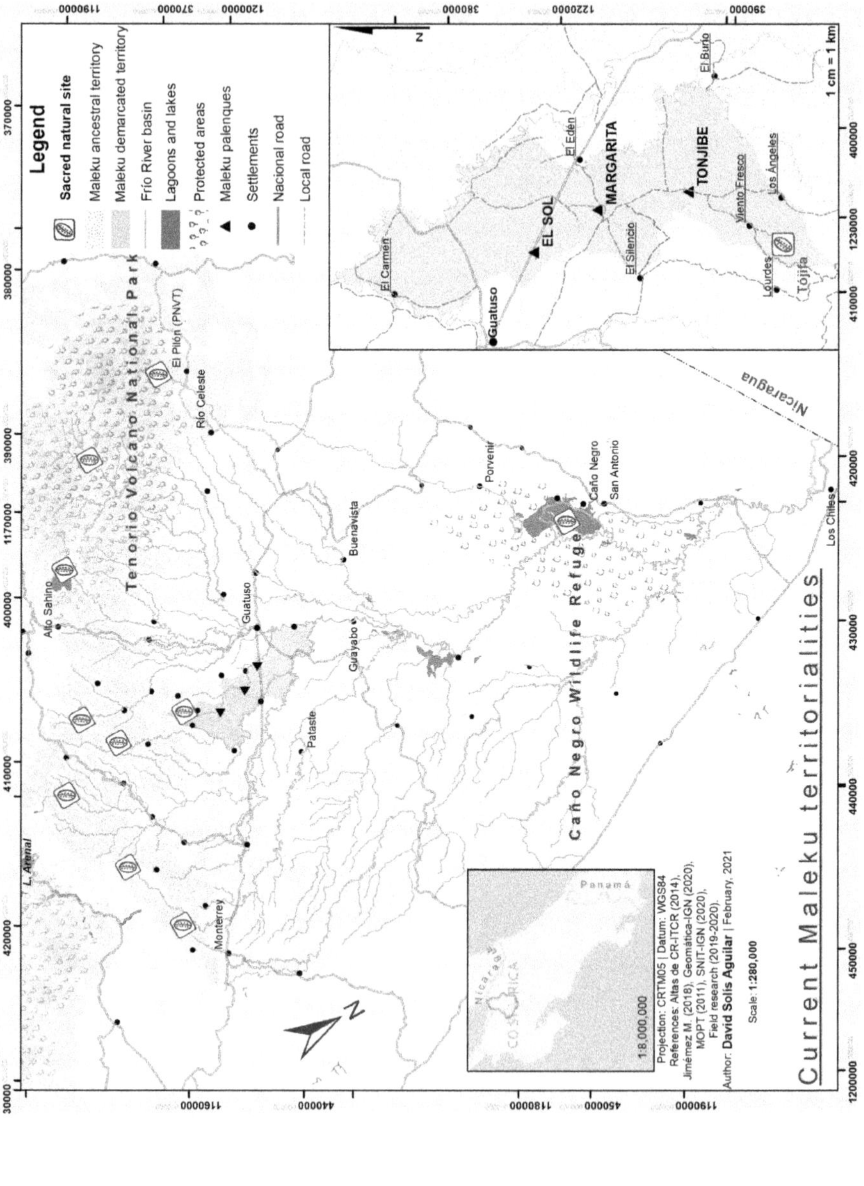


FIGURE 6.1 Maleku territory location. Credits: Map created by David Solis-Aguilar.

(RNVS-CN). In addition, they are currently organising for the recognition of their right to participate in the conservation of the sacred sites *Tióqui riliáca, Piúri facára* and *Ucúriqui chía* in Tenorio Volcano National Park and Protected Zone (PN/ZP-VT), according to Solis-Aguilar (2021).

It is necessary to revise the management of protected areas in Costa Rica in order to implement a rights-based approach to conservation in RNVS-CN and PN/ZP-VT that integrates Maleku's human rights and biocultural heritage. This is based on a culturally appropriate approach that relies on the self-determination and collective agency of communities with effective guarantees of tenure rights over their traditional lands, which are called "community rights-based conservation" (Rights and Resources Initiative, 2020). For this approach, we illustrate the institutional boundaries to making cultural uses of SNSs for Maleku people and how legal recognition of cultural and land rights is inhabited by a focus on species conservation. Therefore, we present the need for community rights-based conservation under the umbrella of the Convention on Biological Diversity through Articles 8j and 10c, and considering the cultural, environmental and social impacts' assessments as per the Akwé: Kon Voluntary Guidelines (Secretariat of the Convention on Biological Diversity, 2004).

## Maleku Ancestral Territory

A historical approximation of the Maleku territory attributes approximately 100,000 hectares to the Maleku, corresponding to the Frío River's middle and upper basin, from the headwaters of its tributaries in Costa Rican Central Range to Caño Negro wetlands (Castillo, 2004). Ancestral Maleku walking and river routes further constitute their territory *Ni maráma ifácfanhéca*, comprising an area of up to 200,000 hectares, extending from the south in the Arenal River, a San Carlos River basin tributary.

This territory included hunting and fishing sites, cultivation areas, resting sites, food gathering and plant extraction areas, as well as sacred sites found along the Frío River basin. The ancestral territory included hunting and fishing routes along the Zapote River basin where the city of Upala is located today, reaching as far as the Solentiname archipelago in Lake Nicaragua through the mouths of the Zapote River, and included the confluence of Frío River (*Ucúrinh*) and San Juan River to the east (see Figure 6.2).

The Maleku culture has been transmitted by elders and considers the continuous existence of the universe since its formation marked by the appearance of the first and male *Tócu*, from whose dream came the second female *Tócu* who accompanied him. After the first *Tócu* dreamed of other *Tócu*, they served as advisory council members and recognised him as their ancestor. This justified his self-proclamation as the creator spirit of the Venado River's headwater called *Nharíne* (Constenla, 2003).

Each Maleku SNS at the ancestral territory river's headwaters is the space of the respective *tócu lhónh maráma*, for Maleku people who had a "good death"

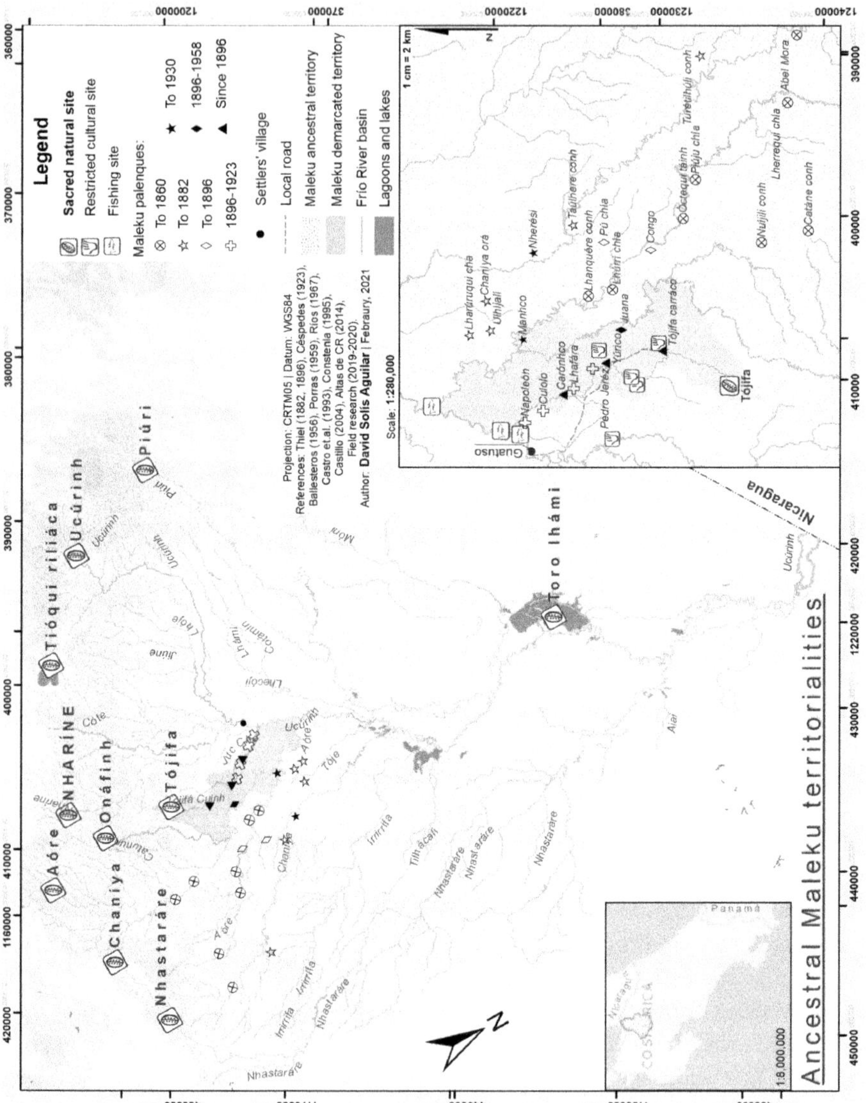


FIGURE 6.2 Maleku ancestral territory, 1860–1950. Credits: Map created by David Solis-Aguilar.

caused by illness or old age. The rivers' headwaters served as an access point for the Maleku to dwell there as *tócu lhónh maráma*. The exact place was determined by their specific clan affiliation, those whom they cared for during their life, as well as their preferred hunting and fishing sites (Bozzoli, 1973; Castro et al., 1993; Constenla, 1982, 1983).

The interaction of the Maleku ancestors in the Frío River basin, based on their practices and beliefs, makes it possible for us to identify their main trajectories across the ancestral territory in approximately 17 settlements or *palenques* on the banks of *Aóre*, *Ulhíjali* and *Chaníya* Rivers prior to colonisation in the 1860s. These *palenques* had adjacent cultivation and food gathering areas and a population of at least 1,050 inhabitants (Borge, 1992; Castillo, 2004, 2005).

Social cohesion among the ancestral Maleku inhabitants of the *palenques* was strengthened by the traditional trip to the sacred site *Toro lhámi*, cared for by the deity *Lhafára*, in the Caño Negro wetlands, through sail on the Frío River at the end of the rain season in March. This trip took about 5 days downstream and up to 8 days upstream and followed strict rule of their culture to obtain valuable proteins from different fish, iguanas and turtles. This seasonal journey culminated in a large celebration at the same site of departure in the Pataste River banks (Castillo, 2015).

In summary, the headwaters of the Frío River tributaries and the *Toro lhámi* wetlands lakes make up the Maleku's sacred landscapes, as a portion of the ancestral territory transformed over time by the relationship with its surrounding nature composed of spiritual beings present there, a sign of Maleku's deep appreciation for their SNSs (Madrigal, 2017). The sacred sites for the Maleku ancestors were the main *Tócu* place at the headwaters of the *Narínhe* River, who had a position of authority over Frío River basin. This area, in turn, constituted the sacred landscape of the spaces walked and rivers that were sailed and navigated by the Maleku ancestors.

## Dispossession by Settlers

Between 1860 and 1899, the Maleku people suffered multiple forms of violence and the proliferation of infectious diseases, which reduced their population by 82% (Castillo, 2005). The brutal acts and events were systematically perpetrated by the trespassing of male groups from Nicaragua into their ancestral territory to extract wild rubber. They established a trade route and enslaved the Maleku children on fort San Carlos fluvial port in Nicaragua (Edelman, 1998).

Because of the Nicaraguan settler's expansion on Maleku ancestral territory, the relationship to cultural spaces was profoundly transformed. They were forced to reorganise the territory's core, consisting of at least 17 settlements on the banks of the *Aóre*, *Ulhíjali* and *Chaníya* Rivers, to survive in the face of the rubber extractors' aggressions. By the 1890s, Maleku people were displaced to the banks of the *Tójifa* and *Onáfinh* Rivers (Zeledón, 2003). This implied the reorganisation of their clans to counteract the decrease in population and possible eradication.

As a consequence, this resulted in the reorganisation of the management of fishing sites, hunting sites, camps and routes for the extraction of plant and timber products.

The transformation of Maleku interrelationships since the 1860s was violent, and after three decades of struggle for survival, their living spaces were relocated to take refuge from settlers' new attacks. The Maleku ancestors had to reorganise agricultural, fishing, hunting and gathering sites among the remaining inhabitants in eight *palenques* by 1900. This is because the forced displacement from the settlements formerly founded on *Aóre* River banks to *Tójifa* River banks established three new settlements named *Carónhco, Tófijá carráco* and *Yúrico* – the *maleku lhaíca* languages toponyms for current *palenque* El Sol, *palenque* Tonjibe and *palenque* Margarita.

In turn, at the beginning of the 20th century, the Maleku relationship to the fishing and hunting trip carried out at the end of the dry season in *Toro lhámi* wetland was temporarily interrupted by the establishment of the Nicaraguan rubber settler camp in current Caño Negro area, which also caused the permanent change of some sites used during that fluvial trip through *Ucúrinh* River (Castillo, 2015). That change involved the development of new spiritual links with their SNSs at Frío River.

The spiritual connection with SNSs wasn't transformed during the first half of the 20th century (Porras, 1959; Ríos, 1967), but the organisation of the eternal abodes for the *tócu lhónh maráma* was redefined. Thus, the Maleku ancestors from the current settlement have other relationships with SNSs. Those from *palenque* Tonjibe are said to have spiritual dwellings in *Nharíne chaconhe* and *Tióqui riliáca*; the people from *palenque* El Sol dwell in *Piúri chaconhe* and *Ucúriqui chía*, and those from *palenque* Margarita dwell in *Tioqui riliáca* and *Piurí chaconhe* (see Figure 6.2). This implied the displacement of many spiritual relationships with SNSs located at other rivers' headwaters.

During the 20th century, logging activities, cattle ranching and growing of crops around the Maleku ancestral territory were driven by Costa Rican colonisation efforts. This took place from the south towards the so-called "agricultural frontier" which clearly implied the commercialisation of the biophysical environment by settlers (Goebel, 2017; Hall, 1984). During the 1950s, the establishment of settlers near the Venado and Purgatorio Rivers' headwaters west of San Carlos County (Flores, 2014) involved a long process of dispossession of Maleku hunting and fishing sites linked to the SNSs of *Nhastaráre* and *Chaníya*.

The Maleku seemed to be cornered by settlers, who only intensified their acts with violent threats and aggressions with firearms and hunting dogs. Eventually, the lands surrounding the rivers' headwaters were seized by settlers (Ballesteros, 2017; Molina, 2017). The Maleku way of life was transformed in the 1960s and 1970s through the loss of land, even in the vicinity of the three *palenques*, by nearby farmers vying to expand their agricultural fields (Ríos, 1967). In face of the colonisers' siege, the Maleku people feared that they would lose their farmland if they were absent for even a few weeks (Bozzoli, 1975).

## A Little (Un)protected Land

The context of agricultural colonisation in northern Costa Rica, particularly in the Frío River basin, and the related violence against the bodies, lands and rivers of the Maleku people, motivated a government agency in 1957 to propose the partial protection of Maleku territory by 14,041.7 hectares, including the river headwaters' sacred sites at *Narínhe chaconhe, Aóre riálica, Onáfiqui chía* and *Tójifa facára* (Arguedas, 1973).

It was not until 1976 that the government decreed the establishment of the "Guatuso Indigenous Reservation", with an estimated extension of 2,993.41 hectares based on official cartographic records. Article 6 of this legislation established that all "The Indigenous Reservations are (...) exclusive for the aboriginal communities that inhabit them (...)" [self-translation] (Presidencia de la República, 1976). The creation of the Maleku demarcated territory in 1976 included only *Tójifa facára* sacred site at Sol River's headwaters, and excluded nine other SNSs, and considered an area of less than 3% of the Maleku ancestral territory.

The Maleku's land dispossession on the surroundings of the three *palenques* continued after the creation of the demarcated territory in 1976 and didn't subside with legal protection. The living conditions for the Maleku people worsened due to the reduction of the legally demarcated territory to 2,743.56 hectares, by a government decree in 1977, an area of which Maleku only owned 15% by 1988 (CONAI, 1984; Tenorio, 1990). In 1990, the Maleku recovered a small portion of their land through the peaceful occupation of farms owned by the non-Indigenous (Guevara & Chacón, 1992), increasing their ownership to 18.9% of the demarcated territory under the second decree in 1977 (Morales, 1996).

Subsequently, through legal actions, the first land delimitation of 1976 was restored (Sala Constitucional, 1999), consolidating the current Maleku demarcated territory. Since 2020, through the peaceful occupation of lands illegally owned by non-Indigenous farmers, the Maleku people have increased their possession to 28.8% of the demarcated territory under the first decree in 1976 (FRENAPI, 2020; INDER, 2019, 2020). The persistence of unequal land occupation by the Maleku within their current demarcated territory affects the SNS *Tójifa facára* at Sol River's waterhead, because this waterbody is owned by non-Indigenous farmers (see Figure 6.1).

The continuum of structural violence shows that the Maleku people survived a cultural genocide since the end of the 19th century within a context of settler colonialism, as native people unprotected by government authorities in the face of agricultural colonisation, like many other native peoples in the Americas (Naccache, 2019).

## Fighting for Conservation

The conservation policies imposed on the Frío River basin by government authorities since the late 1970s have caused a significant impact on Maleku traditional activities and the expression of their cosmological beliefs – the problem is focused in the two protected areas established on their ancestral territory. In 1984, the government decreed the creation of the RNVS-CN, located in the Frío River lower watershed. It is composed of a seasonal lake system, which was included in the Ramsar Convention's list of wetlands in the 1990s. The RNVS-CN is characterised by a great biodiversity, but is heavily impacted by deforestation, wildfires, agrochemical pollution, sedimentation and the dredging of wetlands linked to extensive cattle ranching inside the NPA and pineapple monoculture in its surroundings, as well as illegal hunting and fishing (Brenes, Pérez, Vargas, & Zúñiga, 2016; Corrales, 2018; Fournier et al., 2018).

*Toro lhámi* in the RNVS-CN is still part of the Maleku traditional fishing journey which has been carried out by car since the 1990s, and was authorised previously by the National System of Conservation Areas (SINAC), the government conservation management agency. For the current traditional fishing trip to *Toro lhámi*, the Maleku are prohibited by SINAC officials to hunt species included in their traditional diet like turtles and iguanas; also they have been pressured to fish for species not consumed by their elders such as tilapia (*Oreochromis niloticus*). The rules imposed on the Maleku, despite the inclusion of their culture as a "conservation target" in the current RNVS-CN Management Plan (Cornejo, 2012; SINAC, 2012), fail to guarantee the effective participation of the Maleku in the management of the protected areas.

Controls and restrictions on Maleku cultural practices in the RNVS-CN protected area have been extended with the persecution of fishing in wetlands adjacent to the Frío River. SINAC's persecution of Maleku fishing since the 1990s implied the requirement of permits and frequent confiscations, ignoring the significance of fishing as an activity for family food, with traditional techniques such as handcrafted lines (see Figure 6.3). Nevertheless, since 2011, changes in the public policy of criminal prosecution have prevented accusations against Maleku people based on anthropological expertise that recognises their cultural fishing practices (Fiscalía General, 2011, 2013; Guevara, 2011).

In the upper Frío River watershed, which is part of the Maleku ancestral territory, the PN/ZP-VT protected area is located. It was established as a "forest reserve" in 1976 with 17,450 hectares, subsequently redefined in 1990, and established in 1995 with its current limits and conservation categories. The Tenorio Volcano Protected Zone (ZP-VT) includes the SNS of Lake Cote and its headwaters *Tióqui riliáca*, while the Tenorio Volcano National Park (PN-VT) includes the Frío River headwaters' SNS *Ucúriqui chía*, as well as the Buenavista River's headwaters and waterfall named *Piúri chaconhe* and *Piúri facára* respectively.

The aforementioned sites were visited by Maleku people under strict rules of conduct, according to cosmological belief and faith, for spiritual ceremonies, hunting, fishing, gathering food and medicinal plants, as well as to extract raw materials for the manufacture of clothing, tools and houses (Castillo, 2004). The PN/ZP-VT features important sources of suita palm leaves (*Asterogyne martiana*), used by the Maleku as raw materials for the roofs of traditional houses (see Figure 6.4), which

**FIGURE 6.3** Two Maleku elders fishing on Ucúrinh or Frío River near to Toro lhámi SNS located at Caño Negro Wildlife Refuge. Photo Credits: David Solis-Aguilar.

**FIGURE 6.4** Ethnobotanist explains to his Maleku fellows the suita palm leaves sustainable extraction. Photo Credits: David Solis-Aguilar.

in the last decades have been used for tourism purposes in the currently demarcated Maleku territory. In addition, suita palm leaves are still used for the traditional burial of people who have died of illness or old age (SINAC, 2010).

The Maleku extraction of suita palm leaves in the PN/ZP-VT is prohibited by SINAC officials, who do not grant them the "forest use permits" according to regulations. At the same time, the issue of "subsistence licenses" for the extraction and collection of wild flora requires the Maleku people to submit an application to SINAC, a procedure without consideration of Maleku's living conditions, who would have to travel 55 km to submit a regular application. This situation has been recognised by SINAC in the PN/ZP-VT current Management Plan (SINAC, 2013), but it does not consider mechanisms adjusted to Maleku's traditional practices and their socio-economic conditions of poverty, which means violations of cultural rights related to suita palm uses.

The relationship of the Maleku people with the SNSs in the PN/ZP-VT has been under strain due to the confiscation of two suita palm leaves shipments by police in 2019, which groups of Maleku extracted with non-Indigenous land holders' prior authorisation. These palm leaves were destined for the construction of a Maleku traditional house in the *palenque* Tonjibe elementary school, but part of the leaves was rendered unusable (see Figure 6.5). The suita palm leaves confiscation in 2019 was performed despite the previous endorsement by the Maleku self-governance authority (ADI Maleku, 2015), according to the International Labour Organization Convention No. 169, approved in Costa Rica by 1992. In the end, the Maleku persons involved in this situation experienced no criminal charges, but their territorial rights violations were not acknowledged (Juzgado Penal, 2019).

The Maleku people's memories of ancestral territory include the SNSs *Nharíne*, *Aóre*, *Chaníya*, *Onáfinh* and *Nhastaráre*, respectively, placed in the headwaters of the Venado, La Muerte, Pataste, Cucaracha and Purgatorio Rivers. All these SNSs outside the currently demarcated territory and outside any NPA are located in lands owned by non-Indigenous farmers, and are protected by inapplicable rules related to "forest areas" or "headwater sources" according to environmental law (Asamblea Legislativa, 1992, 1996).

**FIGURE 6.5** Suita palm leaves seized in Guatuso police station, September 2019. Photo Credits: David Solis-Aguilar.

Accordingly, these SNSs have been subjected to deforestation from cattle ranching, and are not accessible to Maleku visitors, much less for traditional hunting or fishing. Therefore, the Maleku people want to build new conservation policies for SNSs in private hands. These still enjoy no specific environmental or cultural protection status, especially the main *Tócu* place in *Nharínhe chacone* at Venado River's headwater.

## Conservation Challenges

As we have shown, Costa Rica's conservation authorities have not yet effectively integrated a policy for legal recognition of the cultural and territorial rights of the Maleku people, according to international legal standards for Indigenous peoples. In spite of this, the Maleku since the 2000s maintain a constant dialogue with SINAC to achieve less restrictive access to the Caño Negro Wildlife Refuge, in order to safeguard the SNS *Toro lhámi*. Maleku people want to continue fishing for species which are culturally important, and to be authorised for turtle and iguana hunting as part of their traditional diet which has spiritual significance.

During the last two decades, Maleku's actions to preserve *Toro lhámi* have been focused on participation in forest fire brigades, and the RNVS-CN's Committee of Surveillance of Natural Resources. In addition, the Malekus have publicly condemned the environmental damage to the wetlands due to sedimentation, mainly caused by the expansion of pineapple cultivation for export and the dredging of nearby wetlands to enable livestock.

A new approach to the conservation of SNSs located in PN/ZP-VT (*Tioqui riliáca, Cóte chía, Piúri chaconhe* and *Ucúriqui chía*) has been demanded by Maleku elders from SINAC officials in 2020 and 2021, based on the legal recognition of territorial rights, specifically to participate in the NPA management and to get unrestricted access to the SNSs. This entails the recognition of suita palm leaves extraction for the construction of traditional roofs, without SINAC's restrictions or police surveillance, regardless of whether these are used for tourism activities in the *palenques* or not.

In addition, the Maleku people have raised the need to participate in the tourism business that SINAC exploits in PN-VT through the trail to their SNS *Piúri facára*. For this purpose, they have proposed to offer their services as cultural guides in public facilities, as well as the sale of their handicrafts.

The COVID-19 pandemic further limited economic opportunities for the Maleku people during 2020, particularly due to the sharp drop in international tourism in Costa Rica, which employed numerous Maleku families. Such circumstances prompted the process of autonomous land recovery in the Maleku demarcated territory, following the demand to government authorities to comply with the IA and the 1976 territorial demarcation decree. The current Maleku land recovery process looks for guarantees for the actual possession of all demarcated land, which implies the compulsory departure of non-Indigenous settlers, according to the IA.

The enforcement of the law in the Maleku demarcated territory would allow for conservation of the SNS *Tójifa facára*, as well as advocacy as a means for improving the integration of rights-based and landscape governance approaches (Blomley & Walters, 2019). The creation in August 2021 of the *Tócu Laca* Intercultural Forum between the Maleku people and SINAC is considered by the community as an opportunity to build a governance and management policy for SNSs.

In order to build a conservation policy for the Maleku SNS, international standards should be considered, such as ILO Convention No. 169 (1989) in its articles on land, territory and resources, the right to ownership and possession of ancestral lands as well as respect for their transmission modalities, protection against land dispossession, protection from being transferred to other lands, the right to participate in the conservation of their lands, and the right to free, prior and informed consent.

In addition, the Convention on Biological Diversity (Secretariat of the Convention on Biological Diversity, 1992) should be considered in Article 8 on In-situ Conservation in relation to respect, preservation and maintenance of knowledge, innovations, and practices of Indigenous and local communities. And the cultural, environmental and social impact assessment criteria proposed in the Akwé: Kon Voluntary Guidelines (2004) should also be included in the *Tócu Laca* Intercultural Forum discussions.

For this goal, the remaining territorial memories of the Maleku elders can help to build an effective rights-based community conservation approach for the Frío River tributaries' headwaters and wetlands. It is necessary to promote the Maleku's sacred landscape governance with community agency because of its ontological link with the rules that *Tócu* left since the beginning of the world, as is expressed by traditional Maleku spirituality, which establishes their ancestral territory in northern Costa Rica.

## References

Arguedas, U. (1973). Algunas notas acerca de los indios Guatusos. In M. E. Bozzoli (Ed.), *Colección Materiales de Estudio. Costa Rica. Patrones culturales de comunidades indígenas.* San José: CEDAL.

Asamblea Legislativa. (1992). *Ley de Conservación de la Vida Silvestre, N° 7317.* San José: SINALEVI.

Asamblea Legislativa. (1996). *Ley Forestal, N° 7575.* San José: SINALEVI.

Asociación de Desarollo Integral de la Reserva Indígena Maleku. (2015). *Normas internas maleku de pesca y extracción de hojas de palma suita: Lista de participantes de pesca colectiva en el RNVS Caño Negro.* Comunicación a la Fuerza Pública y al Ministerio de Ambiente y Energía.

Ballesteros, R. (2017). Mi vida entre los guatusos: 1956. In E. Zeledón (Ed.), *Los aborígenes de Costa Rica: textos históricos, periodísticos y etnográficos* (1° ed., pp. 342–380). San José: EUNED.

Blomley, T., & Walters, G. (Eds.). (2019). *A landscape for everyone: integrating rights-based and landscape governance approaches.* Gland: IUCN.

Borge, C. (1992). *Historia de la colonización de los ríos Frío, Zapote y Pizote en la Zona Norte de Costa Rica* (1° ed.). *Proyecto Regiones Fronterizas.* San José: UCR.

Bozzoli, M. E. (1973). Palenque Margarita y Tonjibe de los indios guatusos. Una visita: Costa Rica: Patrones culturales de comunidades indígenas. In M. E. Bozzoli (Ed.), *Colección Materiales de Estudio. Costa Rica. Patrones culturales de comunidades indígenas.* San José: CEDAL.

Bozzoli, M. E. (1975). *Localidades indígenas costarricenses* (2° ed.). *Colección Aula.* San José: EDUCA.

Brenes, R., Pérez, L., Vargas, Ó., & Zúñiga, E. (2016). *Plan de Control y Protección Refugio Nacional de Vida Silvestre Caño Negro: Tesina para optar por el título de Diplomado en Manejo y Conservación de Áreas Protegidas para Guardaparques.* Heredia: UNA.

Castillo, R. (2004). An ethnogeography of the Maleku indigenous peoples in Northern Costa Rica (Thesis). University of Kansas.

Castillo, R. (2005). Población Indígena Maleku en Costa Rica en el S. XIX. *Anuario de Estudios Centroamericanos, 31,* 115–136.

Castillo, R. (2015). La relación geográfica histórica cultural de los indígenas Maleku con los humedales de Caño Negro. *Vínculos. Revista de Antropología del Museo Nacional, XXXVI*(1), 49–80.

Castro, E., Blanco, A., & Constenla, A. (1993). *Laca majifijica: La transformación de la tierra* (1. ed.). San José: ECUR.

Comisión Nacional de Asuntos Indígenas. (1984). *Censo de tenencia de la tierra en la Reserva Indígena de Guatuso,* Departamento Territorial, CONAI.

Constenla, A. (1982). Algunos aspectos de la etnografía del habla de los indios guatusos. *Estudios de Lingüística Chibcha, 1,* 5–31.

Constenla, A. (1983). Anotaciones sobre la religión tradicional guatusa. *América Indígena, XLIII*(1), 97–123.

Constenla, A. (2003). Dos textos guatusos sobre los profetas del cataclismo. *Estudios de Lingüística Chibcha, XXII,* 61–128.

Cornejo, E. (2012). La integración y legitimidad de los derechos culturales e históricos del pueblo Malecu: En la figura de un comanejo sostenible de los recursos naturales y culturales para la conservación y restauración integral del Refugio de Vida Silvestre Caño Negro (Thesis). UNED.

Corrales, L. (2018). *Ecosistemas Vegetales del Complejo de Humedales de Caño Negro: Los Chiles, Costa Rica* (1 ed). Turrialba: SINAC.

Edelman, M. (1998). A Central American Genocide: Rubber, slavery, nationalism, and the destruction of the Guatusos-Malekus. *Comparative Studies in Society and History, 40*(02), 356–390.

Fiscalía General. (2011). *Política de persecución penal y abordaje de causas indígenas, N° 13-ADM-2011.* San José: Poder Judicial.

Fiscalía General. (2013). *Metodología de trabajo entre la Fiscalía de Asuntos Indígenas y las Fiscalías Adjuntas Territoriales que tramitan causas indígenas, N° 10-ADM-2013.* San José: Poder Judicial.

Flores, C. (2014). *Plan desarrollo Distrital de Monterrey y de Venado* (1 ed.). Ciudad Quesada: Municipalidad de San Carlos.

Fournier, M., Echeverría, S., Mena, F., Arias, M., De La Cruz, E., & Ruepert, C. (2018). Risk assessment of agriculture impact on the Frío River watershed and Caño Negro Ramsar wetland, Costa Rica. *Environmental Science and Pollution Research International, 25*(14), 13347–13359.

Frente Nacional de Pueblos Indígenas. (2020). *Amenazas y asedio de finqueros en conflicto por tierras/territorio en la comunidad indígena Maleku, al Norte de Costa Rica.* FRENAPI. San José, Costa Rica.

Goebel, A. (2017). Bosques, fincas y ciudades. Un acercamiento al proceso socio-metabólico de apropiación en la Región Norte de Costa Rica (1909–1955). *Horizontes, 75*, 13–51.

Guevara, M. (2011). *Informe de peritaje cultural o antropológico sobre el expediente 11–200199-630-PE: 26 de septiembre.* Informe entregado a la Fiscalía de Asuntos Indígenas. UCR.

Guevara, M., & Chacón, R. (1992). *Territorios indios en Costa Rica: Orígenes, situación actual y perspectivas.* San José: García Hermanos S.A.

Hall, C. (1984). *Costa Rica, una interpretación geográfica con perspectiva histórica.* San José: ECR.

Instituto de Desarrollo Rural. (2019). *Levantamiento de la ocupación en Territorio Indígena Maleku: Plan Nacional para la recuperación de Territorios Indígenas*, San José: INDER.

Instituto de Desarrollo Rural. (2020). *Informe sobre el Plan-RTI y sus avances, Oficio, INDER-GG-DRT-FT-474-2020.* San José, Costa Rica: INDER.

Instituto Nacional de Estadísticas y Censos. (Ed.) (2013). *Territorio Indígenas. Principales indicadores demográficos y socioeconómicos: X Censo Nacional de Población y VI de Vivienda 2011.* San José, Costa Rica: INEC.

Juzgado Penal. (2019). *Causa penal por la corta y extracción de palma suita de la Zona Protectora del Volcán Tenorio realizado por personas Maleku,* (190002460630PE). Upala: Poder Judicial.

Madrigal, V. (2017). Cartografía de lo sagrado. Una exploración en la espiritualidad malécu. *Siwo Revista de Teología, 10*(3), 45–59.

Molina, L. A. (2017). Un año con los indios guatusos: 1973. In E. Zeledón (Ed.), *Los aborígenes de Costa Rica: textos históricos, periodísticos y etnográficos* (1° ed., pp. 384–392). San José: EUNED.

Morales, S. (1996). *Lista de poseedores de tierras dentro de la Reserva Indígena de Guatuso.* San José: Fundación Iriria Tsochock.

Naccache, G. (2019). Genocide and Settler Colonialism: How a Lemkinian concept of genocide informs our understanding of the ongoing situation of the Guarani Kaiowá in Mato Grosso do Sul, Brazil. In J. S. Bachman (Ed.), *Studies in genocide and crimes against humanity. Cultural genocide. Law, politics and global manifestations* (pp. 118–139). New York: Routledge.

Porras, Á. (1959). El idioma Guatuso: Fonética y lexicografía (Thesis). Universidad de Costa Rica, San José.

Presidencia de la República. (1976). *Establece las Reservas Indígenas Guatuso, Chirripó, Guaymí de Coto Brus, La Estrella y Talamanca* (5904-G). San José: SINALEVI.

Rights and Resources Initiative. (2020). *Rights-based conservation: The path to preserving Earth's biological and cultural diversity?* Washington, DC: RRI.

Ríos, E. (1967). *Consideraciones sobre los indios Guatusos: Para la Cátedra de Ciencias del Hombre.* San José: Laboratorio de Etnología, UCR.

Sala Constitucional. (1999). *Sentencia al Recurso de Amparo interpuesto por indígenas Maleku con expediente No. 96–007361-007-CO-C, 06229-99.* San José: Poder Judicial.

Secretariat of the Convention on Biological Diversity. (1992). *Convention on Biological Diversity,* (International Agreement). New York: United Nations.

Secretariat of the Convention on Biological Diversity. (2004). *Akwé: Kon Voluntary guidelines for the conduct of cultural, environmental and social impact assessment regarding developments proposed to take place on, or which are likely to impact on, sacred sites and on lands and waters traditionally occupied or used by indigenous and local communities.* CBD Guidelines Series, Montreal.

Sistema Nacional de Áreas de Conservación. (2010). *Estudio etnográfico de la comunidad Maleku: La cosmovisión de la comunidad indígena Maleku y su relación con el Plan de Manejo del Parque Nacional Volcán Tenorio*. Tilarán: SINAC.

Sistema Nacional de Áreas de Conservación. (2012). *Plan de Manejo Refugio Nacional de Vida Silvestre Caño Negro 2012–2020*. Ciudad Quesada: SINAC.

Sistema Nacional de Áreas de Conservación. (2013). *Plan General de Manejo: Parque Nacional Volcán Tenorio y Zona Protectora Tenorio*. Tilarán: SINAC.

Solís Aguilar, D. (2021). Territorialidades del pueblo originario maleku en Costa Rica (Thesis). El Colegio de Michoacán. La Piedad, Mexico. http://colmich.repositorioinstitucional.mx/jspui/handle/1016/1102

Tenorio, L. A. (1990). *Reservas indígenas de Costa Rica* (2° ed.). *Colección indigenista*. San José: CONAI.

Zeledón, E. (2003). *Crónicas de los viajes a guatuso, talamanca del obispo Bernardo Augusto Thiel, 1881–1895* (1° ed.). San José: EUCR.

# PART III

# Examining Themes in Faith-Based Conservation

# 7

# LEGITIMATING AND RESPECTING SACRED GROVES – IMPORTANT FOR HUMAN RIGHTS AND IMPORTANT FOR CONSERVATION

*Edmund Barrow*

## Introduction: Recognizing and Respecting Sacred Groves

Where there is a tree, there is a sacred tree (Barrow, 2019). Sacred Natural Sites (SNSs) are where people and nature meet spiritually (Verschuuren, Wild, McNeely, & Oviedo, 2010). Conservation can help secure sacred groves, empower local custodians and stewards, and protect such groves through the protected area categories. Documenting Indigenous and Community Conserved Areas (ICCAs) helps protect them as conserved areas. But compromise may be needed – recognition in return for greater respect and protection; otherwise, they may be 'hidden and unknown' and so subject to degradation and conversion. ICCAs are natural and/or modified ecosystems containing significant biodiversity values, ecological services, and cultural values, voluntarily conserved by Indigenous Peoples and Local Communities (IPLCs), both sedentary and mobile, through customary laws or other effective means (Kothari, Corrigan, Jonas, Neumann, & Shrumm, 2012). Sacred groves are ICCAs and their custodians are the key owners and managers.

Sacred groves are uninhabited, with strict rules, and are part of landscapes, often forested, with well-defined features, delimited and protected by societies through traditional arrangements (Ramakrishnan, Saxena, & Chandrashekara, 1998). They are created, worshipped in, and protected by local people for cultural and spiritual reasons (Peng, Ning, Zhaoli, & Shengji, 2003). Japanese Shinto and Buddhist sacred groves cover over 110,000 hectares in Japan (Fukamachi & Rackham, 2012). Ghana has over 2,000 groves (Ntiamoa-Baidu, 1995). China has 400 sacred village forests in Yunnan Province alone (Peng et al., 2003). India has over 14,000 described sacred groves but estimates suggest that between 100,000 and 150,000 may exist (Barrow, 2019; Chandrakanth & Romm, 1991).

DOI: 10.4324/9781003246190-10

Though sacred groves occur in most countries, fewer occur in Europe or North America. Many still occur in Africa, India, Asia, and Latin America. In Europe and North America, many sacred groves were destroyed or degraded due to the Industrial Revolution (Zucchelli, 2009), settlement of North America, and religion. However, many still survive and others are being revived. Yet, the relevance of sacred groves remains downplayed by 'modern society' and by conservation (Barrow, 2015).

Threats to sacred and monastic groves include (a) changes in socio-cultural practices and values; (b) erosion of religious beliefs and traditional values; (c) conversion due to land and population pressures combined with in-migration, and a loss of a sense of place; and (d) insecurity of rights. This resulted in degradation, encroachment, conversion, and over-exploitation – all human-induced threats and now there is also climate change (Khan, Khumbongmayum, & Tripathi, 2008; Ormsby & Bhagwat, 2010). This chapter focuses on how to strengthen the security of sacred and monastic groves and their custodians.

## How Sacred Groves Are Being Formally Recognized

There are examples of sacred groves being formally respected, though most have customary recognition. For example, Kaya forests along the coast of Kenya are national monuments (Nyamweru et al., 2008a). There are examples of religions owning the land on which a sacred forest is located (e.g., Montserrat in Europe, or monastic forests in Ethiopia). They may be formally respected as part of a National Park (e.g., Ghana and Europe) or may be declared ICCAs. The following sections provide examples of how sacred trees and groves can be better secured. At the end of each of these sections, I provide a short paragraph on lessons.

### *Sacred Trees*

Fig, Yew, Oak, Olive, and Baobab trees are sacred in most religions and cultures where they grow naturally. Certain trees are gazetted as national monuments and many of these are sacred. Recently, a fig tree in Nairobi was declared a national monument to protect it from being felled for a road.[1] Many Baobab trees in West Africa are national monuments, especially in Senegal (Lewington & Parker, 1999). The 2,000-year-old olive trees of the Garden of Gethsemane in Jerusalem are protected by the three main Abrahamic religions (Hareuveni, 1980; Waisel & Alon, 1980).

Some Buddhists in Asia ordained trees to protect them from logging. For example, a Buddhist Monk in the Karen area of Wat Chan, Thailand ordains trees to protect them from cutting (Burbea, 2015) by wrapping them in the sacred orange cloth, signifying that one cannot take the forest out of Buddhism. He also asked villagers to donate lands of forgiveness, where they forbid tree felling. As a result, they agreed that no trees would be cut, as local people working for illegal loggers felt that this was like killing a monk. In Cambodia, monks wrap monastic robes around trees, which ordain such trees as monks (Burbea, 2015). In Europe, trees are adorned with pieces of cloth and prayers (see Figure 7.1).

***Lessons***: Important trees can be sacred and protected in law – but it requires political will. Where cultural links with trees and forests are strong, tree ordination can be a potent force for securing trees and groves and this is more common in Asia than elsewhere.

**FIGURE 7.1** Prayer Tree adorned with cloth and prayers; Co. Tipperary, Ireland. Photo Credit: Edmund Barrow.

## Sacred Groves in Formal Protected Areas

In South-West China in Yunnan Province, the Dai Holy Hill Forests are protected. By the 1960s, many had been destroyed, but are now being restored. There are at least 447 temple sites in Yunnan, including Tibetan, Chinese, and Taoist Buddhist temples, mosques, and churches (Peng et al., 2003). The Holy Hills demonstrate the role of culture in conservation as cultural and religious values may be more sustainable than legislation (Hongmao, Zaifu, Youkai, & Jinxiu, 2003). The Holy Hills connect nature reserves with other forms of land use, and are often the only remaining rainforest found outside the Xishuangbanna National Nature Reserve. Now they are being integrated into a biodiversity corridor because they are well adapted and found in lower tropical biodiversity-rich foothills (Peng et al., 2003).

Protecting groves is strong in the cultures of some indigenous people in and around some Russian National Parks. Some are in National Parks, e.g., Kenozesky National Park. Sacred groves are part of Zapovedniks, which are nature sanctuaries which can incorporate historical–cultural, historical–archaeological, spiritual, and other types of cultural or natural heritage, and are strictly protected State nature reserves. In north Russia, around Lake Kenozero, there are 45 documented sacred groves (Pungetti & Bhagwat, 2012) which are respected as part of National Parks.

Sri Pada Peak (Adam's Peak) in Sri Lanka is a formal wilderness area and the highest point of Sri Lanka's Central Mountains. The peak is important for conservation, spiritual leaders, and the public, because of its commanding position, and the sacred footprint on the rock summit. This wilderness area safeguards the sacred footprint (*Sri Pada*) of Lord Buddha. Buddhists believe that the sacred footprint is the left foot of Lord Buddha. Muslims believe that it is the footprint of Adam, who stood on the peak for 1,000 years. Christians believe that it is the footprint of St. Thomas, who is said to have brought Christianity to Sri Lanka. Hindus believe that it is the footprint of Hanuman or Shiva (Wickramasinghe, 2006).

***Lessons:*** Respecting the Holy Hill Forests of Yunnan in China as cultural areas and integral parts of village landscapes is now supported by the government, which strengthens their security. These hill areas survived years of degradation caused by political ideology. Where protected areas are gazetted, they can formally respect sacred groves and their custodians but this requires political and protected area managerial effort and support. The combination of the importance of an area for multiple religions serves to better secure the area, for example, Sri Pada Peak. On State-owned lands, sacred groves must be part of equitably negotiated joint management plans. Beliefs can make it possible to embrace diverse ethnic and religious groups as part of common interest to safeguard SNSs.

## Sacred Groves

Kayas are relic forest patches sacred to the Mijikenda people (Githitho, 2006; Nyamweru et al., 2008a). Between 1996 and 1999, 39 Kayas, of an estimated 91, were declared national monuments which protected nearly 2,000 ha. of biodiversity-rich coastal forest. Eleven are now on the World Heritage List.[2] Kayas occur along the Kenyan coast (Robertson, 1987). However, the legislation needs strengthening to better protect Kayas by (a) strengthening local institutions, (b) developing management bodies, and (c) establishing sustainable financing. The biggest problem faced by Kayas is, ironically, tourism, as hotel construction resulted in Kayas being destroyed (Nyangila, 2012). Local communities try to protect their Kayas, but elders said, '*how can we* (the community) *campaign for the preservation of sacred forests or any forest at all, if our leaders grab the only remaining forests*' (Standard Newspaper 27/9/1995).

Ghana formally and legally respects many of its sacred groves, as they conserve biodiversity, protect ecosystems, and regulate the exploitation of nature. Small patches of forest are set aside near settlements, as sacred, and cannot be touched (Ntiamoa-Baidu, 1995). The Boabeng-Fiema Monkey Sanctuary is one example which supports Black and White Colobus Monkeys, sacred to the people there (Ormsby, 2012). In Ghana, the authority for groves varies, but tends to be vested with the King, though functionally authority lies with the village chief and elders.

More sacred groves occur in India than anywhere else in the world (Ramakrishnan et al., 1998). People dedicate sacred groves to local deities or ancestral spirits, and protect them through social traditions (Bhagwat, 2012). But modern pressures weaken these traditions, though people recognize that traditional knowledge and sacred practices are important (Kothari & Das, 1999). Also, British colonialists and post-independent government managers undermined such traditions. Where sacred groves are found on common property lands, legal security is weaker. For example, a plywood industry harvested *Ficus nervosa* as a preferred species and many sacred groves were cut to supply timber (Gadgil, 1992). Many groves are found in Forest Reserves and National Parks (see Figure 7.2) which gives them added security, provided they are part of joint management plans (Malhotra, Gokhala, & Chatterjee, 2001; Ormsby & Bhagwat, 2010). Yet, it is unclear how many sacred groves (or what percentage) are in protected areas. Indeed, some were important for demarcating protected areas, for example, Shai Hills Resource Reserve in Ghana.

Registering sacred groves as ICCAs is important; yet, few are in the ICCA registry (Corrigan et al., 2016). If all sacred groves of the world were recognized, the number of ICCAs would increase by over a quarter of a million (Barrow, 2019)! For ICCAs, local communities are the key decision makers (Kothari & Pathak, 2008), and conservation is achieved directly or as a co-benefit. It is estimated that ICCAs conserve over 12% of the Earth's terrestrial surface, though formally registered ICCAs or OECMs are much fewer (Rights and Resources Initiative, 2020). Unfortunately, official documentation underestimates this scale and does not include most sacred groves. There is a global ICCA registry[3] and ICCAs are an opportunity to recognize and respect thousands of sacred groves. Sacred groves are a great fit with ICCAs and offer opportunities to respect them at national, regional, and global levels as an important conservation strategy, rather than a continued focus on State-owned protected areas (Tauli-Corpuz, Alcorn, & Molnar, 2018).

FIGURE 7.2 Bhimashankar Wildlife Sanctuary, Maharashtra, India. Photo Credits: Ashish Kothari.

***Lessons***: Kaya forests and their cultures are at risk unless better protected in law and politically. Most countries have legislation about national monuments which can be used. But often such legislative support is weak, as the destruction of some 'legally' protected Kaya forests demonstrates. The State can designate sacred groves and respect them – but this usually requires political and customary support. ICCAs can be formally respected at a national ICCA registry level – but this requires political support, and needs careful documentation and FPIC by the custodians.

### *Monastic Forests*

Ethiopia and Eritrea illustrate the importance of monastic forests. These forests provide many daily monastic requirements and are islands of biodiversity in seas of deforested landscapes. Ethiopia has over 35,000 churches and monasteries, and some are over 1,600 years old (Berhane-Selassie, 2008). Monastic forests demonstrate the churches' tenure and are a source for restoration and learning, though the quality and area of many monastic forests are declining.

Monastic forests in Europe are often 'hidden', yet are important. Many European countries established protected areas on sites of former or existing monastic lands, which should be respected as SNSs (Wild & McLeod, 2008). Christian monasteries and their lands thrived in Europe and some are 1,800 years old (Mallarach, 2012). The monastic communities manage these lands, which often have high biodiversity value. While there are no detailed inventories of such monasteries, estimates indicate that thousands still exist. Many forested monasteries contain old growth trees which are important sources of biodiversity for forest restoration. The Santa Creu Hermitage at the Holy Mountain of Montserrat in Catalonia, Spain, has been inhabited for over 1,400 years. Now it is a nature reserve within Montserrat National Park (Mallarach, Corco, & Papayannis, 2016). These monastic communities are self-organized and have continuous conservation records. Most predate formal protected area movements by hundreds of years. Their records show a focus on ecological integrity and landscape diversity.

***Lessons:*** Ancient monastic forests are important repositories for restoration, fostering awareness about conservation, and promoting sustainability. Monastic forests might be well respected by the church, but they also require political and policy support, as well as commitment from the State. Europe needs to respect how important sacred forests on monastic lands are for conservation, and how many occur in National Parks.

## Discussion: How Can We Strengthen Rights and Responsibilities

Sacred groves address shortcomings of formal protected areas (Bhagwat & Rutte, 2006). Different forms of protection can be given, based on the IUCN categories (IUCN & UNEP, 1986) and national regulations (Table 7.1). A global State protected area network may be good for conservation, but can come at a cost to IPLCs as their rights can be ignored (Schreckenberg, Franks, Martin, & Lang, 2016). Custodians of sacred groves require governments to respect, recognize, and have enabling policies so that the threats sacred groves and their custodians face can be addressed (Bhagwat & Rutte, 2006). The scale of sacred groves is testament to their resilience in the face of population and land use pressures. Yet, few sacred groves are recognized as/in formal protected areas. Such SNSs are the world's oldest protected areas, as the Yellowstone model is only 140 years old (Dudley, Higgins-Zogib, & Mansourian, 2009; Verschuuren, 2016), and some are thousands of years old, e.g., Garden of Gethsemane.

Development caused the destruction of many sacred groves, as governments often ignored communities' customary management systems, while allowing commercial forestry, conversion to agriculture, or gazetting National Parks. Yet, conserving sacred groves links nature, culture, and lives. They can be of great value for conserving areas of high biodiversity, sanctuaries for rare species, sites protecting freshwater species, and can be important for restoration (Schaaf, 2003). Other lessons include to:

1 respect sacred groves in terms of the spirituality of people irrespective of religious denomination, spiritual tradition, or land use;
2 appreciate the variety of social institutions responsible for their management;
3 redefine religious responsibilities with respect to the environment;
4 conserve biodiversity, even though conservation may not be a primary goal;
5 integrate spiritual perspectives into conservation, landscape connectivity, and restoration; and
6 respect sacred groves as one form of ICCA.

**TABLE 7.1** Examples of Sacred Groves in IUCN Protected Area Categories

| *IUCN category* | *ICCA type & examples of sacred sites/groves* |
|---|---|
| *Ia & Ib: Strict Nature Reserves* | *Sacred groves, lakes, springs, mountains, islands, etc. Many sacred groves, e.g., Garden of Gethsemane and Cedar Groves of Lebanon.* |
| II: National Park | Community declared sanctuaries, e.g., Ghana and India. |
| III: Natural Monument | Community protected (caves, waterfalls, cliffs, rocks), e.g., Kaya forests as national monuments. |
| IV: Habitat, Species Management Area | Sea turtle nesting sites, watershed forests, community managed wildlife corridors & riparian areas, e.g., Shaman forests, sacred groves in Ghana, and watershed forests in Thailand. |
| V: Protected Landscape, Seascape | Sacred & cultural landscapes & seascapes, e.g., Himalayas, Ganges, Mt. Kenya, and many include sacred groves. |
| VI: Managed Resource Protected Area | Resource reserves, e.g., community forests, grasslands, and waterways, which are under communal rules to assure sustainable harvesting, e.g., Monastic forests, some sacred groves in India, Ghana, and China. |

The protected area movement focuses more on National Parks, less on other categories. And there is a reluctance to recognize protection for cultural values at national levels (Harmon & Putney, 2003). Yet, some protected area authorities appreciate sacred groves in terms of conservation, while spiritual traditions recognize the security protected area status offer (Verschuuren, 2012). Such values should be part of land use planning and formal conservation (Jonas et al., 2017), together with the hundreds of thousands of sacred groves.

There is a focus on expanding the formal protected area network to achieve the goals of the *Convention on Biological Diversity (CBD)*, the Sustainable Development Goals (SDGs), and the Paris Climate Change Accord (Waldron & 87_authors, 2020). But we cannot do this by coercion or as a 'land-grab' from IPLCs (Tauli-Corpuz et al., 2018). Such coercion exacerbates relations between communities and conservation. We would better achieve expansion by respecting, recognizing, and securing the rights and responsibilities of IPLCs. In this way, huge areas can come under the 'flag' of protected areas and conservation, and under different protected area categories.

Sacred groves do just that – communities conserve, at little or no external cost, large areas of land across the globe. These are powerful examples of ICCAs and contribute to conservation, create connectivity, and are locally owned. A recent analysis estimates that communities contribute $3.16–$4.52 billion per year on ICCA management – approximately 16–23% of the total formal investment in protected areas (Tauli-Corpuz et al., 2018). If sacred groves occur in protected areas or are recognized ICCAs, recognition by government authorities can increase protection, if the custodians are supportive.

Sacred groves do not get adequate legal protection. Some are included, and given protection, as a National Park, while others are designated as national monuments. Only four countries in Africa – Lesotho, Tanzania, Mozambique, and Kenya – have legal provisions for local communities to manage sacred groves (Nyamweru & Sheridan, 2008b). Other countries afford greater legal protection for SNSs. For example, many Australian National Parks were handed back to Aboriginal owners (Smith, 1997). Now, in Australia, many sacred groves are part of National Protected Areas, which includes Forest Reserves and World Heritage Sites. This is something we could expand, provided we respect the rights of grove custodians. Canada, the USA, and New Zealand have supportive legislation but it does not appear to make protecting sacred groves any stronger (Thorley & Gunn, 2008).

For communities to better manage sacred groves, policy and law to better respect and secure them need to be supportive. This can be explored with policy support from outside conservation, e.g., in land use, agriculture, and local landscape planning. Legislation for national monuments can be used and implemented for sacred groves even if such legislation is relatively weak.

The conservation category of Other Effective area-based Conservation Measures (OECMs) can help build the necessary respect and recognition of sacred groves (IUCN-WCPA Task Force on OECMs, 2019). Combining this with the importance of sacred groves for religious or spiritual groups with national recognition will better respect their importance on monastic lands, in National Parks, and in landscapes. On State lands, sacred groves should be part of negotiated joint management plans, and part of national ICCA registries. Where protected areas are gazetted, they can respect sacred groves and their custodians. On landscapes, sacred groves can be important culturally and be an integral part of local landscape planning. Where cultural links with trees and forests are strong, tree ordination can be a potent force for securing such trees and groves.

## Conclusion: Sacred Groves Important for People and Conservation

Recognition of, and respect for sacred groves is a necessary first step. Sacred groves are numerous and extensive in most countries and land use systems across the globe. But such respect and recognition should be national in focus, which lays a foundation for incorporating (with FPIC) sacred groves into the IUCN Protected Area Categories, as OECMs and ICCAs, and demonstrate how relevant sacred groves are to conservation, highlighting their importance for culture, human rights, and integrated land use.

Conserving sacred groves through legal recognition is important. Creating greater awareness on the scale, scope, and extent of sacred groves for conservation agencies and religious bodies, nations, and the world is also important. Yet, there is another challenge and paradox: many sacred trees and groves may be 'secret' whose spiritual meaning may be diminished or threatened if they are known and listed (Hamilton, 1998). There are trade-offs – secrecy or risk loss; or recognition and risk abuse of the spiritual. Such choices belong to the custodians.

Strengthening custodian rights and responsibilities, based on FPIC, provides lessons as to how rights and responsibilities can be secured, better respected nationally and globally, and recognized as important cultural and conservation assets. Some of the examples and lessons in this chapter demonstrate how sacred groves can be formally secured for the custodians. This enhances cultural identity, contributes to conservation outcomes, and is important for community livelihoods.

Respecting and recognizing sacred groves, whose scale and extent survives (even thrives) despite declining government budgets, structural adjustment, retrenchment, population and land use pressures, and other macro-economic forces – all of which place pressures on conservation. Yet they survive! This helps secure rural people's rights, which is particularly important for sacred trees and groves to bridge formal conservation with their importance as cultural assets and demonstrate their contribution to conservation. Sacred groves link, practically and spiritually, people and nature and strengthen the argument for the importance of conservation at national and international levels, and how this should contribute to the CBD 30:30 agenda. Formally respecting sacred groves is good for people, good for conservation, good for management, and good financially.

## Notes

1 https://edition.cnn.com/2020/11/11/africa/kenya-president-china-fig-tree/index.html
2 http://whc.unes Countyorg/en/list/1231
3 http://www.iccaregistry.org/

## References

Barrow, E. (2015). The importance of trees and forests for faith based groups and spiritual traditions across the globe. In P. Wit, M. Hassona, & F. Al-Khalaifi (Eds.), *Proceedings of the Second International Forum of the Qur'anic Botanic Garden: Islamic Perspectives on Ecosystem Management* (pp. 220–245). Doha Qatar: Qur'anic Botanic Garden and IUCN Commission on Ecosystem Management.

Barrow, E. (2019). *Our future in nature: Trees, spirituality and ecology*. Bloomington, IN: Balboa Press.

Berhane-Selassie, T. (2008). The socio-politics of ethiopian sacred groves. In M. J. Sheridan & C. Nyamweru (Eds.), *African sacred groves: Ecological dynamics and social change* (pp. 103–116). Oxford: James Currey.

Bhagwat, S. (2012). Sacred groves and biodiversity conservation: A case study form the Western Ghats, India. In G. Pungetti, G. Oviedo, & D. Hooke (Eds.), *Sacred species and sites: Advances in biocultural conservation* (pp. 322–334). Cambridge: Cambridge University Press.

Bhagwat, S., & Rutte, C. (2006). Sacred groves: Potential for biodiversity management. *Frontiers in Ecology and the Environment, 4*(10), 519–524.

Burbea, R. (2015). The Buddha and the Sacred Earth. *Resurgence & Ecologist, 288*, 34–37.

Chandrakanth, M. G., & Romm, J. (1991). Sacred forests, secular forest policies and people's actions. *Natural Resources Journal, 31*(4), 741–755.

Corrigan, C., Bingham, H., Pathak, N., Hay-Edie, T., Tabanao, G., & Kingston, N. (2016). Documenting local contributions to Earth's biodiversity heritage: The global registry. *PARKS, 22*(2), 55–68.

Dudley, N., Higgins-Zogib, L., & Mansourian, S. (2009). The links between protected areas, faiths and sacred natural sites. *Conservation Biology, 23*(3), 568–577.

Fukamachi, K., & Rackham, O. (2012). Sacred groves in Japanese Satoyama landscapes: A case study and prospects for conservation. In G. Pungetti, G. Oviedo, & D. Hooke (Eds.), *Sacred species and sites: Advances in biocultural conservation* (pp. 419–420). Cambridge: Cambridge University Press.

Gadgil, M. (1992). Conserving biodiversity as if people matter: A case study from India. *Ambio, 21*(3), 266–270.

Githitho, A. (2006). The sacred Mijikenda *Kayas* of Coastal Kenya: Evolving management principles and guidelines. In T. Schaaf & C. Lee (Eds.), *Conserving cultural and biological diversity: The role of sacred natural sites and cultural landscapes. Proc. Tokyo Symposium* (pp. 152–157). Paris: UNESCO.

Hamilton, L. S. (1998). Forest and tree conservation through metaphysical constraints. In UNESCO (Ed.), *Natural sacred sites. Cultural diversity and biological diversity. Proc. Int. Symp., Paris 1998* (p. 18). Paris: UNESCO.

Hareuveni, N. (1980). *Nature in our biblical Heritage*. Kiryatone: Neot Kedumin.

Harmon, D., & Putney, A. D. (Eds.). (2003). *The full value of parks – From economics to the intangible*. New York: Rowman and Littlefield.

Hongmao, L., Zaifu, X., Youkai, X., & Jinxiu, W. (2003). Conserving plant biodiversity through traditional beliefs in Xishuangbanna, Southwest China. In C. Lee & T. Schaaf (Eds.), *The importance of sacred natural sites for biodiversity conservation. Proc. Int. Workshop on the importance of sacred natural sites for biodiversity conservation in Kunming and Xishuangbanna biosphere reserve, People's Republic of China* (pp. 126–134). Paris: UNESCO.

IUCN-WCPA Task Force on OECMs. (2019). *Recognising and reporting other effective area-based conservation measures. Gland, Switzerland: IUCN.* Retrieved from Gland, Switzerland. https://portals.iucn.org/library/node/48773

IUCN, & UNEP. (1986). *Review of the protected areas system in the Afrotropical Realm*. Gland: IUCN.

Jonas, H. D., Lee, E., Jonas, H. C., Matallana-Tobon, C., Wright, K., Nelson, F., & Enns, E. (2017). Will 'other effectively areas-based conservation measures' increase recognition and support of ICCAs? *PARKS, 23*(2), 63–78.

Khan, M. L., Khumbongmayum, A., & Tripathi, R. (2008). The sacred groves and their significance in conserving biodiversity – an overview. *International Journal of Ecology and Environmental Sciences, 34*(3), 277–291.

Kothari, A., Corrigan, C., Jonas, H., Neumann, A., & Shrumm, H. (2012). *Recognizing and supporting territories and areas conserved by indigenous peoples and local communities: Overview and National case studies.* Montreal: Secretariat of the Convention on Biological Diversity, ICCA Consortium, Kalpavriksh, and Natural Justice.

Kothari, A., & Das, P. (1999). Local community knowledge and practice in India. In D. A. Posey (Ed.), *Cultural and spiritual values of biodiversity – A complementary contribution to the global biodiversity assessment* (pp. 185–192). London and Nairobi: Intermediate Technology and UNEP.

Kothari, A., & Pathak, N. (2008). Defenders of diversity. *Resurgence, 250*(September/October), 36–37.

Lewington, A., & Parker, E. (1999). *Ancient trees: Trees that live for 1000 years*. London: Collins and Brown.

Malhotra, K., Gokhala, Y., & Chatterjee, S. (2001). *Cultural and ecological dimensions of sacred groves in India*. New Delhi and Bhopal: Indian National Science Academy and New Delhi and Indira Ghandi Rashtriya Manav Sangrahalaya.

Mallarach, J.-M. (2012). Monastic communities and nature conservation: An overview of positive trends and best practices in Europe and the Middle East. In J.-M. Mallarach, T. Papayannis, & R. Vaisanen (Eds.), *The diversity of holy lands in Europe. Proc. 3rd Workshop of the Delos Initiative – Inasi/Aanaar 2010* (pp. 157–173). Gland: IUCN and Metsahallitus Natural Heritage Services Vantaa, Finland.

Mallarach, J.-M., Corco, J., & Papayannis, T. (2016). Christian monastic lands as protected landscapes and community conserved areas: An overview. *PARKS, 22*(1), 63–78.

Ntiamoa-Baidu, Y. (1995). *Indigenous vs. introduced biodiversity conservation strategies: The case of protected area systems in Ghana* (p. 12). Washington, DC: WWF-Biodiversity Support Program.

Nyamweru, C., Kibet, S., Pakia, M., & Cooke, J. A. (2008a). The Kaya forests of coastal Kenya. In M. J. Sheridan & C. Nyamweru (Eds.), *African sacred groves: Ecological dynamics and social change* (pp. 62–86). Oxford: James Currey.

Nyamweru, C., & Sheridan, M. J. (2008b). Introduction. In M. J. Sheridan & C. Nyamweru (Eds.), *African sacred groves: Ecological dynamics and social change* (pp. 1–8). Oxford: James Currey.

Nyangila, J. (2012). Sacred species of Kenyan sacred sites. In G. Pungetti, G. Oviedo, & D. Hooke (Eds.), *Sacred species and sites: Advances in biocultural conservation* (pp. 351–364). Cambridge: Cambridge University Press.

Ormsby, A. (2012). Cultural and conservation values of sacred forests in Ghana. In G. Pungetti, G. Oviedo, & D. Hooke (Eds.), *Sacred species and sites: Advances in biocultural conservation* (pp. 335–350). Cambridge: Cambridge University Press.

Ormsby, A., & Bhagwat, S. (2010). Sacred forests of India: A strong tradition of community-based natural resource management. *Environmental Conservation, 37*(3), 320–326.

Peng, L., Ning, W., Zhaoli, Y., & Shengji, P. (2003). Sacred sites in Northwest Yunnan, China. In C. Lee & T. Schaaf (Eds.), *The importance of sacred natural sites for biodiversity conservation. Proc. Int. Workshop on the importance of sacred natural sites for biodiversity conservation in Kunming and Xishuangbanna Biosphere Reserve, People's Republic of China* (pp. 139–150). Paris: UNESCO.

Pungetti, G., & Bhagwat, S. (2012). Sacred species and biocultural diversity: Applying the principles. In G. Pungetti, G. Oviedo, & D. Hooke (Eds.), *Sacred species and sites: Advances in biocultural conservation* (pp. 367–406). Cambridge: Cambridge University Press.

Ramakrishnan, P. S., Saxena, K. G., & Chandrashekara, U. M. (Eds.). (1998). *Conserving the sacred for biodiversity management.* New Delhi and Oxford: Science Publishing in Association with UNESCO.

Rights and Resources Initiative. (2020). *Rights-based conservation: The path to preserving Earth's biological and cultural diversity?* Washington, DC: Rights and Resources Initiative.

Robertson, S. A. (1987). Preliminary floristic survey of Kaya Forests of Coastal Kenya. A report to the Director of Museums of Kenya (p. 150). Nairobi: National Museums of Kenya.

Schaaf, T. (2003). UNESCO's experience with the protection of sacred natural sites for biodiversity conservation. In C. Lee & T. Schaaf (Eds.), *The importance of sacred natural sites for biodiversity conservation. Proc. Int. Workshop on the Importance of Sacred Natural Sites for Biodiversity conservation in Kunming and Xishuangbanna Biosphere Reserve, People's Republic of China* (pp. 5–12). Paris: UNESCO.

Schreckenberg, K., Franks, P., Martin, A., & Lang, B. (2016). Unpacking equity for protected area conservation. *PARKS, 22*(2), 11–26.

Smith, S. (1997). *Aborigines, land and National Parks in New South Wales.* Retrieved from Sydney: https://www.parliament.nsw.gov.au/researchpapers/Pages/aborigines-land-and-national-parks-in-nsw.aspx

Tauli-Corpuz, V., Alcorn, J., & Molnar, A. (2018). *Cornered by protected areas: Replacing 'fortress' conservation with rights-based approaches helps bring justice for indigenous peoples and local communities, reduces conflict, and enables cost-effective conservation and climate action.* Retrieved from Washington DC: https://rightsandresources.org/publication/cornered-by-protected-areas/

Thorley, A., & Gunn, C. (2008). *Sacred sites: An overview. Report for the Gaia Foundation.* London: The Gaia Foundation.

Verschuuren, B. (2012). Integrating cultural values in nature conservation: Perceptions of culturally significant sites and species in adaptive management. In G. Pungetti, G. Oviedo, & D. Hooke (Eds.), *Sacred species and sites: Advances in biocultural conservation* (pp. 231–246). Cambridge: Cambridge University Press.

Verschuuren, B. (2016). Re-awakening the power of place: Ancient philosophy and practice with relevance for protected areas and conservation in Asia. In B. Verschuuren & N. Furuta (Eds.), *Asian sacred natural sites: Philosophy and practice in protected areas and conservation* (pp. 1–14). London: Routledge.

Verschuuren, B., Wild, R., McNeely, J., & Oviedo, G. (2010). Introduction: Sacred natural sites the foundations for conservation. In B. Verschuuren, R. Wild, J. McNeely, & G. Oviedo (Eds.), *Sacred natural sites – Conserving nature and culture* (pp. 1–13). London: Earthscan.

Waisel, Y., & Alon, A. (1980). *Trees of the land of Israel*. Tel Aviv: Division of Ecology.

Waldron, A., & 87_authors. (2020). Protecting 30% of the planet for nature – Costs, benefits and economic implications. *Working paper analysing the economic implications of the proposed 30% target for area protection in the draft post-2020 Global Biodiversity Framework* (p. 58). Cambridge: Conservation Science Group, University of Cambridge.

Wickramasinghe, A. (2006). Adam's Peak in the cultural landscape of Sri Lanka: Evidence for an eco-cultural basis for conservation. In T. Schaaf & C. Lee (Eds.), *Conserving cultural and biological diversity: The role of sacred natural sites and cultural landscapes. Proc. Tokyo Symposium* (pp. 52–57). Paris: UNESCO.

Wild, R., & McLeod, C. (Eds.). (2008). *Sacred natural sites: Guidelines for protected area managers*. Gland: IUCN.

Zucchelli, C. (2009). *Trees of inspiration: Sacred trees and bushes of Ireland*. Cork: Collins Press.

# 8

# SPIRITUAL VALUES AND ECOSYSTEM SERVICES OF SACRED GROVES IN KARNATAKA, INDIA

*Alison A. Ormsby and Smitha Krishnan*

## Introduction

Sacred forests around the world represent a traditional form of community-based conservation. Also referred to as sacred groves, these are sites that have cultural or spiritual significance to the people who live around them and are primarily conserved for their non-material values (Bhagwat, 2009). Sacred forests have been protected globally for a variety of reasons, including for religious, cultural, and watershed values (Malhotra et al., 2007; Ormsby & Bhagwat, 2010; Ormsby, 2012). These areas are known to provide ecosystem services, such as pollination (Krishnan, Cheppudira, & Ghazoul, 2017), carbon sequestration (Dar et al., 2019; Ramachandra & Bharath, 2019), erosion control, seed banks (Tiwari et al., 1998; Apffel-Marglin & Parajuli, 2000), water quality, and are often the origin of water sources (Burman, 1995).

Services provided by sacred forests include all four major categories of ecosystem services – provisioning, regulating, supporting, and cultural (McNeely & Mainka, 2009), which contribute to ecosystem resilience and stability (Mertz et al., 2007). The types of provisioning and regulating services include biodiversity, clean air, pollination, erosion protection, soil fertility, climate regulation, and water resources. Cultural services provided by ecosystems include their spiritual values. One of the approaches proposed for international biodiversity conservation is payments for ecosystem services. This may be in the form where the beneficiary pays for a service such as carbon sequestration under the Reducing Emissions from Deforestation and Forest Degradation (REDD) programme (Engel et al., 2008).

Sacred forests are often key reservoirs of biodiversity. Relatively little research has been conducted on the ecosystem services provided by community-protected groves. For sacred natural sites in general, Bhagwat (2009) compared the material values represented by quantifying ecosystem services with the non-material

DOI: 10.4324/9781003246190-11

spiritual values of sacred sites and concluded that "the contingent valuation of 'ecosystem services' cannot accurately represent values traditional societies attribute to sacred natural sites" (Bhagwat, 2009, p. 422). Documentation of ecosystem services in 130 sacred natural sites (including groves, forests, meadows, and water bodies) in Uttarakhand, India showed that 24% provided cultural services, 32% supporting services, 12% regulating services, and 32% provisioning services (Gokhale & Pala, 2011).

Globally, India has the highest concentration of groves with an estimate of over 100,000 groves (Malhotra et al., 2007). However, these are disappearing due to loss of cultural beliefs, over-exploitation of natural resources from the groves, and illegal encroachment and conversion to other land-use types (Chandrakanth et al., 2004). Sacred forests traditionally have rules that restrict activities within their boundaries, which can serve a conservation role. Although some groves are small fragments less than one hectare, and others are more extensive, in some cases, these fragments represent the sole remaining natural forests outside of protected areas and are of high conservation value. Sacred forests are known to conserve habitats (e.g. *Myristica* swamps) that are not represented within the current protected area system (Bhagwat & Rutte 2006) and serve as refugia for endemic species including some of which are rarely found in protected areas (Bhagwat et al., 2005b). In addition to trees, lianas, shrubs, epiphytes (Page et al., 2010), macrofungi (Brown, Bhagwat, & Watkinson, 2006), and birds are conserved in groves (Bhagwat et al., 2005a). In regions with human-animal conflict (Bal et al., 2011), fragments could also act as important animal corridors when it is possible to connect multiple fragments to contiguous protected forests (Ramachandra, Bharath, & Vinay, 2019).

It is believed that the sacred grove conservation tradition in India coincided with the start of settled agriculture (Hughes & Chandran, 1998). The motivation behind keeping patches of forest may have included the ecological services that they provide. These include soil conservation, maintenance of water quality, and provision of forest products, including honey and medicinal plants (Tiwari et al., 1998). The communities may have protected groves in honour of traditional gods, animistic deities, or ancestral spirits. In India, many of these original pre-Vedic deities underwent transformation over time into mainstream Hindu gods and goddesses (Apffel-Marglin & Parajuli, 2000; Chandrakanth et al., 2004). Our study focused on how residents near sacred groves in the Kodagu District (also called Coorg) of Karnataka state recognize and value the ecosystem services provided by them. We also quantified pollination services provided by bees nesting in groves, an ecosystem service that is directly perceivable by the coffee farmers within the landscape.

## Methods

### *Study Sites*

Ethnographic and biological research was conducted on sacred groves of Kodagu (Figure 8.1). In Kodagu (area 4,106 Km$^2$), more than 1,200 groves form an informal network of forest reserves with a density of one sacred grove for every 3 Km$^2$

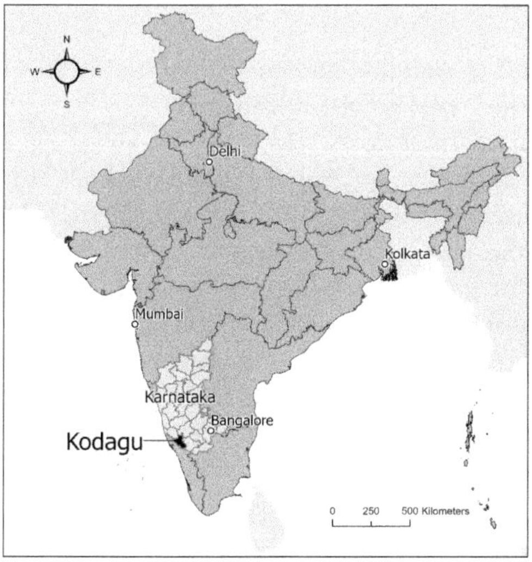


**FIGURE 8.1** Map of India showing Kodagu District study region. Map credit: Elizabeth Forys

of land (Bhagwat et al., 2005a). The groves are interspersed in a coffee and rice paddy matrix. Kodagu is the largest coffee-producing district in India (Garcia et al., 2010).

Our study on pollination services was carried out by Krishnan in 33 groves of Kodagu (2008–2009). The groves were sampled to locate bee colonies and 38 respondents from 21 villages were interviewed to evaluate their knowledge of pollinators and pollination service. Ormsby interviewed 84 residents of 9 villages near approximately 20 groves (2009–2010) to investigate their perceptions of the ecosystem services and spiritual values provided by the groves.

In the study region, most residents practise Hinduism. In addition, there are traditional, non-Hindu representations of gods such as stone monoliths and terracotta offerings that are associated with the sacred groves. The groves are referred to as *devarakadu*, meaning god's forest (Kalam, 1996). The predominant deity associated with the groves is Aiyappa (also spelled Aiappa, Ayappa, Iyyappa, or Ayyappa). Since Aiyappa is considered to be a hunting god, devotees make offerings of small terracotta figurines of a dog or horse, animals that are associated with hunting (Srinivas, 1952). The other deities associated with groves include goddesses such as Chamundi and Bhadrakali and gods such as Eshwara.

Although the sacred forests that were established for older folk deities have been subsumed by Hinduism, the groves are still maintained while Hindu gods are worshipped. Sanskritization, a term used to refer to the replacement of local folk deities with Hindu deities in groves, often results in temples being built within forests (Kalam, 1996; Ormsby & Ismail, 2015). As Tomalin (2004, p. 289) describes, sanskritization is the process by which "regional deities become identified with the pan-Indian Gods and the groves are cleared to make way for temples."

## *Research Approach*

This chapter is the result of combined quantitative and qualitative studies about the ecosystem services provided by the sacred groves of Kodagu.

Ormsby conducted semi-structured interviews with a translator using a 30-question interview guide. A social science, qualitative, mixed methods ethnographic research approach was used (Creswell, 2003; Bernard, 2011). The questionnaire contained questions relating to grove management and resident attitudes, as well as basic demographic information. Interviewees were selected using a variety of approaches – key informant, snowball sampling, and stratified sampling by village neighbourhood and by occupation (Bernard, 2011). In each location, the grove manager, community leader, and/or temple committee member (key informants) were approached for permission to conduct research and to seek a person with whom to visit each grove. To obtain a diversity of perspectives, residents representing a variety of occupations were sought, including farmer, priest, teacher, and herbalist. In particular, people who might use or manage the grove were interviewed. An attempt was made to get an even gender distribution of interviewees, but this was not achieved in every community due to the lack of availability of interviewees.

According to official Forest Department records, each of these villages in the study area harboured at least one grove, with a maximum of 15. Kannada and Kodava languages are spoken locally. The dominant ethnic group is the Kodavas, who were interviewed for this study and speak Kodava (for detailed cultural ethnographies and information, see Srinivas, 1952 and Chinappa, 2003).

In the 33 groves (+1 grove which was clear felled during the study) that were sampled by Krishnan to quantify bee colonies, 38 respondents from 21 villages were interviewed using a structured questionnaire survey to evaluate the knowledge of the community regarding pollinators and pollination services. Coffee farmers living adjacent to the groves and owning coffee plantations located within 500 m from the sacred forests were selected for this study. The knowledge of the respondents regarding the presence of the bee colonies nesting within the sacred forests and adjoining coffee plantations was recorded, including status over time (the previous 10 years) and the farmers' perception of the role of groves in coffee production (as nesting sites for bees that provide pollination services).

Detailed field notes were written after every interview. Interview results were coded for emergent themes. Quotes specifically relating to ecosystem services were analysed and categorized based on the type of ecosystem service mentioned. These descriptions from interviewees are presented as the social science results of this study.

## Results

In our study, the sacred forests were found to be in different states of preservation, including groves that previously existed and are now gone. Results related to each of the major categories of ecosystem services are explained below.

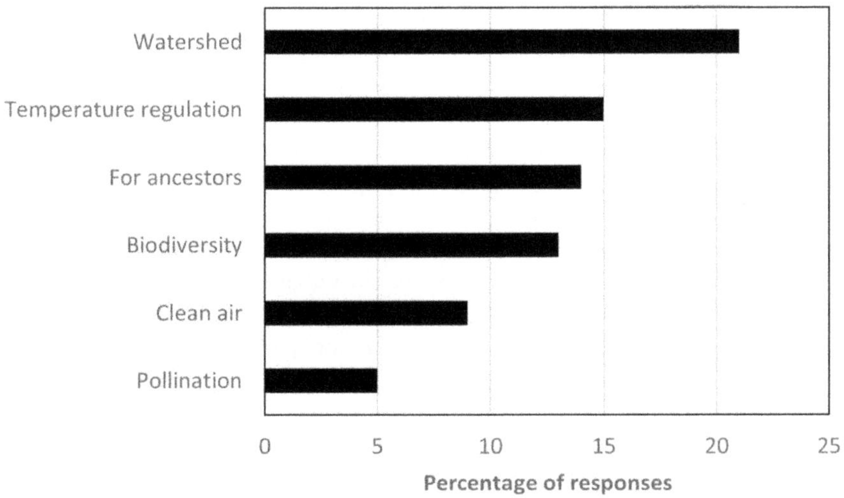


**FIGURE 8.2** Types of ecosystem services identified in study sites. Credits: Created by Alison Ormsby and Smitha Krishnan.

Figure 8.2 summarizes and presents the results of interviews, and the percentage of interviewees mentioning the different types of ecosystem services.

## *Spiritual Values*

The primary reason for protection of the groves studied was due to spiritual reasons (>60% of the respondents). A few specific responses are detailed here. As one family said, "The grove has so much power that if someone does harm or disrespects the grove, they will have a problem." Many residents said about protecting their local grove that, "It is god's grove. If anyone did anything against the grove, there would be harm for the person." Groves in Kodagu often have a core area that is most sacred, which sometimes contains a shrine. As a male farmer said, "The temple is the forest so it should not be destroyed." As one male shopkeeper said about the local grove, "The god resides inside the grove; he needs the cool environment, so cutting of trees should not happen." In addition, a husband and wife shared, "We have faith in god; similarly, we have faith in the grove."

All communities surveyed perform annual festivals associated with the local grove that range in length from one to nine days. Sacred grove festivals are celebrations for the god that resides in the grove and may be conducted at a temple or shrine, often located inside the grove. These festivals are a time when the community renews its connection to the grove and the god. Festivals often involve following strict rules for multiple days before the event, including abstaining from eating meat, hunting, or cutting plants. The festival usually has many aspects, which may involve an annual ritual of taking a statue of the deity out of the temple, typically by a Brahmin (a person considered to be of a superior caste) priest, and taking the deity to a nearby water source where the deity is ritually washed. All festivals normally include several *pujas* – ceremonies where offerings are given, and blessings are sought. Interestingly, some of the festivals involve multiple groves in one community.

As one woman and her brother-in-law explained, "Since this place (the grove) was given for Aiyappa, he has been protecting the village." Furthermore, a female middle-aged resident commented, "The only remaining forest in the village is the grove; if it is destroyed, the younger generation will no longer have respect or faith in god." The grove brings solidarity within the community and gives a sense of protection.

## *Water Regulation*

Many residents also identified the benefits from groves in terms of water resources. One male farmer aged 80 poetically described the value of the grove near his house as "to catch the clouds". The grove is protected in part because it is believed to attract rainfall to the area. Thus, residents are aware of the importance of groves in maintaining perennial streams which are important drinking water sources for the residents. Residents were asked "What do you think should happen to the grove in the future?" A husband and wife, both aged 80 said,

> it should remain as a grove, thicken, and grow – so more moisture is retained in the soil. There is a stream close to the forest that used to flow throughout the year, but now in April or May it dries up, but flows again in June during the monsoon. If the forest is gone, it will dry up.

## *Climate Regulation*

Sacred groves often exist as forest islands in a habitat mosaic – they may be surrounded by non-forest habitats, such as farmland or grassland. In Kodagu, the surrounding areas are typically coffee plantations and/or rice paddies. The groves often stand out as natural habitat areas in contrast to a human-managed landscape. A temperature difference can be perceived when one enters the grove, particularly when walking from an exposed paddy field into a closed-canopy forest, when there is a distinct edge to the grove. The temperature in Kodagu ranges from as low as 9° during the winter nights to as high as 48° during the day in summer months, and is especially hot in the drier, lower elevation areas (unpublished data). As a female labourer aged 45 explained, "each village has a grove for the god. If it is not protected, there would be no rain or cool breeze." According to two middle-aged men interviewed at a temple, "the grove is protected for the environment and the betterment of the village; for cool climate, rainfall, biodiversity, for god." Similarly, a female aged 41 said, "the forest will keep the village cool and provide rainfall for crop cultivation."

## *Pollination Services*

Kodagu is home to four species of social bees, namely, *Apis dorsata* (rock bee or the wild giant honeybee), *Apis cerana indica* (Indian honeybee), *Apis florea* (dwarf honeybee), and *Tetragonula iridipennis* (the stingless bee). Many groves in Kodagu harbour wild bee colonies, specifically *Apis dorsata*, which is migratory. Of the 33 groves that were surveyed by Krishnan, 31 of them harboured colonies of either *Apis dorsata* (in 21) or *Apis cerana indica* (in 18). The region is known for its multi-floral honey and many farmers are also experienced beekeepers. The farmers are also aware of the importance of trees within the groves and the coffee plantations as forage providers to bees. During interviews, there were repeated mentions of *Lagerstroemia microcarpa* (Ben teak), *Mangifera indica* (wild mango), and *Syzygium cumini* (Java plum) as the most important forage-providing trees. About 79% of the respondents reported at least one of the four social bee colonies nesting within their property (Figure 8.3) and more than 50% of respondents reported occurrence of stingless bee colonies nesting in their property. People are aware of and value the pollination services provided by bees to coffee and the related benefits that they reap in terms of higher coffee yields.

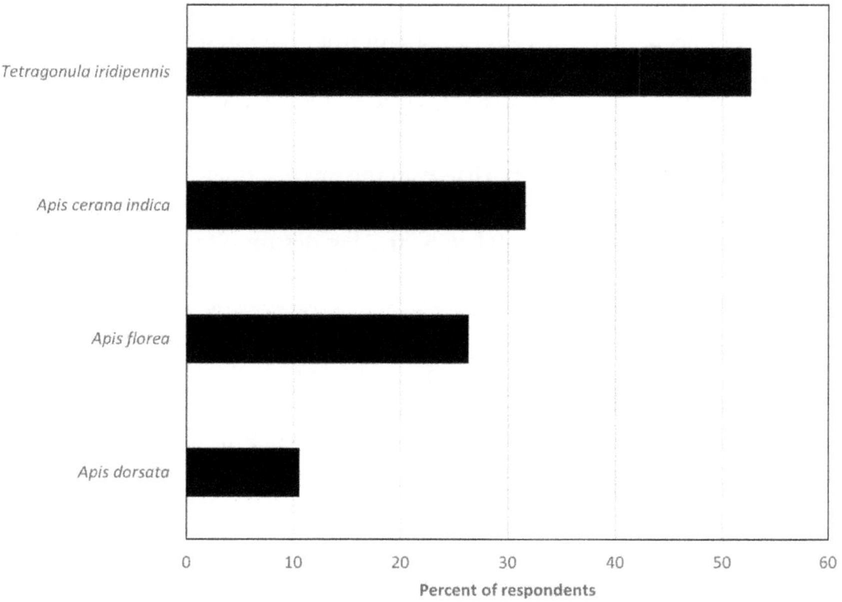


**FIGURE 8.3** Percentage of respondents with bee colonies on their property, by bee species. Credit: Created by Smitha Krishnan.

Pollination was mentioned by some respondents as a value provided by the groves. Of the 38 respondents interviewed by Krishnan, 87% were aware of the fact that the groves are important nesting sites and provide forage resources to wild bees and forage for managed honeybees. Based on Ormsby's interviews, one female resident aged 45 emphasized the importance of groves as providers of refuge, saying "it is good for birds and honeybees – they are there in the forest." In some cases, people traditionally harvest honey from these hives. For example, there are wild bee colonies in many groves, and as one resident said, "it is a natural environment – the honeybees and butterflies help with pollination." In recent years however, farmers have noted a decline in *Apis dorsata* colonies in some regions. Residents living close to a grove noted that colony abundance reduced from about 40 to 50 in the past to just about 10 in recent years. They believe that the reduction in grove size is the main reason for the fall in bee population. Among the 34 study sites that were observed, one of the sacred forests that harboured 17 *Apis dorsata* colonies was cleared during our study period. In addition to encroachment, farmers noted that pesticide usage and insect diseases have led to a decline in bee populations over the years.

Farmers are aware that the pollination services that the bees provide when they visit coffee flowers to collect nectar and pollen have been affected due to the reduction in bee populations, and this is linked to coffee yields. According to a husband and wife, "during the coffee blossom time, the bees used to collect nectar and coffee production was good. Now the population of honeybees has declined, and it has affected coffee production." Residents also benefit from the groves by placing their own boxed hives (for the Indian honeybee) in the grove to benefit from nectar that the bees can obtain from the forests.

In addition, residents believe that the bees act as protectors of the groves. Inappropriate behaviour in the grove is punished by the deity of the grove by bringing them harm such as an attack by bees, or ill health. A female resident aged 39 recounted the oral history that even the mighty ruler Tippu Sultan who was feared by everyone, including the British during their rule in India, was not spared from the wrath of the bees when he tried to destroy a temple in one of the groves. According to legend, Tippu Sultan was trying to destroy all the temples in Kodagu. When he tried to destroy a grove, a fog covered him and reduced his visibility, and honeybees attacked him. Furthermore, regarding a different grove, a male resident aged 79 recounted how an outsider had entered the grove without permission and was attacked by honeybees. He said the person "might not have been clean or appropriate". There are many such accounts from various respondents regarding the role of bees in the protection of groves.

## Discussion

Sacred groves provide many values at a local, regional, and global scale, ranging from intangible, spiritual values to tangible goods and services. Through our research, we documented all types of ecosystem services provided by the groves of Kodagu: provisioning (biodiversity, clean air); regulating (watershed supply, climate regulation, and pollination); supporting (water cycle, rainfall); and cultural/spiritual (beliefs, festivals).

The strict "spiritual fence" provided by the cultural protection given to the groves forbids people from taking advantage of most tangible goods offered by the groves. However, some of this protection is weakening. There is a high demand for the nutrient-rich leaf litter and soil within the groves that can boost production in neighbouring coffee estates and rice paddies. In most cases, collecting leaf litter or soil from the groves is forbidden, but it still often happens. Chandrakanth et al. (1990, p. 209) recommend that "establishing temple forests in the upper and middle stream areas of watersheds would be of great assistance to social forestry programs in reducing erosion and maintaining soil and water conservation structures to improve the soil moisture holding capacity." Locations of groves can affect watershed function, flood control, and microclimates.

Similar to our study, in a study of honey collectors in Kodagu, Demps et al. (2012) found that 89% of respondents report less honey in forests now, compared to 30 years ago. Kodagu is a coffee-growing region and it is known that bees enhance fruitset in coffee. Although *Coffea canephora* (coffee henceforth) is primarily a wind-pollinated species (contributes to 67% of fruitset), fruitset in coffee is enhanced by bees (contributes to 33% of fruitset). The wild honeybee, *Apis dorsata*, is the main pollinator of coffee. They account for 58% of all flower visitors (Krishnan et al., 2012). The migratory *A. dorsata* has very high nesting fidelity and tends to nest on the same tree every year; thus, the region where they nest is assured of their pollination services. They prefer to nest in groves with large trees with a wide crown. They migrate into Kodagu between January and March and stay within the landscape until the onset of the monsoon in June. Coffee flowering occurs between February and April and thus benefits from the pollination services of *A. dorsata*. Wild honeybees prefer to nest in large forests; however, many of these forests have been degraded and some have been cleared or reduced to smaller patches. Sacred forests may be cleared partially or completely to construct a temple, as recorded by another study carried out in one of the groves (with 17 wild bee colonies) that we studied, possibly leading to loss of the pollination services locally due to loss of bee nesting trees. Habitat loss, including reduction in size of groves, could result in a decline in pollination service to crops, including coffee (Krishnan et al., 2017). Pollinators are important in increasing crop yields for many crops across the globe (Klein et al., 2007), and forest patches such as sacred groves can be important in harbouring pollinators that enhance crop yields.

It is evident that these forests provide ecosystem services of substantial value. However, this chapter does not advocate a payment for ecosystem services (PES) scheme to create a financial incentive for protecting the groves (see Engel et al., 2008). Because the groves were established for cultural and spiritual reasons, converting the groves to an economic system based on the ecosystem services they provide risks losing the cultural protection they currently have, which could also change the spiritual significance of a site. While some larger groves may be targeted to be part of a PES programme under REDD, it is advised that these economic arrangements are entered into with caution to avoid disrupting the cultural and spiritual values, and local management of the groves.

A study of eight communities in Madagascar receiving PES found that the payments impacted residents' attitudes but not their behaviours (Sommerville et al., 2010). In addition, the authors noted that in communities with customary land rights, it can be challenging to identify who should receive the payments. In India, Kodagu is known for its coffee and honey production, both of which are affiliated with local forests. Based on a study of honey collectors in Kodagu, Demps et al. (2012, p. 431) noted that "honey collecting seems to be transitioning away from a spiritual and subsistence activity to a profitable economic pursuit, especially for younger men." They concluded that "participation in a market economy has often been observed to drive the loss of traditional ecological knowledge, except in the cases where people have continued access to local resources and cultural reasons for its persistence."

The results of our study demonstrate that local communities already value their community-protected forests primarily for their spiritual, intrinsic values. In addition, many residents recognize and value the ecosystem services provided by groves, without having to receive external payments for those ecosystem services. Thus, cultural beliefs and practices do indeed contribute to biodiversity conservation.

Organizations such as the International Union for the Conservation of Nature (IUCN) and The United Nations Educational, Scientific and Cultural Organization (UNESCO) have created guidelines for management of sacred sites, which include improving global knowledge and understanding of sacred natural sites, protecting these sites while providing appropriate management access and use, and respecting the right of sacred site custodians within an appropriate framework of national policy (Wild & McLeod, 2008). In the future, continued recognition and outside support for the cultural traditions of conserving groves would be helpful, as they will face increased pressure for the use of their natural resources.

## Acknowledgements

We are grateful for the generosity and hospitality of communities in which we conducted research. Thanks to Dr. C.G. Kushalappa at the College of Forestry in Ponnampet, L. Biddanda, S.K. Chengappa, and C.M. Sachin Monnappa. Ormsby thanks the Fulbright-Nehru Research Scholars Program and the Russ Family Fund for financial support. Krishnan thanks Prof. Jaboury Ghazoul for giving her the opportunity to work in Kodagu.

## References

Apffel-Marglin, F., & Parajuli, P. (2000). "Sacred grove" and ecology: Ritual and science. In C. K. Chapple & M. E. Tucker (Eds.), *Hinduism and ecology: The intersection of earth, sky, and water* (pp. 291–316). Cambridge, MA: Harvard University Press.

Bal, P., Nath, C. D., Nanaya, K. M., Kushalappa, C. G., & Garcia, C. (2011). Elephants also like coffee: Trends and drivers of human–elephant conflicts in coffee agroforestry landscapes of Kodagu, Western Ghats, India. *Environmental Management, 47*, 789–801. https://doi.org/10.1007/s00267-011-9636-1LB-Bal2011

Bernard, H. R. (2011). *Research methods in anthropology: Qualitative and quantitative approaches* (5th edn). Lanham, MD: AltaMira Press.

Bhagwat, S. A. (2009). Ecosystem services and sacred natural sites: reconciling material and non-material values in nature conservation. *Environmental Values, 18*, 417–427.

Bhagwat, S. A., Kushalappa, C. G., Williams, P. H., & Brown, N. D. (2005a). A landscape approach to biodiversity conservation of sacred groves in the Western Ghats of India. *Conservation Biology, 19*(6), 1853.

Bhagwat, S. A., Kushalappa, C. G., Williams, P. H., & Brown, N. D. (2005b). The role of informal protected areas in maintaining biodiversity in the Western Ghats of India. *Ecology and Society, 10*, Article No.: 8. Retrieved from http://www.ecologyandsociety.org/vol10/iss1/art8/

Bhagwat, S. A., & Rutte, C. (2006). Sacred groves: Potential for biodiversity management. *Frontiers in Ecology and the Environment, 4*(10), 519–524.

Brown, N., Bhagwat, S., & Watkinson, S. (2006). Macrofungal diversity in fragmented and disturbed forests of the Western Ghats of India. *Journal of Applied Ecology, 43*, 11–17. https://doi.org/10.1111/j.1365-2664.2005.01107.x

Burman, R. J. J. (1995). The dynamics of sacred groves. *Human Ecology, 6*(4), 245–254.

Chandrakanth, M. G., Bhat, M. G., & Accavva, M. S. (2004). Socio-economic changes and sacred groves in South India: Protecting a community-based resource management institution. *Natural Resources Forum, 28*, 102–111.

Chandrakanth, M. G., Gilless, J. K., Gowramma, V., & Nagaraja, M. G. (1990). Temple forests in India's forest development. *Agroforestry Systems, 11*, 199–211.

Chinappa, N. (2003). *Pattole Palame: Kodava culture – Folksongs and traditions*. New Delhi: Rupa Co.

Creswell, J. W. (2003). *Research design: Qualitative, quantitative, and mixed methods approaches*. Thousand Oaks, CA: Sage Publications.

Dar, J. A., Subashree, K., Raha, D., Kumar, A., Khare, P. K., & Khan, M. L. (2019). Tree diversity, biomass and carbon storage in sacred groves of Central India. *Environmental Science and Pollution Research, 26*(36), 37212–37227. https://doi.org/10.1007/s11356-019-06854-9

Demps, K., Zorondo-Rodriguez, F., Garcia, C., & Reyes-Garcia, V. (2012). The selective persistence of local ecological knowledge: Honey collecting with the Jenu Kuruba in South India. *Human Ecology, 40*(3), 427–434.

Engel, S., Pagiola, S., & Wunder, S. (2008). Designing payments for environmental services in theory and practice: An overview of the issues. *Ecological Economics, 65*(4), 663–674.

Garcia, C., Bhagwat, S., Ghazoul, J., Nath, C., Nanaya, K., Kushalappa, C., ... Vaast, P. (2010). Biodiversity conservation in agricultural landscapes: Challenges and opportunities of coffee agroforests in the Western Ghats, India. *Conservation Biology, 24*(2), 479–488.

Gokhale, Y., & Pala, N. A. (2011). Ecosystem services in sacred natural sites (SNSs) of Uttarakhand: A preliminary survey. *Journal of Biodiversity, 2*(2), 107–115.

Hughes, J. D., & Chandran, M. D. S. (1998). Sacred groves around the earth: an overview. In P. S. Ramakrishnan, K. G. Saxena, & U. M. Chandrashekara (Eds.), *Conserving the sacred for biodiversity management* (pp. 69–86). New Delhi: Oxford and India Book House.

Kalam, M. A. (1996). *Sacred groves in Kodagu District of Karnataka (South India): A sociohistorical study*. Pondicherry: Institut Français de Pondicherry.

Klein, A. M., Vaissiere, B. E., Cane, J. H., Steffan-Dewenter, I., Cunningham, S. A., Kremen, C., & Tscharntke, T. (2007). Importance of pollinators in changing landscapes for world crops. *Proceedings of the Royal Society B: Biological Sciences, 274*, 303–313. https://doi.org/10.1098/rspb.2006.3721

Krishnan, S., Cheppudira, K. G., & Ghazoul, J. (2017). Pollinator services in coffee agro-forests of the Western Ghats. In *Agroforestry: Anecdotal to modern science* (pp. 771–795). Singapore: Springer Nature. https://doi.org/10.1007/978-981-10-7650-3_32

Krishnan, S., Kushalappa, C. G., Shaanker, R. U., & Ghazoul, J. (2012). Status of pollinators and their efficiency in coffee fruit set in a fragmented landscape mosaic in South India. *Basic and Applied Ecology, 13*, 277–285. https://doi.org/10.1016/j.baae.2012.03.007

Malhotra, K. C., Gokhale, Y., Chatterjee, S., & Srivastava, S. (2007). *Sacred groves in India.* New Delhi: Aryan Books International.

McNeely, J. A., & Mainka, S. A. (2009). *Conservation for a new era.* Gland: IUCN.

Mertz, O., Ravnborg, H. M., Lovei, G. L., Nielsen, I., & Konijnendijk, C. C. (2007). Ecosystem services and biodiversity in developing countries. *Biodiversity Conservation, 16*, 2729–2737.

Ormsby, A. (2012). Cultural and conservation values of sacred forests in Ghana. In G. Pungetti, G. Oviedo, & D. Hooke (Eds.), *Sacred species and sites: Advances in biocultural conservation* (pp. 335–350). Cambridge: Cambridge University Press.

Ormsby, A., & Bhagwat, S. A. (2010). Sacred forests of India: A strong tradition of community-based natural resource management. *Environmental Conservation, 37*(3), 320–326.

Ormsby, A., & Ismail, S. (2015). Cultural and ecological insights into sacred grove management: Managing timber resources for improved conservation of sacred groves. *Forests, Trees and Livelihoods, 24*(4), 244–258. https://doi.org/10.1080/14728028.2015.1059294

Page, N., Qureshi, Q., Rawat, G., & Kushalappa, C. (2010). Plant diversity in sacred forest fragments of Western Ghats: A comparative study of four life forms. *Plant Ecology, 206*, 237–250. https://doi.org/10.1007/s11258-009-9638-8

Ramachandra, T. V., & Bharath, S. (2019). Global warming mitigation through carbon sequestrations in the Central Western Ghats. *Remote Sensing in Earth Systems Sciences, 2*(1), 39–63. https://doi.org/10.1007/s41976-019-0010-z

Ramachandra, T. V., Bharath, S., & Vinay, S. (2019). Visualisation of impacts due to the proposed developmental projects in the ecologically fragile regions- Kodagu district, Karnataka. *Progress in Disaster Science, 3*, 100038. https://doi.org/10.1016/j.pdisas.2019.100038

Sommerville, M., Milner-Gulland, E. J., Rahajaharison, M., & Jones, J. P. G. (2010). Impact of a community-based payment for environmental services intervention on forest use in Menabe, Madagascar. *Conservation Biology, 24*(6), 1488–1498.

Srinivas, M. N. (1952). *Religion and society among the Coorgs of South India.* Oxford: Oxford University Press.

Tiwari, B. K., Barik, S. K., & Tripathi, R. S. (1998). Biodiversity value, status, and strategies for conservation of sacred groves of Meghalaya, India. *Ecosystem Health, 4*(1), 20–32.

Tomalin, E. (2004). Bio-divinity and biodiversity: Perspectives on religion and environmental conservation in India. *Numen, 51*, 265–295.

Wild, R. W., & McLeod, C. M. (Eds.). (2008). *Sacred natural sites: Guidelines for protected area managers.* Gland: IUCN.

# 9

# ETHIOPIAN ORTHODOX TEWAHEDO CHURCH FORESTS AND POLLINATION SERVICES ACROSS AN AGRICULTURAL LANDSCAPE

*Ashley Massey Marks, Travis W. Reynolds, Sara Hamann, Alemayehu Wassie Eshete, Meg D. Lowman, Sandra Nogué, Kathy J. Willis and Shonil A. Bhagwat*

## Introduction

Protected areas comprise 87% of the Earth's terrestrial surface; yet, there is an increasing awareness of the need for biodiversity conservation beyond protected area boundaries (Bhagwat and Rutte, 2006; Chazdon et al., 2009; Gardner et al., 2009; UNEP-WCMC, 2012). Forest patches have been identified as an important feature in the landscape due to the essential ecosystem services they can provide, which contribute to poverty alleviation, local livelihoods, and human well-being (Howe et al., 2013; Schelhas and Greenberg, 1996; Sheridan and Nyamweru, 2008). Ecosystem services are the 'benefits people obtain from ecosystems' (Millenium Ecosystem Assessment, 2005). In agricultural landscapes, forest fragments can contain nesting habitat and forage resources suitable for pollinators that increase crop yields, in addition to providing other valuable ecosystem services such as pest and weed control and soil retention (Kremen et al., 2007; Nogué et al., 2016; Zhang et al., 2007).

Sacred forests surround places of worship, such as temples, shrines, and churches around the world (Bhagwat and Rutte, 2006; Verschuuren et al., 2010). For example, it has been estimated that in India there are 100,000–150,000 sacred groves, in Japan there are roughly 80,000 Shinto shrine forests, and in Ethiopia there are as many as 35,000 Ethiopian Orthodox Tewahedo Church forests (Jinja Honcho, 2011; Malhotra et al., 2001; Wassie et al., 2005). Sacred natural sites provide cultural ecosystem services, including cultural diversity, spiritual and religious values, recreation and ecotourism, aesthetic values, and knowledge systems (Millenium Ecosystem Assessment, 2005). They also serve as refugia for species and preserve ecological memory in the landscape (Bengtsson et al., 2003; Bhagwat et al., 2005; Mgumia and Oba, 2003; Ntiamoa-Baidu, 2008). They increase landscape connectivity and provide ecosystem services such

DOI: 10.4324/9781003246190-12

as pollination and seed dispersal critical for productive agroecosystems (Bodin et al., 2006; Ishii et al., 2010; Tambat, 2005). Forest fragments, including sacred natural sites, can enhance local livelihoods (Malhotra et al., 2001; Schelhas and Greenberg, 1996; Zhang et al., 2007). This study analyses pollination, an ecosystem service provided by (often small) sacred natural sites within agricultural landscapes.

## Study Area

This study was carried out in the South Gondar administrative zone of the Amhara region in northern Ethiopia. Lying to the east of Lake Tana (11°02'–12°33'N and 37°25'–38°41'E), South Gondar covers an area of 1,461,537 ha with elevation ranging from 1,500 m to 4,231 m. There are two rainy seasons, from March to May and from June to August; average annual rainfall ranges from 700 mm to 1,300 mm and average annual temperatures range from a minimum of 9.3°C to a maximum of 23.7°C (Wassie et al., 2010).

Only 1.2% of the Ethiopian landscape is forest, and thousands of Ethiopian Orthodox Tewahedo Church forests comprise the last remnants of Afromontane forest in the northern highlands (Cardelús et al., 2013; FAO, 2015; Wassie et al., 2005). Afromontane forest has been prioritised for international conservation as part of the Eastern Afromontane biodiversity hotspot and the Ethiopian montane grasslands and woodlands ecoregion (Conservation International, 2011; Mittermeier et al., 2004). The Lake Tana region was also designated as a biosphere reserve in 2014, with complementary goals of conservation and sustainable development (UNESCO, 2019) (see Figure 9.1).

Ethiopian Orthodox Tewahedo Churches, known as *debr* or *geddam*, are powerful local religious and social institutions that are up to 1,500 years old. Churches operate autonomously and their total number is unknown; estimates range from a few thousand to 35,000 (Cardelús et al., 2013; Wassie et al., 2005). Fasting and feasting are regular events in the church calendar. The main feasts include 'nine feasts of the Lord, 33 feasts of our Lady (the Virgin Mary), the feasts of the Apostles, Sunday, Saturday, the feasts of the Angels, the feasts of the righteous (saints) and the feasts of the martyrs' (Ethiopian Orthodox Tewahedo Church, 2003). The fasting regimen is one of the most demanding in Christianity; lay people are required to fast 180 days a year, including two months for Lent, and priests, nuns, monks, and other special groups fast 250 days a year.

Congregants regard the circular church building and surrounding ring of forest as one. The Amharic phrase for the church forests, *atsede betekristian*, includes the entire church compound. Church forests are often several centuries old (McCann, 1997). Associations that serve as social support groups, *mehabirs*, meet and feast in these forests to celebrate their patron saints along with priests and nuns. Festivals in the forests afford the opportunity to invite passers-by, including those in need. Nuns who have lost their husbands are sometimes welcomed to reside in church forests, and are invited to these feasts (Orlowska and Klepeis, 2018).

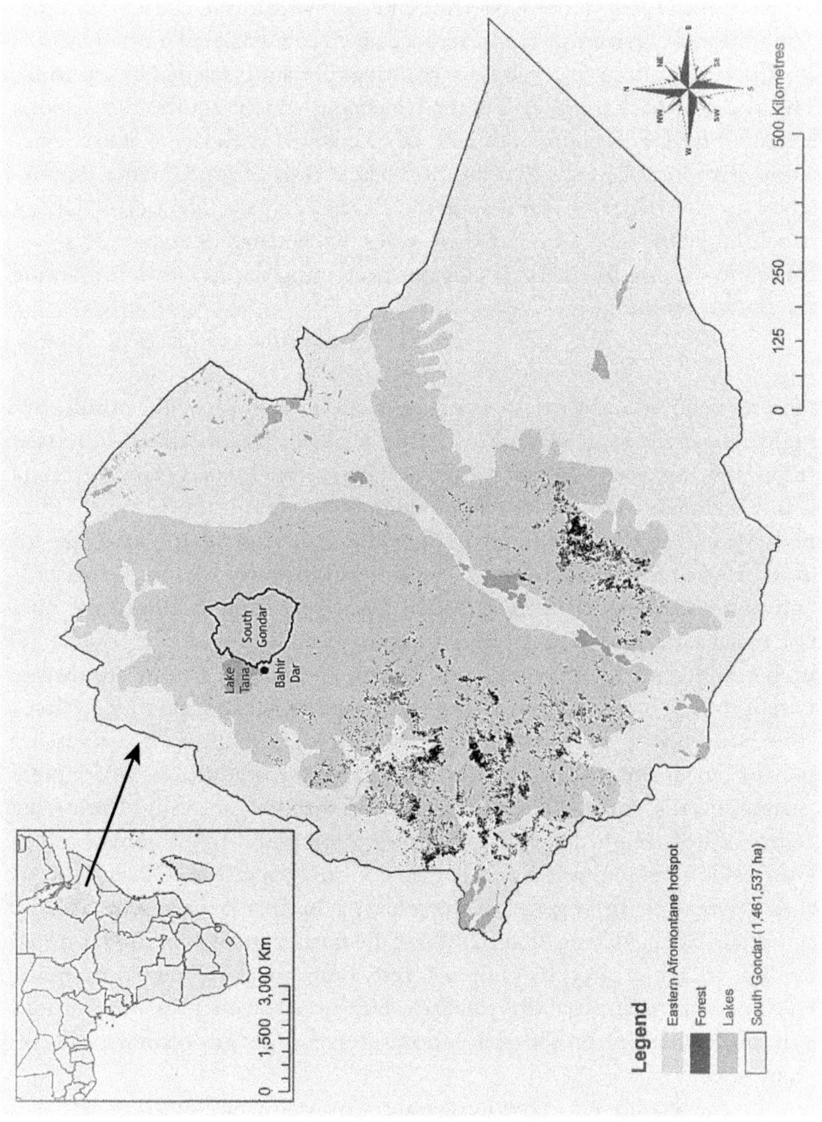


**FIGURE 9.1** Map of Ethiopia including forest cover, the Eastern Afromontane biodiversity hotspot, Lake Tana, and the study area—the South Gondar administrative zone. Credits: Map created by Ashley Massey Marks.

Trees in the church forests range from utilitarian species such as coffee or juniper used for church construction to revered old-age trees believed to house angels guarding the church (Wassie et al., 2005). These forests exist to

> "respectfully cover" and "shelter" the sacred *tabot*, which is deemed to be a replica of the Ark of the Covenant and kept within the church building. The notion of "covering out of respect" is deeply present in other aspects of Ethiopian culture (e.g. religious paintings are always veiled by curtains). The idea goes back to the Ark of the Covenant, which requires the wooden container for God's commandments to be covered with layers of luxurious fabric (Exodus 25:10–22; Exodus 26). The Ark is of fundamental importance for the religious practice of the EOTC and the identity of Christian Ethiopians, who consider themselves descendants of ancient Israel … church forests are like natural curtains protecting the Ark dwelling inside the church building
>
> *(Orlowska and Klepeis, 2018:4).*

Native forests, referred to by priests as *wef zerash* (wild trees planted by birds), are more highly regarded than silviculture. Tree planting within church forests is uncommon; when it does occur, it is catalysed by external actors planting exotic species. In 28 church forests in South Gondar, Wassie et al. (2010) inventoried 168 woody species, 160 of which are indigenous. Functioning as seed banks for native flora, church forests are sources of ecological memory within the agricultural landscape (Wassie and Teketay, 2006). They also serve as refugia for flora and fauna that have largely disappeared from northern Ethiopia.

Church communities are dynamic, however, and church forests are threatened as religious values change. Threats include unchecked grazing by livestock, the building of 'second churches' within the compounds, increasing proximity of burial sites to the church building, and the additions of concrete headstones and walls, trails, and clearings for community gatherings (Orlowska and Klepeis, 2018). High demand for fuelwood and land in Ethiopia renders them vulnerable to exploitation and conversion to agriculture. Church forests have been decreasing in vegetation density and in area over recent decades (Cardelús et al., 2013; Wassie et al., 2005). One farmer in South Gondar noted that after his village cleared its church forest, crop yields decreased dramatically. His observation inspired this research. Here, we consider the implications of church forest distribution for pollination services and agricultural yields on a landscape scale.

Ethiopia is the second most populous country in Africa, and subsistence-based livelihoods are dependent on agriculture and livestock. Private land holdings for subsistence farming make up over 95% of Ethiopia's agriculture and less than 5% is commercial agriculture (Central Statistical Agency of Ethiopia, 2011). In Amhara region, 43% of people live in extreme poverty (zur Heide, 2012) and in the South Gondar administrative zone, land holdings average only 1.38 ha per household (Central Statistical Agency of Ethiopia, 2012c).

Ethiopia is Africa's top exporter of honey and wax and Amhara's honey production accounts for 22% of national production (Central Statistical Agency of Ethiopia, 2013; Pauly and Hora, 2013).

Most beehives in South Gondar (95%) are traditional cylindrical hives with fixed frames. Fifty-four per cent of honey is sold and 44% is used for household consumption (Central Statistical Agency of Ethiopia, 2012a). Honey is an important complement to local diets, especially during periods of food scarcity. It is served as a medicine or tonic and as a fermented festival drink of honey wine, or *tej*. It is made by fermenting honey and adding water and Gesho (*Rhamnus prinoides*, a shrub used like hops), then flavouring it with a variety of spices. Tej is offered to nuns and priests, passers-by, and the disadvantaged during the aforementioned festivals in church forests, fostering social cohesion. Beeswax is used to make votive candles for religious ceremonies (Ejigu et al., 2009).

Honeybees require nectar and pollen from plants. From the nectar, they make honey to give them energy, and pollen provides them with protein, vitamins, fatty substances, and other nutrients (Amssalu et al., 2003). There are over 6,000 flowering plants in Ethiopia, most of which provide forage resources for honeybees. The high diversity of honeybee flora is one reason Ethiopia has the highest bee density in Africa (Fichtl & Adi, 1994). The spring rains, or *belg*, from March to May bring a flowering period dominated by trees. The *kiremt*, or big rains, from June to August make honeybee forage resources plentiful from September to November. This coincides with the *meher* season of crops harvested from September to February. After the dry season from November to February, forage resources are scarce from January to March (Bezabeh, 2006). Honeybees are central-place foragers, which means that they seek out floral resources in the landscape and then return to fixed nest sites in habitat patches (Kremen et al., 2007). *Apis mellifera* frequently nests in tree hollows, which are still found in church forests, thanks to prohibitions against tree clearance. Some communities practise traditional as well as modern beekeeping (using hives with movable frames) within church forests (Roubik, 1989; Wassie et al., 2005; zur Heide, 2012).

While cereal crops like rice and wheat are self-pollinating, many global food crops rely on insects for pollination. Pollination by honeybees has been shown to improve the yields of some global food crops, including pulses and oilseeds, which are grown by farmers in South Gondar (Admassu and Nuru, 1999; Central Statistical Agency of Ethiopia, 2012b; Klein et al., 2007; Matiur et al., 1995). Of the total area of crops planted for subsistence farming during the meher season in South Gondar, 12–20% of crops are pulses and 4–6% are oilseeds (Central Statistical Agency of Ethiopia, 2012b) (see Table 9.1).

Despite accounting for <20% of crop production in South Gondar, pulses and oilseeds are economically higher value crops than cereals and serve as critical local sources of dietary protein, animal fodder, and nitrogen fixation (Rashid et al., 2010). After coffee, oilseeds and pulses are the second and third largest export crops in Ethiopia, and the production of pulses has surged over the last decade. Ethiopia is the second largest producer of faba beans after China and one of the top ten honey and pulse-producing nations in the world (FAO, 2015; Rashid et al., 2010).

**TABLE 9.1** Crops of South Gondar and Level of Yield Increases from Pollination by Bees and Other Insects. Credits: Ashley Massey Marks

| | *Insect pollination* | *Pollinators (bees in bold)* |
|---|---|---|
| **Cereals** | Does not increase yields (Klein et al., 2007) | Self-pollinating |
| **Pulses** | | |
| *Vicia faba* (Faba Beans) | 10–40% increase in yields (Klein et al., 2007) | ***Apis mellifera L., Bombus lapidarius L., Bombus pascuorum Scopoli, Bombus hortorum L.,*** *Anthophora plumipes Pallas, Eucera spp.,* ***Megachile rotundata Fabricius*** |
| *Pisum sativum* (Field Peas) | 0–10% increase in yields (Klein et al., 2007) | *Apis spp., Bombus spp.* |
| *Phaseolus vulgaris* (Haricot Beans) | 0–10% increase in yields (Klein et al., 2007) | ***Apis florea Fabricius, Bombus spp.,*** *Thysanoptera* |
| *Cicer arietinum* (Chickpeas) | Does not increase yields (Klein et al., 2007) | *Apis mellifera L., Megachilidae, Halictidae* |
| *Lens culinaris* (Lentils) | Does not increase yields (Klein et al., 2007) | *Apidae* |
| *Lathyrus sativus* (Grass Peas) | Insect pollination observed (Matiur et al., 1995) | *Apidae* |
| **Oilseeds** | | |
| *Guizotia abyssinica* (Neug or Niger Seed) | ~43% increase in yields (Admassu and Nuru, 1999) | *Apis mellifera bandasii* later named *Apis mellifera simensis* (Pauly and Hora, 2013) |
| *Linum usitatissimum* (Linseed) | 0–10% increase in yields (Klein et al., 2007) | *Apis mellifera L., Bombus spp.* |
| *Brassica napus* (Rapeseed) | 10–40% increase in yields (Klein et al., 2007) | ***Apis mellifera L., Bombus spp.,*** *Andrena spp., Osmia cornifrons Radoszkowski, Osmia lignaria Say,* ***Halictus spp.,*** *Syrphidae* |
| **Root crops** | | |
| *Solanum tuberosum* (Potatoes) | Increases seed production (Klein et al., 2007) | Unknown |

## Methods

We estimated the extent of pollination zones surrounding Ethiopian church forests in South Gondar, Amhara, a geographical area prioritised for forest conservation and sustaining local livelihoods (zur Heide, 2012). We focused on pollination services provided by honeybees (*Apis mellifera*), because these insects (1) are ubiquitous in Ethiopia (Pauly and Hora, 2013); (2) serve as indicator species in many studies investigating pollination service delivery and distance from natural and semi-natural habitat patches (Nogué et al., 2016; Ricketts et al., 2008); (3) are key pollinators for all crops grown in South Gondar that benefit from insect pollination (Admassu and Nuru, 1999; Central Statistical Agency of Ethiopia, 2012b; Klein et al., 2007); and (4) as aforementioned, contribute directly to local livelihoods and the economy.

We asked, what is the extent of South Gondar cropland that falls within the pollination zones of church forests? How might pollination services provided by church forests contribute to local livelihoods?

To identify church forests in South Gondar, we systematically scanned Google Earth and Bing Maps. The aerial view of silver or brightly painted church domes in patches of green Afromontane forest contrasted with the surrounding brown agricultural landscape (see Figure 9.2).

We created a point layer of church domes in ArcGIS v.10.0 and overlaid a vegetative cover layer we derived from an April 2013 Landsat 8 satellite image (30 m, <2% cloud cover [<1% in the study area], WGS 84, Africa Albers Equal Area Conic projection) by calculating the Normalized Difference Vegetation Index (NDVI) as (near-infrared − red)/(near-infrared + red) where near-infrared is band 5 (0.845–0.885 μm) and red is band 4 (0.630–0.680 μm) (Tucker, 1979). In April the contrast between agricultural land and forests was pronounced. The locations of over 60 Afromontane forests in South Gondar and neighbouring zones were validated via ground-truthing. We mapped the perimeters of 15 forests using handheld GPS units and cross-referenced these areas with our NDVI layer. We performed a supervised classification, first training the software by identifying sample pixels in the image to represent different land use classes. It applied this information to map land use classes across the South Gondar satellite image. Pixels with NDVI>0.25 were identified as likely forest. Next, to map the church forests, we used the point layer of church domes and selected contiguous forest pixels within 1,000 m of the church points. Ground-truthing indicated that forested pixels in South Gondar that were not identified as church forests were predominantly eucalyptus plantings. We calculated the area of church forests and performed spatial statistics on their distribution (Average Nearest Neighbour, NNRatio: 0.80, NNZScore: −12.42, $p = 0.00$). We mapped cropland using GlobCover data (ESA GlobCover Project, 2009).

We catalogued bee species observed in the church forests in geolocated photographs by Phillip Harpootlian and specimens collected by Erica McAlister in January 2012. Next, we assessed the suitability of church forests as honeybee habitat. We assessed the floral resources for pollinators by consulting vegetation lists recorded in church forest plots and identifying honeybee forage (Bezabeh, 2006; Wassie et al., 2005). We mapped the seasonal availability of forage using flowering period data, available for 39 of the 61 species of honeybee flora in church forests (Fichtl & Adi, 1994).

We created pollination 'donuts' surrounding the church forests using the foraging distance of honeybees, which we derived via two methods. First, we considered a global meta-analysis of 23 studies by Ricketts et al. (2008) that employs

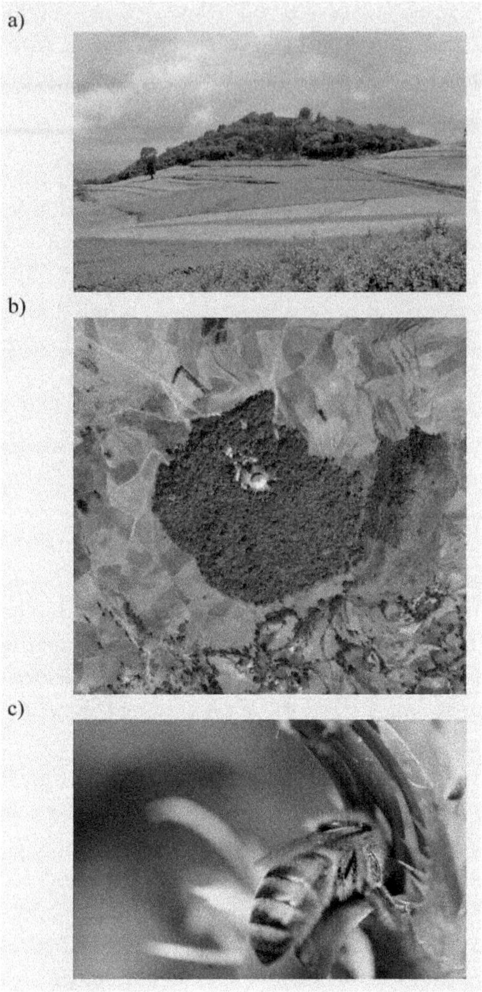


**FIGURE 9.2** Debresena Ethiopian Orthodox Tewahedo Church forest, South Gondar; church dome surrounded by forest and an agricultural landscape, (a) from the agricultural matrix, and (b) aerial view. (c) Honeybee (*Apis mellifera*) in Gelawdios church forest in South Gondar. Photo Credits: Google Earth and Phillip Harpootlian.

Bayesian techniques to quantify the distance of 50% decay of pollinator visitation rates from habitat patches. Twenty of the studies included visitation rates of the pollinator *Apis mellifera*. Ricketts et al. found a distance of 2,170 m for the 50% decline of pollinator visitation. We applied this distance (2,170 m) to the edge of church forests to create the donuts.

We applied a second method to test whether the distance of 2,170 m derived from the metadata was appropriate for South Gondar. We calculated the foraging distance of the local honeybee subspecies *Apis mellifera simensis* using the relationship between bee body size and foraging distance found by Greenleaf et al. (2007). We applied their logistical regression to the intertegular span measurement of *Apis mellifera simensis* (3 mm) (Pauly and Hora, 2013), and calculated a predicted maximum feeding distance of 2,206 m. With this distance only 36 m greater than the distance derived from the global metadata on pollinator visitation rates, we are confident that global metadata is applicable in our case, if not slightly conservative.

Having determined the areas that honeybees could pollinate from church forests, next we assessed which of those areas were croplands. We overlaid the church forest pollination donuts with the croplands layer derived from GlobCover data with the forested pixels removed (ESA GlobCover Project, 2009). We then calculated the area of church forest pollination zones that cover croplands.

## Limitations

While the pollination zones indicate where pollination services could occur in the agricultural landscape of South Gondar, pollination service delivery is also strongly dependent on landscape structure, pollinator abundance, microclimate, and the presence of competitors, predators, pathogens, and parasites. Therefore, pollinator services will be heterogeneous in these landscapes (Kremen et al., 2007). South Gondar's landscape structure is quite homogenous and predominantly composed of a sea of agricultural land with islands of church forests. Performing this type of analysis with finer scale data like crop distribution maps, which were unavailable for our study area, could lend additional insights into pollination services. This could support agricultural extension. For example, farmers could be encouraged to plant their self-pollinating cereal crops in areas too far to be reached by pollinators from church forests, while prioritising the planting of pollinator-dependent crops closer to the forest.

The vegetative species richness of church forests varies with altitude and grazing intensity (Wassie et al., 2010). The number of honeybee flora species in a given church forest will, thus, vary. As previously mentioned, honeybee hives in South Gondar are predominantly (95%) traditional cylindrical hives, but traditional and modern hives are both used for beekeeping in church forests. We assumed the church forests are the only location for hives; however, there may be modern hives kept by farmers outside of the church forests within the agricultural matrix. Our pollination services map only covers potential pollination zones from hives in forest patches.

## Results

We recorded 1,173 church forests in South Gondar, ranging in size from 0.09 to 254.94 ha (see Figure 9.3).

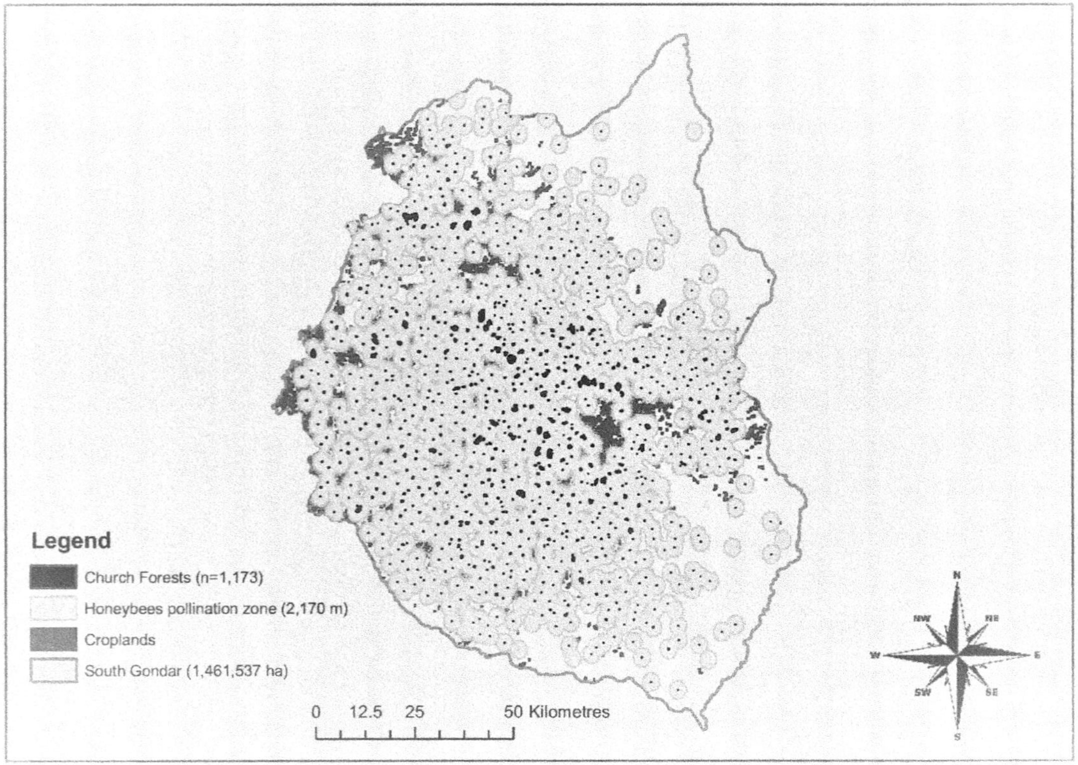


**FIGURE 9.3** South Gondar church forests cover 7,235 ha (0.5% of South Gondar). Pollination zones with radii of 2,170 m in South Gondar, Amhara, Ethiopia, were overlaid with croplands to determine that pollination zones of honeybees cover 282,866 ha (~33%) of croplands (19% of South Gondar). Photo Credits: Figure created by Ashley Massey Marks (data derived from ESA GlobCover Project, 2009; Ricketts et al., 2008).

The largest forests likely reflect the capture of national forest, such as Alem Saga in central South Gondar, or eucalyptus plantations within 1,000 m of church domes. Most church forests are quite small, however, with an average size of 6.16 (±0.54) ha. In total the 1,173 church forests cover 7,235 ha, or 0.5% of South Gondar. Spatial statistics show that the observed mean distance between church forests in South Gondar is 1,503 m and there is less than 1% likelihood that the clustered pattern of church forests in South Gondar is due to random chance.

Ten honeybee and wild bee species were observed in South Gondar, including *Apis mellifera* and species in the *Lasioglossum* and *Megachile* genuses. Honeybee flora comprise 64% of vegetation species surveyed in church forests. Many of these species provide resources outside of the crop flowering period of the year when such resources are otherwise scarce. In addition to church forests, eucalyptus plantations in the landscape provide honeybee forage throughout the year, but particularly after the belg when trees are the predominant flowering vegetation (Bezabeh, 2006).

## Discussion

In agricultural landscapes where the pressure to convert forest to agricultural land is intense, small forest fragments are conserved due to spiritual and religious values. Sacred natural sites provide resources for direct use by local communities, such as fuelwood, medicinal plants, and harvest of non-timber forest products. This study shows how sacred natural sites can also provide ecosystem services to a much greater area of the agricultural landscape beyond their footprints. In this case, Ethiopian church forests provide pollination services for croplands and contribute significantly to local livelihoods in the process.

We found that church forests in South Gondar are quite small, covering a total of 7,235 ha—only 0.5% of South Gondar. However, croplands are nearby, within the honeybee foraging distance of 2,170 m for 88% of the church forests. The pollination zones surrounding the forests extend to cover 282,866 ha (~83%) of South Gondar's croplands. Overall, pollination services extend to an area nearly 42 times the footprint of these sacred natural sites.

How do small forest fragments provide ecosystem services across such a large area? The church forests are clustered, with a mean observed distance between them of 1,503 m—well within the honeybee foraging distance of 2,170 m. The clustered distribution of church forests in South Gondar is most likely due to their occurring within densely populated human landscapes which follow agricultural settlement patterns. This is what accounts for the widespread pollination coverage of croplands in South Gondar. Our findings support the assertion of Bodin et al. (2006) that the locations of sacred forests are key determinants of pollination service delivery in the landscape. In the future, cropland beyond pollination zones should be targeted by agricultural extension agents for the planting of pollination reservoirs to create additional habitat for pollinators (Brosi et al., 2008). Additionally, future work could build on this study to test pollination frequency, geolocate hives, and apply this approach in other landscapes where pollination science and crop mapping are more advanced.

Our comparison of church forest vegetative species with honeybee flora flowering periods indicates that the forests provide honeybee forage resources throughout the year, including during dearth periods when resources are scarce. This is an important contribution to local livelihoods. Amhara beekeepers report the lack of suitable honeybee forage as their main constraint (Ejigu et al., 2009).

The bulk of the literature on the ecological role of sacred natural sites focuses on biodiversity conservation within the sites' boundaries. However, sacred natural sites may also provide ecosystem services in the surrounding landscape. Identifying ecosystem services such as pollination for agricultural crops can support local farmers' efforts to secure land tenure for forest fragments. While such small fragments may initially escape the notice of policy makers, spatial analyses such as this one demonstrate the outsized impact that small church forests can have on local agricultural livelihoods. We recommend conserving these forest fragments, securing their land tenure regimes, and planting pollinator forage corridors where supplemental pollinator forage is required. Local policy makers would be wise to assess ecosystem services beyond the footprint of forest fragments when determining conservation and land use priorities.

## Acknowledgements

We would like to thank the many priests of the Ethiopian Orthodox Tewahedo Church who provided information and access to their church forests and communities. This material is in part based upon work supported by the National Science Foundation under Grant No. SMA-1359367. Any opinions, findings, and conclusions or recommendations expressed in this material are those of the authors and do not necessarily reflect the views of the National Science Foundation.

## Works Cited

Admassu, A., & Nuru, A. (1999, June 7–8). Effect of honeybee pollination on seed yield and oil content of Niger (Guizotia abyssinica). *1st National Conference of Ethiopian Beekeepers Association* (pp. 67–73) Ethiopia: Addis Ababa.

Amssalu, B., Nuru, A., Radloff, S. E., & Hepburn, H. R. (2004). Multivariate morphometric analysis of honeybees (Apis mellifera) in the Ethiopian region. *Apidologie, 35*(1), 71–81.

Bengtsson, J., Angelstam, P., Elmqvist, T., Emanuelsson, U., Folke, C., Ihse, M., … Nyström, M. (2003). Reserves, resilience and dynamic landscapes. *AMBIO: A Journal of the Human Environment, 32*, 389–396.

Bezabeh, A. (2006). Seasonal intensity of flowering and pollen-forage selectivity by honeybees, *Apis mellifera bandasii* in central highlands of Ethiopia. In T. Degefa F. Feyissa (Eds.), *Institutional arrangements and challenges in market-oriented livestock agriculture in Ethiopia: Proceedings of the 14th annual conference of the Ethiopian Society of Animal Production* (ESAP), 2006, Addis Ababa. 230.

Bhagwat, S. A., Kushalappa, C. G., Williams, P. H., & Brown, N. D. (2005). A landscape approach to biodiversity conservation of sacred groves in the Western Ghats of India. *Conservation Biology, 19*, 1853–1862.

Bhagwat, S. A., & Rutte, C. (2006). Sacred groves: Potential for biodiversity management. *Frontiers in Ecology and the Environment, 4*, 519–524.

Bodin, Ö., Tengö, M., Norman, A., Lundberg, J., & Elmqvist, T. (2006). The value of small size: Loss of forest patches and ecological thresholds in Southern Madagascar. *Ecological Applications, 16*, 440–451.

Cardelús, C. L., Scull, P., Hair, J., Baimas-George, M., Lowman, M. D., & Wassie Eshete, A. (2013). A preliminary assessment of Ethiopian sacred grove status at the landscape and ecosystem scales. *Diversity, 5*, 320–334.

Central Statistical Agency of Ethiopia. (2011). *Large and medium scale commercial farms sample survey, 2010/11 (2003 E.C.). Statistical Bulletin.* Addis Ababa: The Federal Democratic Republic of Ethiopia.

Central Statistical Agency of Ethiopia. (2012a). Crop and livestock product utilization (Private Peasant Holdings, Meher Season). *Agricultural Sample Survey 2011/2012 (2004 E.C.).* Addis Ababa: The Federal Democratic Republic of Ethiopia.

Central Statistical Agency of Ethiopia. (2012b). *Crop production forecast sample survey, 2012/13 (2005 E.C.).* Addis Ababa: Federal Democratic Republic of Ethiopia.

Central Statistical Agency of Ethiopia. (2012c). Land utilization (Private Peasant Holdings, Meher Season). *Agricultural Sample Survey 2011/12 (2004 E.C.).* Addis Ababa: The Federal Democratic Republic of Ethiopia.

Central Statistical Agency of Ethiopia. (2013). Livestock and livestock characteristics. *Agricultural Sample Survey 2012/13 (2005 E.C.).* Addis Ababa.

Chazdon, R. L., Harvey, C. A., Komar, O., Griffith, D. M., Ferguson, B. G., Martínez-Ramos, M., ... Philpott, S. M. (2009). Beyond reserves: A research agenda for conserving biodiversity in human-modified tropical landscapes. *Biotropica, 41*, 142–153.

Ejigu, K., Gebey, T., & Preston, T. R. (2009). Constraints and prospects for apiculture research and development in Amhara region, Ethiopia. *Livestock Research for Rural Development, 21*(10): Article #172.

ESA GlobCover Project. (2009). *GlobCover* [Data Set]. European Space Agency and UCLouvain. http://due.esrin.esa.int/page_globcover.php.

Ethiopian Orthodox Tewahedo Church. (2003). The Ethiopian Orthodox Tewahedo Church Faith and Order. Retrieved 16 November 2021 from https://www.ethiopianorthodox.org/english/calendar.html.

FAO. (2015). FAOSTAT. Rome, Italy. Retrieved from http://www.fao.org/faostat/

Fichtl, R., & Adi, A. (1994). *Honeybee flora of Ethiopia.* Weikersheim: Margraf Verlag.

Gardner, T. A., Barlow, J., Chazdon, R., Ewers, R. M., Harvey, C. A., Peres, C. A., & Sodhi, N. S. (2009). Prospects for tropical forest biodiversity in a human-modified world. *Ecology Letters, 12*, 561–582.

Greenleaf, S. S., Williams, N. M., Winfree, R., & Kremen, C. (2007). Bee foraging ranges and their relationship to body size. *Oecologia, 153*, 589–596.

Howe, C., Suich, H., Van Gardingen, P., Rahman, A., & Mace, G. M. (2013). Elucidating the pathways between climate change, ecosystem services and poverty alleviation. *Current Opinion in Environmental Sustainability, 5*, 102–107.

Ishii, H. T., Manabe, T., Ito, K., Fujita, N., Imanishi, A., Hashimoto, D., & Iwasaki, A. (2010). Integrating ecological and cultural values toward conservation and utilization of shrine/temple forests as urban green space in Japanese cities. *Landscape and Ecological Engineering, 6*(2), 307–315. https://doi.org/10.1007/s11355-010-0104-5

Jinja Honcho. (2011). *Jinja Honcho: Association of Shinto Shrines*. http://www.jinjahoncho.or.jp/en/.
Klein, A.-M., Vaissière, B. E., Cane, J. H., Steffan-Dewenter, I., Cunningham, S. A., Kremen, C., & Tscharntke, T. (2007). Importance of pollinators in changing landscapes for world crops. *Proceedings of the Royal Society B: Biological Sciences, 274*, 303–313.
Kremen, C., Williams, N. M., Aizen, M. A., Gemmill-Herren, B., LeBuhn, G., Minckley, R., ... Ricketts, T. H. (2007). Pollination and other ecosystem services produced by mobile organisms: A conceptual framework for the effects of land-use change. *Ecology Letters, 10*, 299–314.
Malhotra, K. C., Chatterjee, S., Gokhale, Y., & Srivastava, S. (2001). *Cultural and ecological dimensions of sacred groves in India*. New Delhi & Bhopal: Indian National Science Academy & Indira Gandhi Rashtriya Manav Sangrahalaya.
Matiur, R. M., Jagdish, K., Anisur, R. M., & Ali, A. M. (1995). Natural outcrossing in *Lathyrus satlvus* L. *The Indian Journal of Genetics and Plant Breeding, 55*, 204–207.
McCann, J. C. (1997). The plow and the forest: Narratives of deforestation in Ethiopia, 1840–1992. *Environmental History, 2*, 138–159.
Mgumia, F., & Oba, G. (2003). Potential role of sacred groves in biodiversity conservation in Tanzania. *Environmental Conservation, 30*, 259–265.
Millenium Ecosystem Assessment. (2005). *Ecosystems and human well-being: Synthesis: A report for the Millennium Ecosystem Assessment*. Washington, DC: Island Press.
Mittermeier, R. A., Gil, P. R., Hoffman, M., Pilgrim, J., Brooks, T., Goettsch, C., ... Da Fonseca, G. A. B. (2004). *Hotspots revisited: Earth's biologically richest and most threatened terrestrial ecoregions*. Chicago: The University of Chicago Press.
Nogué, S., Long, P. R., Eycott, A. E., De Nascimento, L., Fernández-Palacios, J. M., Petrokofsky, G., ... Willis, K. J. (2016). Pollination service delivery for European crops: Challenges and opportunities. *Ecological Economics, 128*, 1–7.
Ntiamoa-Baidu, Y. (2008). Indigenous beliefs and biodiversity conservation: the effectiveness of sacred groves, taboos and totems in Ghana for habitat and species conservation. *Journal for the Study of Religion, Nature and Culture, 2*, 309–326.
Orlowska, I., & Klepeis, P. (2018). Ethiopian church forests: a socio-religious conservation model under change. *Journal of Eastern African Studies, 12(4)*, 674–695. https://doi.org/10.1080/17531055.2018.1519659
Pauly, A., & Hora, Z. A. (2013). Apini and Meliponini from Ethiopia (Hymenoptera Apoidea Apidae Apinae).
Belgian Journal of Entomology, 16, 1–35.
Rashid, S., Yirga, C., Behute, B., & Lemma, S. (2010). *Pulses value chain in Ethiopia: Constraints and opportunities for enhancing exports*. Washington, DC: International Food Policy Research Institute (IFPRI).
Ricketts, T. H., Regetz, J., Steffan-Dewenter, I., Cunningham, S. A., Kremen, C., Bogdanski, A., ... Viana, B. F. (2008). Landscape effects on crop pollination services: Are there general patterns? *Ecology Letters, 11*, 499–515.
Roubik, D. W. (1989). *Ecology and natural history of tropical bees*. Cambridge: Cambridge University Press.
Schelhas, J., & Greenberg, R. S. (1996). *Forest patches in tropical landscapes*. Washington, DC: Island Press.
Sheridan, M., & Nyamweru, C. (Eds.) (2008). *African sacred groves: Ecological dynamics & social change*. Oxford: Oxford University Press.
Tambat, B. (2005). Seedling mortality in two vulnerable tree species in the sacred groves of Western Ghats, South India. *Current Science, 88*, 350.

UNEP-WCMC. (2012). *World database on protected areas.* Retrieved August 2014 from http://www.wdpa.org/.

UNESCO. 2019. *Lake Tana biosphere reserve, Ethiopia.* Retrieved 15 January 2021 from https://en.unesco.org/biosphere/africa/lake-tana.

Verschuuren, B., McNeely, J., Wild, R., & Oviedo, G. (2010). *Sacred natural sites: Conserving nature and culture.* London and Washington, DC: Earthscan.

Wassie, A., & Teketay, D. (2006). Soil seed banks in church forests of northern Ethiopia: Implications for the conservation of woody plants. *Flora – Morphology Distribution Functional Ecology of Plants, 201*(1), 32–43. https://doi.org/10.1016/j.flora.2005.04.002

Wassie, A., Teketay, D., & Powell, N. (2005). Church forests in North Gonder Administrative Zone, Northern Ethiopia. *Forests, Trees and Livelihoods, 15*, 349–373.

Wassie, A., Sterck, F. J., & Bongers, F. (2010). Species and structural diversity of church forests in a fragmented Ethiopian Highland landscape. *Journal of Vegetation Science, 21*, 938–948.

Zhang, W., Ricketts, T. H., Kremen, C., Carney, K., & Swinton, S. M. (2007). Ecosystem services and dis-services to agriculture. *Ecological Economics, 64*, 253–260.

zur Heide, F. (2012). Feasibility study for a Lake Tana Biosphere Reserve, Ethiopia. https://www.nabu.de/imperia/md/content/nabude/international/machbarkeitsstudie_lake_tana.pdf

# 10

# CULTIVATING FAITH

## Exploring the Role of Faith-Based Relationships in a Christian Conservation Agriculture Programme in the Dakatcha Woodland, Kenya

*Peter Rowe and Stanley Baya*

## Introduction

'It is a blessing to keep the environment clean'. Peter had just arrived in Malindi, Kenya from Edinburgh, Scotland when he heard those words. Having set out to explore the nexus of faith and conservation for his MSc degree, he was in Kenya to learn about the work of A Rocha Kenya (ARK), a Christian environmental conservation organisation. 'Where are you going?', asked a taxi driver after his exit from the airport. 'A Rocha' Peter responded. 'A Rocha, I know it', the taxi driver said, and with that they were headed towards ARK's field centre, *Mwamba* (Swahili for 'rock'). Shortly into the 30-minute journey, Peter began speaking with the driver, Samuel,[1] about what brought him to Kenya. After hearing about Peter's research and Christian faith, Samuel, who informed Peter he was a devout Catholic, readily grasped the connection between the two often-disparate spheres of Christianity and conservation. 'It is a blessing to keep the environment clean', Samuel said as they drove past people working their *mashamba* (Swahili for 'fields'). Kenya, perhaps more than other places, provides a unique setting for everyday encounters between Christianity and conservation, given the influence the two have had in shaping the nation's social and physical landscapes.

Although the role of faith in conservation is being increasingly acknowledged in mainstream conservation agendas (UNEP, 2020), much of the scholarship that has been produced at this nexus is instrumental in nature, concerned largely with how faith can motivate conservation (Mcleod and Palmer, 2015). Indeed, this work is foundationally important for establishing the notion that conservation can be motivated by religious conviction. To move beyond, however, the aim of this book, and of this chapter, to explore the conservation of nature *through* faith, implies that faith serves as not just a motivator, but indeed as a mechanism that shapes the way conservation initiatives are inspired, implemented, and sustained.

DOI: 10.4324/9781003246190-13

With this in mind, this chapter explores the role of faith in shaping a Christian conservation initiative led by ARK in the Dakatcha Woodland, an unprotected forest in Kilifi County recognised by BirdLife International and the International Union for Conservation of Nature (IUCN) as a Key Biodiversity Area (KBA) (see Figure 10.1). Specifically, we demonstrate how relationships founded

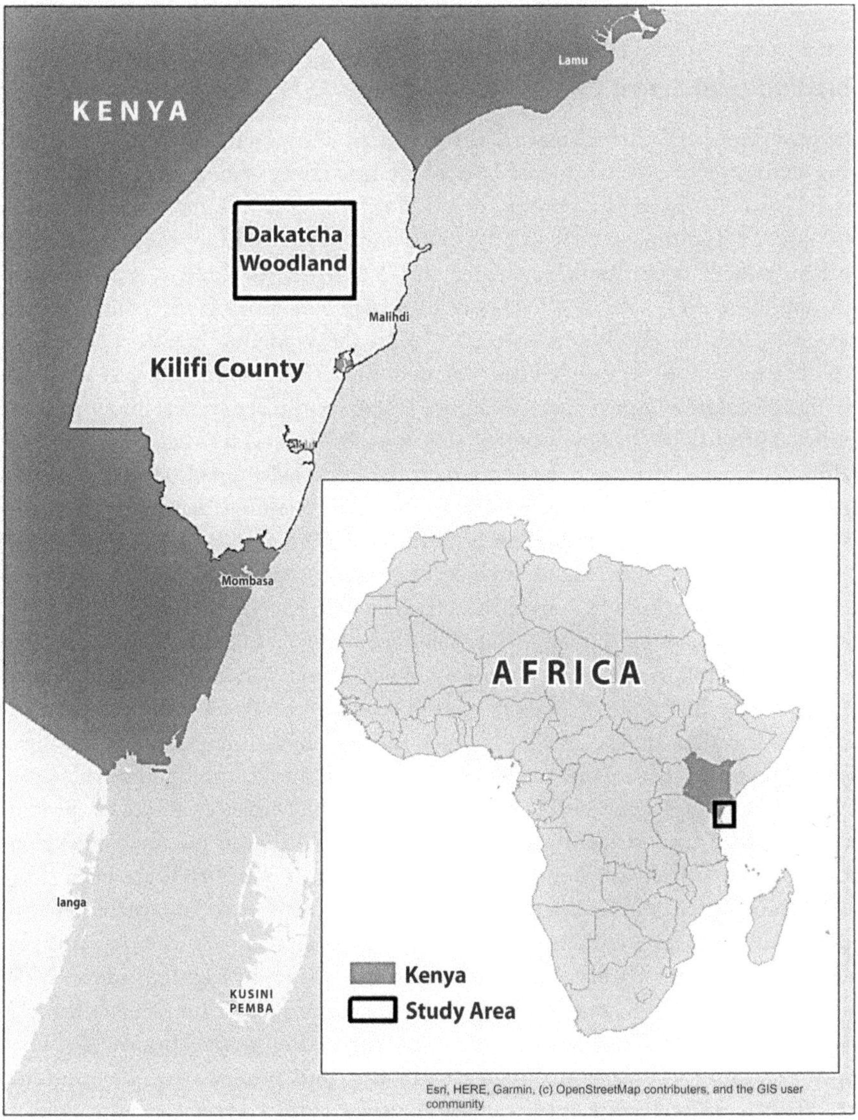


**FIGURE 10.1** Map showing Kenya, Kilifi County, and the approximate boundary of the Dakatcha Woodland. Photo Credits: Produced by Peter Rowe with Esri ArcMap software.

on shared Christian faith between ARK staff and programme participants serve as the fabric through which a food security and biodiversity conservation programme, Farming God's Way (FGW), is implemented. Additionally, this chapter serves to problematise what we call *contractual conservation*, and we suggest that notions of Biblical covenant may be helpful for conceptualising conservation relationships. Before delving into the case study, we outline the prominence of Christianity in Kenya, and the trajectory of conservation in the country, with an eye towards conservation relationships.

## Christianity in Kenya

The prominence of Christianity in Kenya is hard to overstate. Even when considering statistics alone, the influential place of Christianity is evident. Though Kenya's religious landscape is certainly diverse, according to the most recent census, over 80% of Kenyans identify as Christian, with 70% attending religious services weekly (KNBS, 2020; Berkley Center, 2017). Further, in 2014, it was estimated that approximately half of all schools in Kenya had a religious affiliation, the majority Christian (Berkley Center, 2017). In the academic literature, the place of Christianity in Kenya is well documented. From its introduction and influence during colonisation (Spear and Kimambo, 1999), to its place in post-independence politics (Gifford, 2009), development (Parsitau, 2011), and beyond (Van Klinken, 2019), Christianity has come to be perhaps the 'most salient social force' in Kenya (Gifford, 2015, p. 12). Given this, that Christianity itself has many diverse manifestations in the country should not be surprising. Ranging from mainline Christian denominations introduced during colonisation (Anglican, Baptist, Catholic, etc.) to African Instituted Churches,[2] Christianity in Kenya cannot be generalised. Thus, ARK's theology is not generalisable across Kenya, but should rather be seen as a thread within the variegated tapestry of Christianity throughout the country.

Despite the influence of Kenya's many *Christianities* (see Gifford, 2009) on the country's social landscape, their influence on the physical landscape is less clear. However, some recent work has begun to provide initial insights into the nexus of Christianity and environmentalism more broadly. Moyer et al. (2012) provide an overview of faith-based organisations (FBOs) working on sustainable development projects in Kenya. They demonstrate that a variety of FBOs are engaged in sustainability programmes, ranging from waste reduction to sustainable energy. Additionally, Spaling and Vander Kooy (2019) provide an account of the entanglements between Christianity and conservation agriculture by highlighting an FGW programme run by a Canadian non-governmental organisation (NGO) through a church in Kenya. Their work, while more focused on unpacking the claims of FGW regarding soil moisture, increased labour, and productivity, demonstrates that 'faith is an important and animating element of...conservation agriculture' (p. 423), and deserves more attention in conservation and development scholarship. To understand better the conservation context into which the work of ARK enters into, we now turn to a brief overview of conservation in Kenya.

## *Kenyan Conservation and Conservation Relationships*

Within conservation practice in Kenya, three interrelated trajectories can be discerned: 'fortress' or 'fences and fines' conservation, emphasising strictly enforced protected areas (Brockington, 2002), community-based conservation, which, in theory, aims to develop and enact conservation strategies with, and for the benefit of, local people (Western et al., 2015), and neoliberal or 'for profit' conservation (Dempsey, 2016; Fletcher, 2020), a newer wave in conservation practice in which large corporate organisations seek to harness 'economic markets as putative mechanisms for financing nature conservation' (Fletcher, 2020, p. 2). Though each rose to prominence at different moments in Kenya's conservation history, varieties of each remain in practice. While this corpus of literature is expansive, and a recap of it will not be attempted (for in-depth accounts, see Matheka, 2008; Chongwa, 2012; Cockerill and Hagerman, 2020), what is crucial to establish is that conservation in Kenya, much like Christianity in Kenya, is not static, but characterised by diversity and dynamism. Despite this extensive body of work, we contend that the link between what we term *contractual conservation* and conservation relationships, that is, relationships between governing body/international institution/NGO staff and conservation programme participants, has yet to be explored. Here, contractual conservation can be thought of as conservation initiatives based on legal agreements between a governing body/institution/NGO and participants, where the benefit to be gained from the conservation initiative is *conditional*. While contracts for conservation in Kenya have been employed mainly in payment for ecosystem services schemes (Andeltová et al., 2019), they have also become common in wildlife conservation initiatives (Muthiani et al., 2011). Thus, contractual conservation enters into the complex space of conservation relationships.

While much work has been produced on community perceptions of conservation initiatives (see, for example, Matseketsa et al., 2018), less work has been done exploring everyday relationships between staff and participants. However, two important takeaways can be gleaned from the small yet growing conservation relationships literature. First, relationships are paramount to implementing conservation schemes (Thondhlana and Cundill, 2017). Generally accepted is that transparent relationships between staff and local communities lead to better achievement of conservation goals (Thondhlana and Cundill, 2017). Second, however, is that these relationships are often sites of conflict (Mutanga et al., 2015). Despite the growing literature around contracts and relationships in conservation, to our knowledge, no attention has been given to faith in this space. This is surprising, given the overwhelming presence of faith adherents in key biodiversity hotspots where many conservation and development programmes are carried out (Bhagwat et al., 2011; Mikusiński et al., 2014). In Kenya, where both Christianity and conservation are hugely influential, we believe that it is essential to understand better the linkages between these two spheres.

## Case Study: Farming God's Way in the Dakatcha Woodland: Faith-Based Relationships at Work

### *Methodology*

The data for this chapter was collected in June 2019 at ARK's field centre and four FGW plots in the Dakatcha Woodland using semi-structured interviews and participant observation. In total, seven interviews with seven ARK staff were conducted, averaging 59 minutes. Like all researchers, our positionalities, including our Christian faiths, were brought into the research environment. Indeed, despite the surge of literature within human geography and political ecology on various aspects of researcher positionality (Sultana, 2007; Kohl and McCutcheon, 2015), faith appears to be a relatively 'forgotten factor' (Selinger, 2004). While an in-depth exploration into how our faiths shaped this research is beyond the scope of this chapter, it is worth noting briefly two points as to how Peter's faith shaped the research process that led to this chapter.

First, his Christian faith motivated him to study the work of ARK. Like any researcher, he chose to do research in his field because he is passionate about it. Thus, this research is inherently not neutral. Peter chose to explore the work of ARK because he connected to it, not only intellectually, but spiritually and emotionally as well. This resonates with Gold (2002), who writes that, as a Christian doing research on the Focolare Movement in Italy, her academic work reflected her religious convictions so that there was 'nothing impartial' about her choice of research (p. 228). Second, his faith meant that he was able to participate as a believer in faith-based activities held at ARK's office, including communal worship and Bible studies. On participating (as opposed to observation alone), Shah (2017) notes that, 'It is the participation…that has the potential to reveal unique new insights' (p. 54). This notion came to life in Peter's research experience, as participating in these activities with ARK staff as a spiritual insider provided him with a rich understanding of staff and organisational theology.[3]

### *A Rocha International and A Rocha Kenya*

The origins of A Rocha International (ARI – 'A Rocha' is Portuguese for 'The Rock') can be traced to the 1980s when Anglican church leaders Peter and Miranda Harris relocated from England to Portugal with the aim of establishing a field study and bird observation centre 'with a clear Christian character' (Harris, 1993, p. 9). In Portugal, the Harris' envisioned a centre where their Christian convictions would be equally apparent in their conservation work as it was in the more typical ministry they left behind (Harris, 1993). Today, the A Rocha network consists of locally led initiatives in 19 countries around the world (A Rocha, 2021). One of ARI's most recent initiatives is their African Forest Programme that 'aims to conserve 50,000 hectares of biodiversity' across Ghana,

Nigeria, Uganda, and Kenya (A Rocha, 2021). Essential to this effort is ARK's work in the Dakatcha Woodland.

ARK was established as an NGO in Kenya in 1998. Since then, ARK has been working to carry out their mission to 'make God's love for His world known by demonstrating how to practically care for it...' (A Rocha Kenya, 2021). Active in the Dakatcha Woodland for nearly 20 years, ARK began to work with farmers on conservation agriculture techniques in 2014 (A Rocha, 2021). Today, the primary community conservation scheme employed by ARK in the Dakatcha Woodland is FGW, which applies the three pillars of conservation agriculture (no tillage, mulching, and crop rotation) alongside Biblical principles to 'train... community members...to be able to obtain better productivity from their farms and in so doing learn to...care for the rest of God's creation...' (A Rocha, 2021; see also Dryden, 2009). Given this, in addition to improving food security, ARK envisions FGW as a biodiversity conservation initiative, as it is ARK's hope that through FGW, farmers will gain a deeper appreciation for God's creation. Currently, ARK works with 12 churches of a variety of denominations across the Dakatcha Woodland.

## *The Dakatcha Woodland*

Covering over 188,000 hectares, the Dakatcha Woodland is home to 13 IUCN Red List species, including four classified as endangered: Clarke's Weaver (*Ploceus golandi*), the Sokoke Scops Owl (*Otus ireneae*), the Sokoke Pipit (*Anthus sokokensis*), and the Golden-rumped Sengi (*Rhynchocyon chrysopygus*) (A Rocha Kenya, 2020). One of the last remaining fragments of the East African Coastal Forest which historically spanned from Somalia to Mozambique (A Rocha Kenya, 2020), the Dakatcha Woodland is suffering increasing habitat fragmentation due to a number of activities, including unregulated charcoal production, increasing commercial agriculture, and shifting cultivation (A Rocha Kenya, 2020, 2021). Indeed, since 2001, the Dakatcha Woodland has seen over 16% tree cover loss (A Rocha Kenya, 2020).

Of course, the Dakatcha Woodland is home to more than nonhuman life. Located within the Magarini sub-county of Kilifi County, the Dakatcha Woodland has a sparsely distributed population of approximately 66,000 people (KNBS, 2020). In Magarini sub-county, and certainly in the Dakatcha Woodland, most residents (78%) are engaged in farming, the vast majority for subsistence (97%) (KNBS, 2020). Apart from subsistence farming, other employment avenues include commercial agriculture and charcoal production (KNBS, 2020). In terms of the area's religious landscape, unfortunately data is not available at the sub-county level. However, for Kilifi County as a whole, approximately 67% identify as Christian (Catholic, Protestant, Evangelical, etc.), 17% as Muslim, 2% as practising African Traditional Religions, and the remaining 14% claiming another or no religion (KNBS, 2020).

## *Speaking 'in a Language They Understand': Manifesting Faith-Based Relationships*

Despite the relative remoteness of the Dakatcha Woodland, the area is home to dozens of churches. As a non-denominational Christian conservation organisation, ARK seeks to collaborate with churches across denominational lines. Currently, ARK is working with Baptist, Pentecostal Evangelistic Fellowship of Africa, Seventh Day Adventists, and African Instituted Churches. Once a potential church partnership is identified, the process of building a relationship between ARK (the organisation and the staff) and church leadership begins by reaching out to churches in the Dakatcha Woodland.

Rather than presenting themselves as representatives from an NGO, one ARK staff member, a male in his 30s, noted the importance of '*presenting ourselves as brothers and sisters in Christ…*'. If the pastors and *wazee* (Swahili for elders) are interested in what ARK is sharing, they are invited to a one-day workshop at ARK's Watamu field centre to learn more about the Biblical foundations for environmental stewardship and FGW. According to interviews with ARK staff, during this workshop, value is constructed, in part, by speaking '*to them* [pastors and *wazee*] *in a language they understand*'. Broadly, this means that ARK staff focus on communicating their conviction that deforestation and environmental degradation in the Dakatcha Woodland '*is not a scientific problem*' but '*a spiritual problem*'. With this, ARK situates the locus of environmental degradation not in poor practice or management, but in the heart. Going further, according to one ARK staff member, a male in his 30s, the main goal that they attempt to accomplish with pastors and *wazee* throughout the seminar is to help them '*understand the value of God's creation…to God Himself, and your value as man and woman to God himself and this triangular relationship between God, man, and the creation*'. Following this workshop, if the interest remains, ARK staff ask the pastors and *wazee* if they can present four environmentally themed sermons over four consecutive weeks in their respective churches, culminating with a sermon about FGW.

## *Maintaining Relationships: Secular, Faith-Based, or Somewhere in Between?*

Relationships between ARK, partner churches, and farmers are maintained in a variety of ways, across a variety of timescales, using explicitly faith-based activities and what could more broadly be called 'friendship work' (Girgis, 2007, p. 357). Currently, ARK carries out two explicitly faith-based activities in an effort to maintain relationships. First is the aforementioned sermon series in which ARK staff preach sermons that focus on communicating to congregants (and thus potential FGW farmers) themes such as God's love for creation and the role of humans in environmental stewardship. During the fourth sermon, ARK staff introduce FGW not only as a method of farming laid out by God (Dryden, 2009, p. 7), but also as a means of biodiversity conservation in the Dakatcha

Woodland. Indeed, ARK staff were clear in communicating that though the primary goal of FGW is to increase food security, a hoped-for by-product of participation in FGW is the cultivation of Christian conservation ethic which shapes how congregants think about and act towards the Dakatcha Woodland and nonhuman life therein, including the endangered species ARK works to protect. Once the sermon series concludes, the second activity, a Bible study, is led by ARK staff. According to several interviews, the Bible studies are meant to build on the '*foundation*' laid down during the sermon series, and '*reinforce the concept of...Biblical environmental stewardship*'.

ARK staff also engage in a number of activities that could fall under the umbrella of 'friendship work' (Girgis, 2007, p. 357). Such activities include weekly check-in visits by ARK staff to FGW farmers (see Figure 10.2), as well as more personal activities such as eating together, attending weddings, funerals, and, according to one ARK staff member, a male in his 50s, '*simply spending time with them*'. This, however, begins to diverge from the 'friendship work' described by Girgis, who defines it as 'the constructive, empowering work that practitioners do in order to build capacity with others' (Girgis, 2007, p. 357). While this seems to align with the work of ARK staff, Girgis goes on to note that 'Friendship work is not emotional, nor is it done to make friends with another person' (Girgis, 2007, p. 357). This represents a significant departure from the work of ARK, as some staff were consistent in reporting that actually befriending FGW participants is crucial

**FIGURE 10.2** ARK staff member (right) talking with an FGW farmer in the Dakatcha Woodland. Photo Credits: Peter Rowe.

for successful work. Before discussing how these relationships are mobilised for environmental action, it is worth noting that, though some maintenance activities (eating together, attending weddings and funerals) are not explicitly faith-based, it is difficult at best to fully separate them from 'faith-based' work. Though ARK staff and FGW participants share a host of potential commonalities (language, ethnicity, nationality), because they are interacting within relationships founded on a diverse yet shared Christian faith, the work they do is inherently under the umbrella of Christianity. Thus, though ARK staff are interacting with churches and farmers as NGO workers, they are also (and perhaps more importantly) interacting as '*...brothers and sisters in Christ...*'.

## *From Maintenance to Mobilisation*

Like maintenance, the mobilisation of relationships occurs over several timeframes. Primarily, relationships are mobilised in the short term to encourage farmers to adopt FGW. As outlined above, ARK staff spend much time cultivating relationships based on shared faith, hoping that individuals will gain a deeper understanding of and appreciation for God's love of creation, and out of that, consider adopting FGW. Since ARK began recruiting FGW participants by building relationships with local churches in 2018, ARK has seen participation steadily grow. Currently, ARK works with approximately 60 farmers across 12 churches throughout the Dakatcha Woodland. In addition to improving food security, the smaller scale at which FGW is practised translates to practical conservation impact by reducing encroachment into the Dakatcha Woodland and habitats of endangered species therein. In the long term, ARK's Christian faith shapes the duration of relationships with churches and farmers in two distinct ways that combine to produce relationships intended to span well beyond relationships typical in many conservation initiatives.

First is the notion that environmental degradation generally, and deforestation in the Dakatcha Woodland specifically, is '*a spiritual problem*'. Alongside this is the belief expressed by several ARK staff that '*spiritual*' change is a slow process, often playing out over a span of years, even decades. Given this, in several interviews, ARK staff revealed a commitment to '*walking with communities for the long term...*'. Given their extended duration, relationships, according to a female ARK staff member in her 20s, are seen as '*almost a lifelong relationship that you can't really let go*'. Second, in practising FGW, ARK is committed to using informal, faith-based relationships. An emergent theme expressed by ARK staff was captured through an interview with a staff member, a female in her 20s, who noted the desire to see change [in FGW farmers] as a result not of contractual obligation, but as a result of change '*totally from your heart*'. Indeed, the idea of using contracts was viewed as unbiblical by at least one staff member, a male in his 50s, who saw the consequences of not fulfilling a contract as '*effectively a punishment*' which '*doesn't do a relationship much good...*'. Explaining further, the staff member noted that '*God isn't about contracts, He's about covenant*'.

### *Towards a Covenant Conservation*

As alluded to above, contracts for conservation are becoming increasingly prevalent. So, the question we would like to pose is: what comes beyond contractual conservation? Based on our case study above, it is clear that faith at times has a central role to play in the relationships which make practical conservation action possible. The faith-based relationships cultivated between ARK staff and programme participants serve as the fabric through which conservation initiatives such as FGW are carried out. Indeed, ARK's Christian faith shapes the way relationships with churches and farmers manifest, are maintained, and mobilised, demonstrating a viable faith-based conservation relationship framework. Further, the relationships ARK staff cultivate with FGW participants can be viewed not in terms of contractual obligation, but rather in the context of Biblical covenant.

Of the themes that tie together the Bible, covenant is one of the most important (Mackie, 2017). While covenants can be very personal (a marriage covenant), Biblical covenant can also be quite large in scale, such as the Noahic covenant (*Genesis*, 9: 9–11). Though different types of covenants exist throughout the Bible, it can generally be conceptualised as a personal and relational partnership and space to accomplish a goal together within the context of God's unconditional love (Mackie, 2017). Brinig and Nock (1999) note that covenants depart in three ways from 'secular, legal contracts': permanence, unconditional love, and involvement, or at least recognition, of God (p. 11). Thus, while a contract is a 'legal relationship in which...parties bind themselves together on mutual *conditions*' (Torrance, 1970, p. 54, italics in original), a covenant, in theological terms, is a 'promise binding two people or two parties to love one another *unconditionally*' (Torrance, 1970, p. 54, italics in original). Taking these together, a conservation covenant can be thought of as a long-term relationship between parties in which they work together towards mutually accomplishing conservation goals within the context of God's unconditional love.

This notion of covenant, though not part of formal conservation strategy used by ARK, begins to manifest, in part, in their work in the Dakatcha Woodland. According to numerous interviews, several ARK staff noted that relationships with communities would not terminate, even if the FGW project did. Indeed, when asked how long relationships with farmers were intended to last, one staff member, a female in her 30s, responded with a puzzled look and replied, '*until infinity*'. Regarding unconditional love, one staff member, a female in her 20s, stated that the importance ARK places on developing relationships comes from '*valuing love, because you really have to love these people*'. The notion of involving God in relationships between ARK and FGW farmers is most clear, as these relationships materialise in churches throughout the Dakatcha Woodland and are maintained through various faith-based activities such as sermons and Bible studies. Additionally, several staff members discussed in interviews the importance of praying about the work they undertake and with churches and farmers in the Dakatcha Woodland. Further, one staff member, a male in his 50s, noted that relationships with churches and farmers in the Dakatcha Woodland were about '*trust, and...being known and knowing the other party and sharing and being open*' about conservation and development hopes. While this model of covenant conservation may be applicable to the work of ARK and potentially other FBOs, we recognise that the vast majority of conservation organisations in Kenya have no formal faith basis. Nonetheless, we believe that conceptualising conservation relationships as more akin to Biblical covenant rather than legal contract could be impactful in helping conservation organisations, both faith-based and secular, think about the relationships they cultivate.

## Conclusion

This chapter is inherently incomplete. While our case study offers initial insights into the realities of faith-based relationships in conservation, we recognise the nuance of relationship building, especially in a context as dynamic and diverse as Kenyan Christianities and conservation. Despite this, by exploring the role of faith-based relationships, we have presented a practical example of how ARK is working to conserve nature *through* faith. In doing so, we have shown three things. First, ARK's Christian faith moves beyond motivation, fundamentally shaping their conservation programmes, whereby faith-based relationships become the very fabric through which conservation manifests. Second, we have shown that far from being on the periphery, ARK's Christian faith is paramount in actively constructing value in their work in the Dakatcha Woodland. By speaking to church leaders '*in a language they understand*', ARK staff work to build relationships on a shared conviction that both environmental degradation and indeed environmental conservation in the Dakatcha Woodland are a '*spiritual issue*'. Third, this chapter has shown that the relationships ARK staff seek to cultivate with church leaders and FGW farmers may be reflective more of Biblical covenant than legal contract. While far greater research is needed to explore this possibility, its potentials and its pitfalls, our work demonstrates the centrality of faith in conserving nature.

## Notes

1 This name has been changed in an effort to protect anonymity.
2 African Instituted Churches are churches whose 'establishment and growth has taken place on African soil, under the initiatives of Africans' (Mwaura, 2004, p. 161).
3 We use the term 'spiritual insider' cautiously, as while there may be significant overlaps between Peter's personal theology and the theologies of ARK staff, there may be significant departures as well (Snellman, 2020).

## References

A Rocha International. (2021, April 21). *A Rocha International*. http://www.arocha.org
A Rocha Kenya. (2020, December 15). *A Rocha Dakatcha nature reserve concept note*. https://www.arocha.or.ke/wp-content/uploads/sites/16/2020/10/202009-Dakatcha_Land_Purchase_Acres-min.pdf

A Rocha Kenya. (2021, January 7). *Farming God's way*. https://www.arocha.or.ke/farming-gods-way/

Andeltová, L., Catacutan, D. C., Wünscher, T., & Holm-Müller, K. (2019). Gender aspects in action-and outcome-based payments for ecosystem services—A tree planting field trial in Kenya. *Ecosystem Services, 35*, 13–22.

Anonymous Interviews 1–7 (2019, June 7–18). Interviewed by: Peter Rowe.

Berkley Center. (2017, March 28). *Faith and development in focus: Kenya*. https://berkleycenter.georgetown.edu/publications/faith-and-development-in-focus-kenya

Bhagwat, S. A., Dudley, N., & Harrop, S. R. (2011). Religious following in biodiversity hotspots: Challenges and opportunities for conservation and development. *Conservation Letters, 4*(3), 234–240.

Brinig, M., & Nock, S. (1999). Covenant and contract. *Regent University Law Review, 12*, 9.

Brockington, D. (2002). *Fortress conservation: The preservation of the Mkomazi Game Reserve, Tanzania*. Bloomington: Indiana University Press.

Chongwa, M. B. (2012). The history and evolution of national parks in Kenya. In *The George Wright Forum* (Vol. 29, No. 1, pp. 39–42). Hancock, MI: George Wright Society.

Cockerill, K., & Hagerman, S. (2020). Historical insights for understanding the emergence of community-based conservation in Kenya: International agendas, colonial legacies, and contested worldviews. *Ecology and Society, 25*(2), 1–19.

Dempsey, J. (2016). *Enterprising nature: Economics, markets, and finance in global biodiversity politics*. Hoboken, NJ: John Wiley & Sons.

Dryden, G. (2009, May 22). *Farming God's Way trainer's reference guide*. http://media.virb-cdn.com/files/c5/FileItem-279260-FGW_Trainers_Reference_Guide.pdf

Fletcher, R. (2020). Neoliberal conservation. In M. Aldenderfer (Ed.), *Oxford research encyclopedia of anthropology*. New York: Oxford University Press.

Gifford, P. (2009). *Christianity, politics, and public life in Kenya*. New York: Columbia University Press.

Gifford, P. (2015). *Christianity, development and modernity in Africa*. London: Hurst.

Girgis, M. (2007). The capacity-building paradox: Using friendship to build capacity in the South. *Development in Practice, 17*(3), 353–366.

Gold, L. (2002). Positionality, worldview and geographical research: A personal account of a research journey. *Ethics, Place & Environment, 5*(3), 223–237.

Harris, P. (1993). *Under the bright wings*. Vancouver, BC: Regent College Publishing.

KNBS. (2020, October 21). *2019 population and housing census*. Retrieved from https://www.knbs.or.ke/?page_id=3142

Kohl, E., & McCutcheon, P. (2015). Kitchen table reflexivity: Negotiating positionality through everyday talk. *Gender, Place & Culture, 22*(6), 747–763.

Mackie, T. (2017, August 19). *Covenant – The language of faith*. Retrieved from https://www.youtube.com/watch?v=AyVEEZDsKdA

Matheka, R. M. (2008). Decolonisation and wildlife conservation in Kenya, 1958–68. *The Journal of Imperial and Commonwealth History, 36*(4), 615–639.

Matseketsa, G., Chibememe, G., Muboko, N., Gandiwa, E., & Takarinda, K. (2018). Towards an understanding of conservation-based costs, benefits, and attitudes of local people living adjacent to Save Valley conservancy, Zimbabwe. *Scientifica*.

Mcleod, E., & Palmer, M. (2015). Why conservation needs religion. *Coastal Management, 43*(3), 238–252.

Mikusiński, G., Possingham, H. P., & Blicharska, M. (2014). Biodiversity priority areas and religions—A global analysis of spatial overlap. *Oryx, 48*(1), 17–22.

Moyer, J. M., Sinclair, A. J., & Spaling, H. (2012). Working for God and sustainability: The activities of faith-based organizations in Kenya. *Voluntas: International Journal of Voluntary and Nonprofit Organizations, 23*(4), 959–992.

Mutanga, C. N., Vengesayi, S., Muboko, N., & Gandiwa, E. (2015). Towards harmonious conservation relationships: A framework for understanding protected area staff-local community relationships in developing countries. *Journal for Nature Conservation, 25*, 8–16.

Muthiani, E. N., Njoka, J. T., Kinyua, P., & Gitau, G. (2011). Partnership challenges of community wildlife sanctuaries in Laikipia County, Kenya. In *KASAL Program Conference, Kenya Agricultural Research Institute Headquarters* (pp. 9–12). Nairobi: Kenya Arid and Semi-Arid Lands Research Programme.

Mwaura, P. N. (2004). African instituted churches in East Africa. *Studies in World Christianity, 10*(2), 160–184.

New International Version Bible. (2011). *Genesis, 9*, 9–11.

Parsitau, D. S. (2011). The role of faith and faith-based organizations among internally displaced persons in Kenya. *Journal of Refugee Studies, 24*(3), 493–512.

Selinger, L. (2004). The forgotten factor: The uneasy relationship between religion and development. *Social Compass, 51*(4), 523–543.

Shah, A. (2017). Ethnography? Participant observation, a potentially revolutionary praxis. *HAU: Journal of Ethnographic Theory, 7*(1), 45–59.

Snellman, G. (2020). Approaching Laestadianism as an insider researcher. *Approaching Religion, 10*(1), 40–54.

Spaling, H., & Vander Kooy, K. (2019). Farming God's way: Agronomy and faith contested. *Agriculture and Human Values, 36*(3), 411–426.

Spear, T., & Kimambo, I. (1999). *East African expressions of Christianity.* Eastern African Studies Series, Oxford: James Currey. Athens, OH: University Press.

Sultana, F. (2007). Reflexivity, positionality and participatory ethics: Negotiating fieldwork dilemmas in international research. *ACME: An International Journal for Critical Geographies, 6*(3), 374–385.

Thondhlana, G., & Cundill, G. (2017). Local people and conservation officials' perceptions on relationships and conflicts in South African protected areas. *International Journal of Biodiversity Science, Ecosystem Services & Management, 13*(1), 204–215.

Torrance, J. B. (1970). Covenant or contract?: A study of the theological background of worship in seventeenth-century Scotland. *Scottish Journal of Theology, 23*(1), 51–76.

UNEP. (2020, November 3). *Faith for Earth initiative.* United Nations Environment Programme. Retrieved from https://www.unenvironment.org/about-un-environment/faith-earth-initiative

Van Klinken, A. (2019). *Kenyan, Christian, queer: Religion, LGBT activism, and arts of resistance in Africa* (Vol. 3). University Park: Penn State Press.

Western, D., Waithaka, J., & Kamanga, J. (2015). Finding space for wildlife beyond national parks and reducing conflict through community-based conservation: The Kenya experience. *Parks, 21*(1), 51–62.

# 11

# SHINTO SHRINE FORESTS AND HAPPINESS IN JAPAN

*Ashley Massey Marks, Ken Kitatani and Shonil A. Bhagwat*

## Introduction

Over half of the world's population lives in urban areas and 67.2% of the world's population is projected to live in urban areas by 2050. In Europe, North America, Australia, New Zealand, and Japan, 77.7% of people live in cities with projections of 85.9% by 2050 (United Nations, 2011, 2012). Green spaces in urban areas decrease stress and improve well-being; research to date focuses on high-income countries: The United States, the United Kingdom, European countries, and Japan (Wolf et al., 2012). Li (2009) notes that residents of Japanese prefectures with less forest cover have higher standardised mortality ratios from cancer than those living in prefectures with greater forest cover.

In addition to health benefits, research indicates that green space can improve happiness. Over 20,000 participants in a United Kingdom study responded to a smartphone survey to report their happiness. Increased happiness was reported in green spaces as opposed to urban environments (MacKerron & Mourato, 2013). Following the Kingdom of Bhutan's reporting of 'Gross National Happiness' (GNH) as an alternative to Gross Domestic Product (GDP), measuring happiness on a national scale has gained traction in numerous countries, including China, France, Austria, the United Kingdom, and Japan (Atkisson, 2012). In East Asian cultural contexts, happiness is defined in terms of social interpersonal connectedness (Uchida et al., 2004).

Sacred natural sites are 'areas of land or water having special spiritual significance to peoples and communities' (Oviedo & Jeanrenaud, 2007:79) and range in size from small cemeteries to large urban forests. Undertaking inventories of urban sacred natural sites has been identified as a top research priority with the aim of improving their management and protection (Jackson & Ormsby, 2017). Urban sacred natural sites face a multitude of challenges, including 'urban sprawl, air

DOI: 10.4324/9781003246190-14

pollution, and the clearing and fragmentation of sites' (Ormsby, 2021:1). This study focuses on sacred natural sites in the highly urbanised nation of Japan; 91.3% of Japan's population lives in urban areas (United Nations, 2011). Large swaths of forest in Japan have been cleared for urban development. The World Health Organization recommends that cities have a minimum of 9 m2 green space per capita (Kuchelmeister, 1998). Asian cities average 39 m2 green space per capita, African cities 79 m2 green space per capita, and Latin American cities 255 m2 green space per capita (Economist Intelligence Unit, 2012; Kuchelmeister, 1998). In Japan, Tokyo has 5.8 m2 green space per capita, Osaka has 4.5 m2 green space per capita, and in Aichi prefecture, 26 of 32 cities have green space per capita below the 9 m2 green space per capita WHO threshold (Economist Intelligence Unit, 2012; Tokyo Metropolitan Government Bureau of Construction, 2012; Yamamoto, 2010).

Shinto shrines, or '*jinja*', persist despite urban growth. The surrounding Shinto shrine forests, or '*chinju no mori*', comprise one form of green space in Japanese cities. During the Meiji Era (1868–1912), the government decreed Shintoism the national religion and claimed that the divine rights of the Emperor were rooted in Shinto. By restricting state funds to one *jinja*, or 'shrine', per village, the government attempted to consolidate and nationalise Shinto practice. During this period, the number of Shinto shrines in Japan decreased from 200,000 to 120,000 (Shimazono, 2009). After the Second World War, Japan's rapid industrialisation spurred a wave of migration to cities, effectively depopulating the countryside. Although Shintoism was once regarded as a rural Japanese religion, Shinto shrine forests persisted. Their legal status as state land, in the same classification as Japanese Imperial Court land, served as a bulwark against land use change. Today, Shintoism is practised by 79.2% of Japan's population, many of whom also practise Buddhism (Central Intelligence Agency, 2013).

While State Shinto was heavy-handed in its attempt to financially influence autonomous shrines run by Shinto priests and priestesses, it also linked Shintoism with Japanese nationalism. Another essentialist view is described as 'ethnic', and conceives that a connection with nature is part of being Japanese:

> One could say that the basis of Japan's culture and spirituality lies in these forests. The deep woods, sacred places into which men do not lightly tread, evoked reverence and stirred the imagination. As if it were embedded in our DNA, awe of sacred trees still dwells in the hearts of modern Japanese
> *(Japan National Tourism Organization, 2005:1).*

Essentialist conceptions of Shintoism have been extensively critiqued; yet, Shinto is still described through a variety of alternative paradigms, including ahistoric, an imperial non-religion, environmental, as asserting otherness, local, universal, and spiritual (Rots, 2013). Thus, it is important to note that this paper is not claiming Shinto to be an 'environmental' religion, even though we focus on the distribution of Shinto shrine forests in the landscape.

Shinto shrine forests persist across the country to welcome and honour *kami*, or 'Shinto gods', and to lend a 'dignified' atmosphere to shrine buildings

(Domenig, 1997). Shinto shrine forests range in size from less than a hectare, to the 5,400 ha Grand Shrine of Ise, which is 2,000 years old and includes stands of primeval forest (Ishii et al., 2010; Manabe et al., 2008; Public Affairs Headquarters for Shikinen-Sengu, 2010). The concept of sacred forests in Japan as the meeting place for men and gods dates back to Jomon times (10,000 to 300 BC) (Tsukada, 2005). Masaaki Ueda, former chairman of the Shaso Gakkai, a society for sacred natural sites in Japan, said:

> The Japanese viewed giant trees and pillars as yoshiro– meaning "a place where gods draw near." Even as they cut down the forests to develop rice paddies and towns, there was a sense of taboo, that if you felled a tree in a sacred forest, you would be cursed. As a result they preserved woods with original flora. Although established as forbidden forests that people were not supposed to enter, the sacred woods became the center of village community because people could gather around them and offer performances of dance and music to the gods
>
> *(Tsukada, 2005:1).*

Festivals, or '*matsuri*', with dance and music still regularly occur at Shinto shrines, and, in the modern era, are scheduled for Saturdays and Sundays to align with the Japanese work week (Nelson, 2006). These large gatherings bring people together and promote a sense of social cohesion. People visit shrines on many special days, including to celebrate New Year, bless their newborns a few weeks after birth, for '*setsubun*' – chasing away evil spirits at the start of spring, celebrating the harvest, '*obon*' – visits from the ancestors, and for '*shichi-go-san*' – girls aged three and seven and boys aged three and five dress in kimonos and receive prayers for their health and development (Jinja Honcho, 2011).

Entering a Shinto shrine on a typical day is a markedly different experience, with a quiet and peaceful atmosphere. One enters through wooden torii gates – open archways that separate the secular world from the *kami*, but allow the public to access the shrine at any time. After passing through the gates, visitors walk down an entrance path lined with trees and into shrine precincts with sacred trees. Visitors wash their hands and mouths at the purification trough before making offerings and prayers at the offering hall. There are places to leave wishes on wooden plates, '*ema*', and in some shrines, receive fortune telling paper slips, '*omikuji*'. Sacred trees, stones, etc., are marked with a straw rope with zigzag paper strips, '*shimenawa*' (Figure 11.1).

Shintoism provides a feeling of connection with friends, family, Japanese identity, and the natural world. Happiness is a relative concept, and in Japan, happiness is constructed as the realisation of social harmony (Kitayama et al., 2000). Evans (2001:ii) notes:

> In Shinto, each individual stands on a vertical line connecting the kami, [their] ancestors and [their] descendants, past and future. Additionally, each person is also on a horizontal line that connects [them] with [their] neighbour, [their] friends, society, country, and with foreign nations.

FIGURE 11.1 Meiji shrine, Tokyo from top left, clockwise: Entering through torii gate, three women in festival wear walking down the entrance path lined with trees, sacred trees marked with 'shimenawa', purification trough. Photo Credits: Ashley Massey Marks.

Social trust has been found to have significant positive effects on subjective assessments of happiness in Japan, as reported by the 14,538 respondents of the Japanese General Social Survey (JGSS) (Kuroki, 2011).

Shintoism emphasises practice, or *jissen*, over theology or doctrine, and the 'practice of Shinto leads [its practitioners] to live in harmony with *Dai Shizen no Meguri*, the ceaseless cycle of Great Nature' (Evans, 2001:ii). Kishimoto (1968:126) describes how Shinto shrine forests enrich the lives of urban residents:

> At a corner of a busy district of a city a Jinja is quietly situated. The precinct of the sanctuary is full of green covered by huge old trees… At due time and season, rituals are observed as they were in ancient days. The performance produces in the participants a certain kind of sacred feeling… Stepping into the precinct, into its quietude, a visitor rediscovers his inner self having been lost all the while in the busy life of the city… To have a brief moment like this a few times a week for a lay member of a Jinja, means an immeasurable enriching of his life. The Jinja becomes the oasis of urban life.

Empirical research in Japan has investigated the physiological benefits of *shinrin-yoku*, or 'taking in the forest atmosphere or forest bathing' (Li, 2009; Park et al., 2009; Tsunetsugu et al., 2009). Over 25% of 498 respondents in a 2003 survey engaged in *shinrin-yoku* in the past year and over 66% of respondents were interested in *shinrin-yoku* for health benefits (Cabinet Office of Government of Japan, 2004).

*Shinrin-yoku* may reduce the risk of psychosocial stress-related diseases (Morita et al., 2007). Diabetic patients who engaged in *shinrin-yoku* nine times over a six-year period decreased their blood glucose levels after 30 minutes of walking in the forest, more than diabetic subjects using cycle ergometers, walking on tread-mills, or swimming in hot spring pools (Ohtsuka et al., 1998). Li (2009) found that office workers in Nagano prefecture who stayed in the forest for three-day/two-night retreats decreased stress levels and exhibited improved immune activity up to a week, and in some cases over a month, later. Morita et al. (2007) found that stress relief occurred upon arriving at the forest, and the type of forest, size, and time spent in the forest did not determine health benefits. This is especially important considering the high rates of stress in the Japanese workforce, coupled with a relatively low incidence of green space in Japanese urban areas. According to government surveys, 58% of workers in Japan report stress at work, and there is even a Japanese term, *karoshi*, for 'death from overwork' (Ministry of Health Labour and Welfare, 2008).

## Methodology

We created a forest map layer for Japan in ArcMap GIS using GlobCover satellite data (ESA GlobCover Project, 2009), a map layer of core urban areas, using GlobCover's designation of artificial surfaces and associated areas (ESA GlobCover Project, 2009), and a map layer of surrounding urban areas (areas of dense human habitation beyond the core urban areas), using 2002–2003 MODIS satellite data at 1 km resolution (Schneider et al., 2003). We overlapped our layers of forest and core urban areas with the MODIS urban areas layer in ArcMap and removed any intersecting pixels to create our surrounding urban areas layer. We calculated the area of forest, core urban areas, and surrounding urban areas for each of Japan's 47 prefectures, and the population density of each prefecture (GADM, 2012; Statistics Bureau of Japan, 2005). Geographically referenced point data of 80,000 Shinto shrines (Geographic coordinate system WGS-1984) were downloaded from a Shinto shrine database (White 曜社 Co., n.d.), mapped in ArcMap v.10.0 and geographical duplicates were removed. While extensive (n=57,742) and geographically expansive, our analysis is constrained as point data. The future availability of boundary data for sacred natural sites will enable additional analyses.

First, we calculated the Shinto shrine forest density per prefecture, and then we assessed the average observed distance between Shinto shrine forests using the Average Nearest Neighbour tool in ArcMap. For each prefecture, we calculated green space per capita – the preferred metric of urban green space guidelines to ensure that green space provision keeps pace with population growth (Kuchelmeister, 1998). We calculated the Shinto shrine forest density in each prefecture's forest, core urban areas, and surrounding urban areas.

Finally, we employed a Japanese national happiness index as a coarse non-monetary method of measuring cultural ecosystem services on a national scale. We chose to use Sakamoto's 'Happiness Ranking' as it is a national happiness index developed by Japanese researchers who selected indicators for happiness in a Japanese context. Statistical tests were performed in SPSS v. 21. Sakamoto's 'Happiness Ranking' comprises 40 objective indicators within the categories of (1) life and family issues, (2) labour issues, (3) safety and security issues, and (4) medical care and health issues (Sakamoto, 2011). We compared the number of Shinto shrine forests per 1,000 people and Shinto shrine forest density with Sakamoto's well-being index for the 47 prefectures of Japan.

## Results

We identified 57,742 Shinto shrine forests across the 47 prefectures of Japan; there is an observed mean distance of 913.82 m between Shinto shrine forests (Average Nearest Neighbour, NNRatio: 0.23, NNZScore: −351.91, p = 0.00). We found that 66% of Japan comprises forest, 1% is core urban areas, and 14% is surrounding urban areas. The remaining 19% of Japan is dominated by other land cover types (ESA GlobCover Project, 2009; Kitayama et al., 2000; Schneider et al. 2003; White 曜社 Co., n.d.). We found that 12,935 (~22%) of Shinto shrine forests are located inside larger forested areas, 30,423 (~53%) are located in surrounding urban areas, and 1,512 (~3%) are located in core urban areas. The remaining 12,872 Shinto shrine forests (~22%) are located in the other land cover types. The density of Shinto shrine forests in forested areas across Japan is 0.05 Shinto shrine forests/km2, in core urban areas, there are 0.75 Shinto shrine forests/km2, and in surrounding urban areas, there are 0.57 Shinto shrine forests/km2. We found significant differences between the Shinto shrine forest density in forest versus core urban areas (Related Samples Wilcoxon Signed Rank Test, $n = 44$, $p = 0.000$) and the Shinto shrine forest density in forest versus surrounding urban areas (Related Samples Wilcoxon Signed Rank Test, $n = 47$, $p = 0.000$).

In the prefecture of Tokyo, home to the most populated city in Japan (36.5 million people), core urban areas comprise 43.73% of the prefecture; however, 63.96% of Shinto shrine forests in the prefecture are found in core urban areas, with another 27.55% found in surrounding urban areas (Central Intelligence Agency, 2013). At 889, Tokyo has the most Shinto shrine forests in core urban areas of any prefecture, 689 more than the second-most, Saitama at 200. However, Tokyo's core urban areas are more expansive than other prefectures' core urban areas. Thus in eight other prefectures, the Shinto shrine forest density is higher than Tokyo's 1.13 Shinto shrine forests/km2, the highest being Shimane

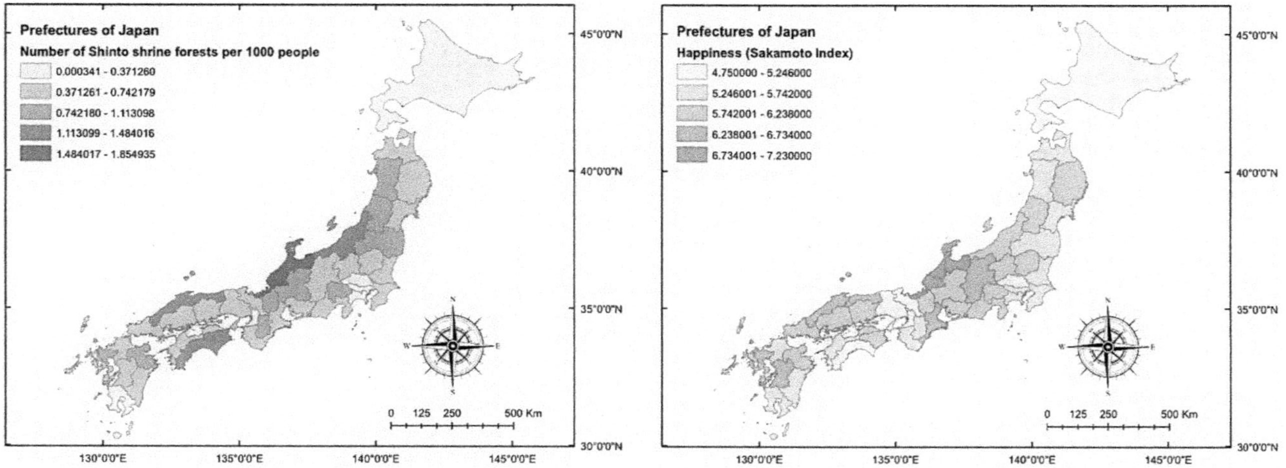


**FIGURE 11.2** Number of Shinto shrine forests per 1,000 people and happiness (Sakamoto Index) for the 47 prefectures of Japan. Photo Credits: Figure created by Ashley Massey Marks – data derived from GADM, 2012; Sakamoto, 2011; Statistics Bureau of Japan, 2005; White 曜社 Co., n.d.

with 5.45 Shinto shrine forests/km2. When the number of Shinto shrine forests in Tokyo is added to the total number of 'secular' green spaces reported by the Tokyo Metropolitan Government Bureau of Construction (2012) (e.g. metropolitan and municipal parks, national gardens, and nature parks), Shinto shrine forests in Tokyo prefecture's 53 districts comprise on average 14.15 ± 1.85% (range: 2.65–86.36%, median 9.95%) of overall green space.

Of the 44 prefectures with core urban areas, 11 prefectures have a Shinto shrine forest density of >1 Shinto shrine forest/km2 and 23 have a Shinto shrine forest density of >0.5 Shinto shrine forest/km2. Of the 47 prefectures with surrounding urban areas, 4 have a Shinto shrine forest density of >1 Shinto shrine forest/km2 and 27 have a Shinto shrine forest density of >0.5 Shinto shrine forest/km2.

The number of Shinto shrine forests per 1,000 people is positively correlated with Sakamoto's 'Happiness Ranking' for the 47 prefectures of Japan (Spearman correlation, $n = 47$, $p = 0.000$, two-tailed, correlation coefficient = 0.591) (Sakamoto, 2011) (Figures 11.2 and 11.3; Table 11.1).

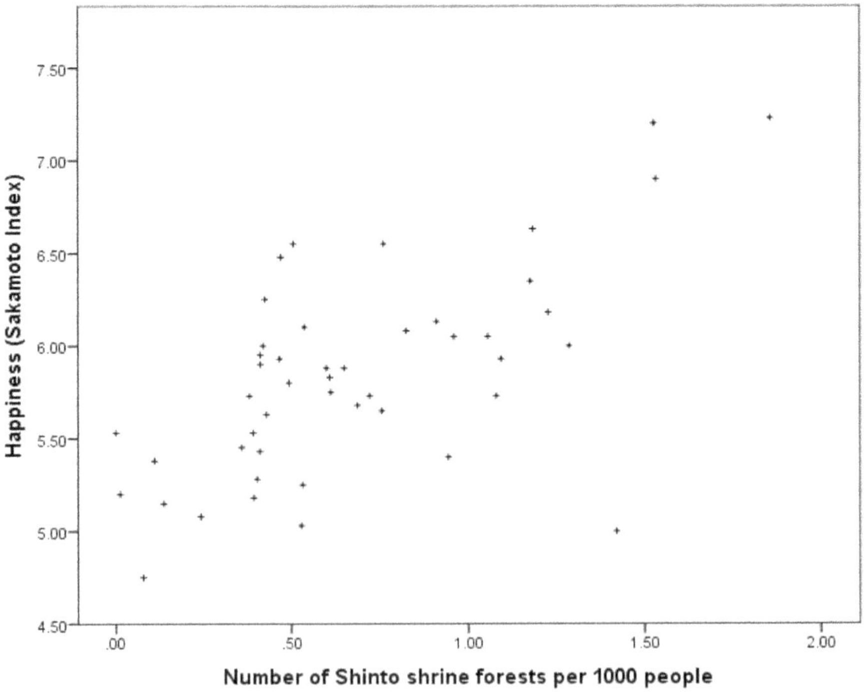


**FIGURE 11.3** Happiness (Sakamoto Index) is positively correlated with the number of Shinto shrine forests per 1,000 people, Spearman correlation ($n = 47$, $p = 0.000$, two-tailed, correlation coefficient = 0.591). Photo Credits: Figure created by Ashley Massey Marks – data derived from Sakamoto, 2011; Statistics Bureau of Japan, 2005; White 曜社 Co., n.d.

**TABLE 11.1** Number of Shinto shrine forests per 1,000 people and happiness (Sakamoto Index) for the 47 prefectures of Japan. Photo Credits: Table created by Ashley Massey Marks – data derived from Sakamoto, 2011; Statistics Bureau of Japan, 2005; White 曜社 Co., n.d.

| *Prefecture* | *Number of Shinto shrine forests* | *Number of Shinto shrine forests/1,000 people and rank* | | *Happiness (Sakamoto Index) and rank* | |
|---|---|---|---|---|---|
| Fukui-ken | 1,524 | 1.85 | 1 | 7.23 | 1 |
| Ishikawa-ken | 1,799 | 1.53 | 2 | 6.9 | 3 |
| Toyama-ken | 1,697 | 1.53 | 3 | 7.2 | 2 |
| Kōchi-ken | 1,132 | 1.42 | 4 | 5 | 46 |
| Tokushima-ken | 1,042 | 1.29 | 5 | 6 | 17 |
| Niigata-ken | 2,982 | 1.23 | 6 | 6.18 | 10 |
| Tottori-ken | 718 | 1.18 | 7 | 6.63 | 4 |
| Shimane-ken | 873 | 1.18 | 8 | 6.35 | 8 |
| Yamagata-ken | 1,330 | 1.09 | 9 | 5.93 | 19 |
| Fukushima-ken | 2,258 | 1.08 | 10 | 5.73 | 27 |
| Yamanashi-ken | 934 | 1.06 | 11 | 6.05 | 14 |
| Ōita-ken | 1,160 | 0.96 | 12 | 6.05 | 15 |
| Akita-ken | 1,080 | 0.94 | 13 | 5.4 | 37 |
| Shiga-ken | 1,256 | 0.91 | 14 | 6.13 | 11 |
| Gifu-ken | 1,738 | 0.82 | 15 | 6.08 | 13 |
| Saga-ken | 659 | 0.76 | 16 | 6.55 | 5 |
| Nara-ken | 1,076 | 0.76 | 17 | 5.65 | 31 |
| Ehime-ken | 1,060 | 0.72 | 18 | 5.73 | 28 |
| Ibaraki-ken | 2,047 | 0.69 | 19 | 5.68 | 30 |
| Nagasaki-ken | 962 | 0.65 | 20 | 5.88 | 23 |
| Tochigi-ken | 1,236 | 0.61 | 21 | 5.75 | 26 |
| Okayama-ken | 1,194 | 0.61 | 22 | 5.83 | 24 |
| Iwate-ken | 832 | 0.60 | 23 | 5.88 | 22 |
| Kagawa-ken | 544 | 0.54 | 24 | 6.1 | 12 |
| Aomori-ken | 765 | 0.53 | 25 | 5.25 | 40 |
| Hyōgo-ken | 2,959 | 0.53 | 26 | 5.03 | 45 |
| Kumamoto-ken | 931 | 0.51 | 27 | 6.55 | 6 |
| Gunma-ken | 999 | 0.49 | 28 | 5.8 | 25 |
| Nagano-ken | 1,032 | 0.47 | 29 | 6.48 | 7 |
| Shizuoka-ken | 1,769 | 0.47 | 30 | 5.93 | 20 |
| Wakayama-ken | 446 | 0.43 | 31 | 5.63 | 32 |
| Mie-ken | 795 | 0.43 | 32 | 6.25 | 9 |
| Yamaguchi-ken | 627 | 0.42 | 33 | 6 | 16 |
| Aichi-ken | 2,998 | 0.41 | 34 | 5.9 | 21 |
| Hiroshima-ken | 1,187 | 0.41 | 35 | 5.95 | 18 |
| Miyagi-ken | 972 | 0.41 | 36 | 5.43 | 36 |
| Fukuoka-ken | 2,041 | 0.40 | 37 | 5.28 | 39 |
| Kyōto-fu | 1,045 | 0.39 | 38 | 5.18 | 42 |
| Chiba-ken | 2,378 | 0.39 | 39 | 5.53 | 33 |

(*Continued*)

| *Prefecture* | *Number of Shinto shrine forests* | *Number of Shinto shrine forests/1,000 people and rank* | | *Happiness (Sakamoto Index) and rank* | |
|---|---|---|---|---|---|
| Miyazaki-ken | 441 | 0.38 | 40 | 5.73 | 29 |
| Kagoshima-ken | 632 | 0.36 | 41 | 5.45 | 35 |
| Saitama-ken | 1,719 | 0.24 | 42 | 5.08 | 44 |
| Hokkaidō | 770 | 0.14 | 43 | 5.15 | 43 |
| Tōkyō-to | 1,390 | 0.11 | 44 | 5.38 | 38 |
| Ōsaka-fu | 696 | 0.08 | 45 | 4.75 | 47 |
| Okinawa-ken | 17 | 0.01 | 46 | 5.2 | 41 |
| Kanagawa-ken | 3 | 0.00 | 47 | 5.53 | 34 |

Sakamoto's 'Happiness Ranking' (Sakamoto, 2011) is negatively correlated with population size of the 47 Japanese prefectures (2005) (Spearman correlation ($n$ = 47, $p$ = 0.001, two-tailed, correlation coefficient = −0.476)), but is not correlated with the population density of the prefectures nor the density of Shinto shrines.

## Discussion

Cultural ecosystem services provided by forest patches, including sacred natural sites such as Japanese Shinto shrine forests, are difficult to quantify and incorporate in ecosystem service assessments, especially at the landscape scale. The Millennium Ecosystem Assessment defines ecosystem services as the 'benefits people obtain from ecosystems', and includes cultural diversity, spiritual and religious values, recreation and ecotourism, aesthetic values, and knowledge systems as cultural ecosystem services (Millenium Ecosystem Assessment, 2005).

Urban green space and forests can improve the well-being of urban populations, including decreasing stress and improving health (Wolf et al., 2012). We hypothesised that sacred natural sites, such as Japanese Shinto shrine forests, provide cultural ecosystem services to urban residents who comprise 91.3% of Japan's population, including social cohesion and stress relief (Central Intelligence Agency, 2013). The development of non-monetary methods of valuing cultural ecosystem services and mapping the distribution of urban sacred natural sites have been identified as research priorities (Daily et al., 2009; Jackson & Ormsby, 2017). First we asked: What is the prevalence and distribution of Shinto shrine forests across the 47 prefectures of Japan? We found Shinto shrine forests to be prevalent ($n$ = 57,742) across the 47 prefectures of Japan. When we explored the distribution of Shinto shrine forests in different land cover types (forest: $n$ = 12,935, 22%; surrounding urban areas: $n$ = 30,423, 53%; core urban areas: $n$ = 1,512, 3%),

Shinto shrine forests were more prevalent in urban areas than in forest areas and also more densely distributed in urban areas than in forest areas (forest: 0.05 Shinto shrine forests/km2; surrounding urban areas: 0.57 Shinto shrine forests/km2; core urban areas: 0.75 Shinto shrine forests/km2).

The distribution of Shinto shrine forests can perhaps be more easily understood as the distribution of religious institutions near the people who use them. Nelson (2006:162) writes, 'As recently as twenty years ago, the household to which a child belonged was considered automatically affiliated with a local shrine' and if they have migrated away from their ancestral area, 'a local sense of belonging guides the choice of the shrine in which one's descendants will be dedicated'. In more densely populated urban areas, there are more religious institutions to meet demand. The land tenure afforded to Shinto shrines helps them stave off the pressures of urban development, which is especially important in highly urbanised Japan with its densely populated cities, because green space is extremely limited compared to other urban areas in the world (Economist Intelligence Unit, 2012; Kuchelmeister, 1998; Tokyo Metropolitan Government Bureau of Construction, 2012; Yamamoto, 2010). Over half of the prefectures had densities of at least one Shinto shrine forest for every 2 km2 in both core and surrounding urban areas.

Our finding that Sakamoto's 'Happiness Ranking' is negatively correlated with the population size of the 47 Japanese prefectures supports the notion in the literature that the negative aspects of urban living (e.g. overcrowding, pollution, and crime) can affect happiness. Japanese critiques of the Sakamoto Index note that subjective indicators of happiness and objective indicators usually match, whereas the Sakamoto Index's prefectural ranking is significantly different than subjective indicators of happiness (i.e. happiness surveys such as the JGSS). Kuroki (2011) analyses subjective indicators of the JGSS versus the Sakamoto Index and concludes that Sakamoto's indicators are more accurately described as a 'quality-of-life' ranking rather than Sakamoto's 'Happiness Ranking'. When we compared the distribution of Shinto shrine forests with the national happiness index, we found greater happiness in prefectures with more Shinto shrine forests per capita. The three prefectures with the fewest Shinto shrine forests relative to population were Kanagawa (3 shrines for 8,791,597 people), Okinawa (17 shrines for 1,361,594 people), and Osaka (696 shrines for 8,817,166 people). These three prefectures also placed on the low end of the Sakamoto Index: Kanagawa (34th of 47), Okinawa (47th of 47), and Osaka (41st of 47). Okinawa is the southernmost prefecture in Japan and was part of the Ryukyu Kingdom until 1879, with a culture and language different from mainland Japanese. The introduction of Shinto shrines was meant to assimilate Okinawans into mainstream Japanese culture, but Shintoism was not embraced by Okinawans (Grayson, 2005).

The three prefectures with the highest number of Shinto shrine forests relative to population are all located in the Hokuriku region in the northeastern part of the main island of Japan: Fukui (1.855), Ishikawa (1.532), and Toyama (1.526).

The same three prefectures have the highest happiness ranking according to the Sakamoto Index: Fukui (7.23 points), Toyama (7.20 points), and Ishikawa (6.90 points). Fukui is known for its child-friendly environment, good welfare and social programmes, and low rates of fire, crime, and accidents (Hosei University, 2011). While these factors may be responsible for Fukui's high happiness ranking, they may also be linked to Shinto shrine forests and the sense of community and social cohesion that they provide.

The prefecture of Kochi is an anomaly: It has a high number of Shinto shrine forests per 1,000 people (4th of 47) but a low happiness ranking (46th of 47); it is a visible outlier in the lower right hand quadrant of the correlation graph. In Kochi, typhoons cause frequent landslides and floods leading to huge losses of life and property, which lead to low rankings in the safety and security indicators in the Sakamoto Index.

It is important to note that there are numerous factors influencing a prefecture's happiness and the number of Shinto shrine forests per capita, that is to say, correlation does not mean causation. For example, this study also found that the population of a prefecture negatively influenced its happiness, which may be due to the negative aspects of urban living, e.g. pollution, crowding, obesity, etc., and/or the challenges associated with governing more people, e.g. the stretching of prefectural resources. The number of Shinto shrine forests per capita may be the result of factors that affect happiness, for example, good local governance, as in the case of Fukui, or historical political processes, as in the case of Okinawa.

With countries like Bhutan reporting 'Gross National Happiness', there is an increasing focus on how well-being should be determined in a nationally relevant context. As research into happiness is prioritised, happiness data should become available at a finer scale than overall rankings. This will enable future research to perform regression analyses to more fully explain the relationship between happiness and sacred natural sites at the national scale. As sacred natural sites are often small, autonomously managed by religious institutions, and seemingly scattered across the landscape, the extent of the cultural ecosystem services they provide can be overlooked. Our study demonstrates the potential for spatial analysis to contribute to this emerging research area. Future work may also determine the qualities of sacred natural sites that increase well-being. For example, Shinto shrine forests are in many cases old stands of forest with more biodiversity than other types of secondary forest and may provide different cultural ecosystem services than newer green spaces (Dean et al., 2011). Furthermore, the religious and spiritual values underpinning sacred natural sites may provide different benefits than 'secular' green space, such as improving social trust and relatedness, as seen in the Japanese case (Kuroki, 2011; Zelenski & Nisbet, 2012). Calls in the literature for non-monetary valuations of cultural ecosystem services on a landscape scale may soon be answered. This paper is a first look at such an approach, and finds that finer scales of data hold additional promise for more definitively linking sacred forests with ecosystem services assessments.

## Acknowledgements

We would like to acknowledge the tremendous contribution that Shinto shrine priests and priestesses have made and continue to make towards the well-being of Japan, both its land and its people. We are grateful for the work of the former Alliance of Religions and Conservation, especially Martin Palmer and Chantal Elkin, for raising awareness globally about the green tenets of Shinto. Thanks to Simon Abele at the University of Oxford for his assistance with the Shinto shrine forest point data and prefectural boundaries, and for the financial support of the University of Oxford's Fell Fund and James Martin Biodiversity Institute.

## References

Atkisson, A. (2012). *Life beyond growth: Alternatives and complements to GDP- measured growth as a framing concept for social progress, 2012 Annual Survey Report*. Institute for Studies in Happiness, Economy, and Society. https://www.greengrowthknowledge.org/sites/default/files/downloads/resource/Life_Beyond_Growth_ISHES.pdf.

Cabinet Office of Government of Japan. (2004). The opinion poll on forests and life 2003. In E. Morita, S. Fukuda, J. Nagano, N. Hamajima, H. Yamamoto, Y. Iwai, T. Nakashima, H. Ohira, & T. Shirakawa (2007). Psychological effects of forest environments on healthy adults: Shinrin-yoku (forest-air bathing, walking) as a possible method of stress reduction. *Public Health, 121*(1), 54–63. https://doi.org/10.1016/j.puhe.2006.05.024

Central Intelligence Agency. (2013). *Japan: People and society*. The World Factbook 2013–2014. https://www.cia.gov/the-world-factbook/about/archives/.

Daily, G. C., Polasky, S., Goldstein, J., Kareiva, P. M., Mooney, H. A., Pejchar, L., … Shallenberger, R. (2009). Ecosystem services in decision making: time to deliver. *Frontiers in Ecology and the Environment, 7*, 21–28.

Dean, J., Van Dooren, K., & Weinstein, P. (2011). Does biodiversity improve mental health in urban settings? *Medical Hypotheses, 76*, 877–880.

Domenig, G. (1997). Sacred groves in modern Japan: Notes on the variety and history of Shintō shrine forests. *Asiatische Studien: Zeitschrift der Schweizerischen Asiengesellschaft, 51*, 91–121.

Economist Intelligence Unit. (2012). *The Green City index*. Siemens AG. https://assets.new.siemens.com/siemens/assets/api/uuid:cf26889b-3254-4dcb-bc50-fef7e99cb3c7/gci-report-summary.pdf.

ESA GlobCover Project. (2009). *GlobCover* [Data Set]. European Space Agency and UCLouvain. http://due.esrin.esa.int/page_globcover.php.

Evans, A. L. (2001). *Shinto¯norito: A book of prayers: in the tradition of Tsubaki Grand Shrine*. Victoria: Trafford.

GADM. (2012). *GADM database of Global Administrative Areas* (Version 2.0) [Data Set]. https://gadm.org.

Grayson, J. H. (2005). 'Shinto' and Japanese popular religion: Case studies of multivariant practice from Kyushu and Okinawa. *Japan Forum, 17*(3), 347–367. https://doi.org/10.1080/09555800500283885

Hosei University. (2011). *Hosei University releases prefectural ranking of well-being*. Hosei Graduate School of Regional Policy and Design. http://ishes.org/en/happy_news/2011/hpy_id000254.html.

Ishii, H. T., Manabe, T., Ito, K., Fujita, N., Imanishi, A., Hashimoto, D., & Iwasaki, A. (2010). Integrating ecological and cultural values toward conservation and utilization of shrine/temple forests as urban green space in Japanese cities. *Landscape and Ecological Engineering, 6*(2), 307–315. https://doi.org/10.1007/s11355-010-0104-5

Jackson, W., & Ormsby, A. (2017). Urban sacred natural sites – A call for research. *Urban Ecosystems, 20*, 675–681. https://doi.org/10.1007/s11252-016-0623-4

Japan National Tourism Organization. (2005). *Forests of Japan's Holy Heartland- Sacred trees that framed a culture.* Japan's Arts and Culture. http://www.jnto.go.jp/eng/indepth/cultural/kie/forests/index.html.

Jinja Honcho. (2011). *Jinja Honcho: Association of Shinto Shrines.* http://www.jinjahoncho.or.jp/en/.

Kishimoto, H. (1968). Urbanization and Jinja Shinto in Japan. In C.J. Bleeker (Ed.), *Proceedings of the XIth International Congress of the International Association for the History of Religions, Claremont, CA* (pp. 117–126). Leiden: Brill.

Kitayama, S., Markus, H. R., & Kurokawa, M. (2000). Culture, emotion, and well-being: Good feelings in Japan and the United States. *Cognition & Emotion, 14*(1), 93–124. https://doi.org/10.1080/026999300379003

Kuchelmeister, G. (1998). *Urban forestry: Present situation and prospects in the Asia and Pacific region, FAO Asia-Pacific Forestry Sector Outlook Study.* Food and Agriculture Organization of the United Nations. http://www.fao.org/3/x1577e/X1577E00.htm.

Kuroki, M. (2011). Does social trust increase individual happiness in Japan? *Japanese Economic Review, 62*(4), 444–459. https://doi.org/10.1111/j.1468-5876.2011.00533.x

Li, Q. (2009). Effect of forest bathing trips on human immune function. *Environmental Health and Preventive Medicine, 15*(1), 9–17. https://doi.org/10.1007/s12199-008-0068-3

MacKerron, G., & Mourato, S. (2013). Happiness is greater in natural environments. *Global Environmental Change, 23*(5), 992–1000. https://doi.org/10.1016/j.gloenvcha.2013.03.010

Manabe, T., Ito, K., Isono, D., & Umeno, T. (2008). The effects of the regulation system on the structure and dynamics of green space in an urban landscape: The case of Kitakyushu city. In S.-K. Hong, N. Nakagoshi, B. Fu, & Y. Morimoto (Eds.), *Landscape ecological applications in man-influenced areas: Linking man and nature systems.* Spring Science + Business Media B.V. 10.1007/1-4020-5488-2

Millenium Ecosystem Assessment. (2005). *Ecosystems and human well-being: Synthesis: A report for the Millennium Ecosystem Assessment.* Washington, DC: Island Press.

Ministry of Health Labour and Welfare. (2008). *Overview of labour health survey results 2007.* https://www.mhlw.go.jp/.

Morita, E., Fukuda, S., Nagano, J., Hamajima, N., Yamamoto, H., Iwai, Y., … Shirakawa, T. (2007). Psychological effects of forest environments on healthy adults: Shinrin-yoku (forest-air bathing, walking) as a possible method of stress reduction. *Public Health, 121*(1), 54–63. https://doi.org/10.1016/j.puhe.2006.05.024

Nelson, J. K. (2006). *A year in the life of a Shinto shrine.* Seattle: University of Washington Press.

Ohtsuka, Y., Yabunaka, N., & Takayama, S. (1998). Shinrin-yoku (forest-air bathing and walking) effectively decreases blood glucose levels in diabetic patients. *International Journal of Biometeorology, 41*(3), 125–127. https://doi.org/10.1007/s004840050064

Ormsby, A. (2021). Diverse values and benefits of sacred natural sites. *Trees, Forests and People, 6*, 100136. https://doi.org/10.1016/j.tfp.2021.100136

Oviedo, G., & Jeanrenaud, S. (2007). Protecting sacred natural sites of indigenous and traditional peoples. In J. M. Mallarach & T. Papayannis (Eds.), *Protected areas and spirituality.* IUCN and Publicacions de l'Abadia de Montserrat.

Park, B. J., Tsunetsugu, Y., Kasetani, T., Kagawa, T., & Miyazaki, Y. (2009). The physiological effects of Shinrin-yoku (taking in the forest atmosphere or forest bathing): evidence from field experiments in 24 forests across Japan. *Environmental Health and Preventive Medicine, 15*(1), 18–26. https://doi.org/10.1007/s12199-009-0086-9

Rots, A. P. (2013). *Forest of the gods: Shinto, nature, and sacred space in contemporary Japan* (Publication No. 44540) [Doctoral dissertation, University of Oslo]. DUO Research Archive.

Sakamoto, K. (2011). *Happiness ranking.* Hosei Graduate School of Regional Policy Design. http://www.hosei.ac.jp/documents/koho/photo/2011/11/20111110.pdf.

Schneider, A., Friedl, M. A., McIver, D. K., & Woodcock, C. E. (2003). Mapping urban areas by fusing multiple sources of coarse resolution remotely sensed data. *Photogrammetric Engineering & Remote Sensing, 69*(12), 1377–1386. https://doi.org/10.14358/pers.69.12.1377

Public Affairs Headquarters for Shikinen-Sengu. (2010). *The Grand Shrine of Ise: Spiritual Home of the Japanese People.* Tokyo: Public Affairs Headquarters for Shikinen-Sengu.

Statistics Bureau of Japan. (2005). *Population Census of Japan.* https://www.stat.go.jp/english/data/kokusei/2005/poj/mokuji.html.

Shimazono, S. (2009). State Shinto in the lives of the people: The establishment of emperor worship, modern nationalism, and Shrine Shinto in Late Meiji. *Japanese Journal of Religious Studies, 36*(1), 93–124. https://doi.org/10.18874/jjrs.36.1.2009.93-124

Tokyo Metropolitan Government Bureau of Construction. (2012). *Parks of Tokyo.* http://www.kensetsu.metro.tokyo.jp/english/kouen/gaiyo/01.html.

Tsukada, K. (2005). *100,000 sacred woods remain in Japan- do they still evoke reverence and awe?* Japan's Arts and Culture. http://www.jnto.go.jp/eng/indepth/cultural/kie/forests/kie_forests_02.html.

Tsunetsugu, Y., Park, B.-J., & Miyazaki, Y. (2009). Trends in research related to "Shinrin-yoku" (taking in the forest atmosphere or forest bathing) in Japan. *Environmental Health and Preventive Medicine, 15*(1), 27–37. https://doi.org/10.1007/s12199-009-0091-z

Uchida, Y., Norasakkunkit, V., & Kitayama, S. (2004). Cultural constructions of happiness: Theory and empirical evidence. *Journal of Happiness Studies, 5*(3), 223–239. https://doi.org/10.1007/s10902-004-8785-9

United Nations. (2011). *World urbanization prospects: The 2011 revision.* Population Division of the Department of Economic and Social Affairs of the United Nations Secretariat. https://www.un.org/en/development/desa/population/publications/pdf/urbanization/WUP2011_Report.pdf.

United Nations. (2012). *World population prospects: The 2010 revision.* Population Division of the Department of Economic and Social Affairs of the United Nations Secretariat. https://doi.org/10.18356/08b807d4-en

White 曜社 Co., Ltd. (n.d.). *Japan Shrine Portal Site.* http://www.jinja.in/.

Wolf, K., Flora, K., & Housley, E. (2012). *Research on the beneficial aspects of the experience of nature in cities: A literature review.* Open Spaces Sacred Places National Awards Initiative. TKF Foundation. https://your.kingcounty.gov/dnrp/library/water-and-land/land-conservation/meeting-09-29-2016/benefits-of-nature-in-cities-lit-review-02-12.pdf.

Yamamoto, K. (2010). Evaluation of the degree of the sufficiency of public green spaces as an indicator of urban density in the Chubu Metropolitan Area in Japan. *World Academy of Science, Engineering and Technology, 4*, 586–594.

Zelenski, J. M., & Nisbet, E. K. (2012). Happiness and feeling connected. *Environment and Behavior, 46*(1), 3–23. https://doi.org/10.1177/0013916512451901

# 12

# SPIRITUAL LEADERS BUILD COMMON GROUND

## Enacting Worldviews through Legal Pluralism in the Face of Neoliberalism

*Bas Verschuuren and Felipe Gomez*

### Introduction

In Indigenous Guatemalan societies, sacred natural sites (SNS) are central to people's spiritual life and well-being and play an integral role in natural resource management and governance (Delgado & Mariscal, 2010; Ybarra, 2011). The spiritual significance of SNS extends to, and is characteristic of, Indigenous peoples' relationship with the wider landscape. We propose an understanding of this wider landscape as a *spiritscape* because it emphasises the importance of spiritual values and spirituality in different worldviews when working with SNS. According to McNiven (2004), Studley (2010), and Verschuuren (2022), the landscape becomes a spiritscape because it is animated with ancestors, spirits, creator beings and other mythological or symbolic figures that imbue it with spiritual energies, life, and sentience. Callicott et al. (2007) characterise the spiritscape paradigm by psycho-spiritual connections which are enacted through ritual and ceremony and preceded over by spiritual guides, shamans, or custodians who maintain harmony and good relations between all the elements – human, natural and spiritual – of their cosmologies. Spiritscapes can be understood as elements of biocultural landscapes, landscapes defined by their human-nature interactions which have co-evolved and developed over many generations (Verschuuren et al., 2014). These biocultural entanglements make spirituality part of human well-being as well as the well-being of ecosystems (Caillon et al., 2017).

In this chapter, SNS are brought into being by engagement, attachment, and enactment of indigenous ontologies (Woolgar & Lezaun, 2013; Dwiartama & Rosin, 2014; Blaser, 2012; Viveiros de Castro, 2008). We investigate the role of SNS from the perspective of Guatemalan spiritual leaders (hereafter named *Ajq*

DOI: 10.4324/9781003246190-15

*Ijab*) who view SNS as part of their cosmologies (Conz, 2014; Ybarra, 2011). We explore the creation of common grounds in Guatemala where *Ajq Ijab* and their communities defend their SNS and livelihoods through activism, legal processes, public political dialogues, and by building bridges with various layers of government and private companies. Common grounds may be defined as:

> consisting of the worldviews and realities of Indigenous people as well as nonindigenous conservation and development actors that affect sacred natural sites. Such common ground is commonly created out of the resolution of contestation, violence, political opposition, participation, representation, reconciliation and diplomacy
>
> *(Verschuuren, 2017, p. 21).*

In our exploration of how common grounds are created, we follow the *Ajq Ijab* of *Maya* communities which have united in *Oxlajuj Ajpop*, the National Council for Maya Spiritual Leaders in Guatemala, which works with Maya K'iche and other Indigenous and ethnic communities in Guatemala. The *Ajq Ijab* that perform the rituals of healing during the ceremony explained that they perform other healing activities with individuals, even over long distances:

> A sacred natural site is important to a person's well-being. It makes a person integer and whole and at peace with himself so he grows in his spirituality and thinking and can have a positive influence on a situation
>
> *(Ajq Jiab, senior male person who identifies as Maya and guides in the local community).*

The role of the *Ajq Jiab* is to resolve conflicts and speak justice, and can be understood to be advising and guiding the community as well as individuals. One Ajq'ij explained his role as follows:

> Maya Ajq Ijab are specialists who guide marriages, guide the spiritual life of the community and community members as well as healers, midwives, and specialists for the treatment of bone. The symbol of our authority is the staff that stands for justice, righteousness, good behaviour and the wise advice passed onto us by our ancestors.
>
> *(Ajq'ij, a senior male person who identifies as Maya and holds a leadership role in Oxlajuj Ajpop).*

We investigated disjunctures and contestations between different ontologies that form part of a mutually created common ground constructed through the development of a law proposal, community research, and public dialogues. While doing so, we discuss the challenges posed to the Mayan communities in creating a politics that is conducive to their worldviews.

## Legal Recognition for Guatemala's Sacred Sites

In Guatemala, Indigenous spiritual leaders named *Ajq Ijab* organised themselves in Oxlajuj Ajpop, an organisation whose name means 'the thirteen spiritual leaders'. It was conceived in 1991 by seven organisations of spiritual leaders from the Maya, Xinca, and Garífuna ethnic groups and includes over 1,100 spiritual leaders from 12 provinces throughout Guatemala (Gomez & Caal, 2003). Since its inception, Oxlajuj Ajpop has united spiritual leaders into a movement in support of regaining ownership and control over sacred places during and after the Guatemalan Civil War (Ybarra, 2010). The *Ajq Ijab* protect their sacred sites as affirmation of their identity, a spiritual responsibility to the ancestors, the cosmos and the land. The civil war lasted from 1960 to 1996 between various guerrilla groups (supported by ethnic Maya Indigenous peoples and mestizo peasant communities) and the government of Guatemala (Kubota, 2017). During the civil war, care for many sacred sites had been disrupted and many spiritual leaders lived in exile. After the civil war, peace negotiations between Indigenous peoples and the government took place and agreements were made that became known as the 'Peace Accords' (Frank, 2013). The peace accords state the ambition to restore sacred sites and bring them under indigenous control and ownership (Conciliation Resources, 1997, p. 44).

As part of the agreements made under the Peace Accords, the *Ajq Ijab* stress the need to protect, restore, and re-dignify many of the traditional sacred sites that are located around the country. As a first step towards achieving this, the *Ajq Ijab*, united in Oxlajuj Ajpop, developed a law proposal (hereafter referred to as 'the law proposal') entitled: '*Initiative in Support of the Law on Sacred Natural Sites of Indigenous Peoples No. 3835, On the Occasion of the New Mayan Cycle Oxlajuj B'aqtun*' (see Gomez et al., 2010, Figure 12.1). This law proposal could potentially bring ownership, management and governance of indigenous sacred sites back into the hands of Indigenous people. Such developments would also provide the legal backing for a possible conservation strategy for the numerous SNS across Guatemala. Not only are many of these SNS located in unique ecosystems, harbour endangered species, and help protect biodiversity; they are also connected as a network central to the spirituality and well-being of the Indigenous communities that have looked after them for generations – a social-spiritual conservation network (Verschuuren, 2019). In this chapter, we analyse the development of the law proposal and support activities as manifestations of the *Ajq Ijab* ability to conserve their SNS in the context of neoliberal politics.

Oxlajuj Ajpop undertook much work on the legal aspects of the law proposal, including a legal review and sectoral analysis across administrative fields (i.e. forestry, water and education) and set out recommendations to enable the implementation of the law proposal (see Oxlajuj Ajpop, 2008b). Besides the need for a cross-sectoral implementation of the law proposal, there were challenges posed by land ownership and religious movements. According to some of the *Ajq Ijab*, several sacred sites that are located on lands owned by individuals, rather

**FIGURE 12.1** Cover of the 'Initiative in support of the law on Sacred Natural Sites of Indigenous Peoples No. 3835'. Source (Photo Credits): Oxlajuj Ajpop.

than by the community, have lost the protection of the landowners because they converted to evangelicalism:

> Having the management and governance of sacred natural sites legally recognised is important for the continuation of indigenous worldviews. There are problems with sacred sites such as companies that want to develop constructions on top of them or the catholic church who replace the sacred places with images of its own
>
> *(Ajq Ijab, senior male person who identifies as Maya).*

In several communities, private land ownership has been a conscious strategy that the communities chose during the land reform after the civil war (Ybarra, 2013). The main objective was for the community to maintain self-governance over its ancestral land and sacred sites to prevent the possibility of selling large tracts of land to private companies or leasing them out for economic gain. In one particular site, Oxlajuj Ajpop was able to acquire ownership of the most central part of the sacred site and hold it in trust for the *Ajq Ijab*. Oxlajuj Ajpop and the *Ajq Ijab* came to this decision out of fear of mining companies that were known to undertake prospecting missions throughout the country (Costanza, 2016). Such prospecting, as well as most mining activities, would damage SNS, and the fear existed that the government would lease these lands to mining companies because it legally owns all minerals below the subsoil.

Indigenous people in Guatemala know a long and violent history of persecution and civil war and are struggling to rebuild their communities and get their interests represented in national politics (Monterroso & Bravo, 2008). Guatemalan politics are dominated by *Ladino* (non-Indigenous, i.e. Guatemala's dominant social group) proponents of a neoliberal economy where the rights to exploit the countries' natural resources are sold to private businesses. The impact of mining, forestry, and hydro-electric projects on the territories and resources of Indigenous communities is significant (Costanza, 2016). This also leads to misappropriation, violence and abuse of Indigenous peoples' rights (Kubota, 2017). The work on the law proposal and the parallel strategy to reclaim the administration of sacred sites across Guatemala can also be interpreted as part of a broader struggle for the expression of Maya identity, its cosmovision, and a claim for what Ybarra calls 'Maya spiritual rights' (2013, p. 549). This term is also used by Oxlajuj Ajpop in relation to a provision of the law proposal which states that:

> The Government assumes the commitment to promote, jointly with indigenous spiritual organisations, regulation of access to these ceremonial centres, guaranteeing the free practice of indigenous spirituality within conditions that are respectful of spiritual leaders
>
> *(Iniciativa de Ley de Los Lugares Sagrados de Los Pueblos Indigenas No. 3835, En El Marco Del Nuevo Ciclo Maya Oxlajuj B'aqtun, 2008, p. 12).*

The above of course is a legal expression – an outcome of the struggle for recognition of Maya identity and an affirmation of the importance of the free practice of Maya spirituality and the role of the *Ajq Ijab* in Mayan societies. In daily life, SNS help maintain this spirituality, not only through communal relationships with the cosmos, but also in individual well-being.

Throughout our research, many SNS have shown to be spaces for retreat and strengthening of indigenous identity based on the practice of ceremony that connects individuals and communities with Maya cosmology (see Figure 12.2). Ceremony is typically guided by one or more spiritual leaders who create a process for healing individuals as well as groups of people based on specific days indicated by the *Chol'ij*, the Maya solar calendar. Such processes can be seen as therapeutic counselling, and also as strengthening indigenous identity through Maya spirituality and cosmology. Quang et al. (2013) already found that sacred sites can be used by Indigenous people to retreat and resist the oppression of mainstream politics and serve as places of learning while aspiring to new visions of Indigeneity and law. As such, SNS are places of power that, through their numinous character, possess an agency that affects Indigenous peoples' relations to the state and possibly the ability of Indigenous peoples to resist neoliberal transnational and state politics.

**FIGURE 12.2** The local community gathers for ceremony performed by the Ajq Jiab at Chusaqrib'Al in Santa Cruz del Quiché. The ceremony is used to discuss issues pertinent to the community, to perform spiritual healings and to plan for strengthening the conservation of this and other sacred sites in the region. Photo Credits: Bas Verschuuren.

## *Ajq Jiab* Working with the Government

The *Ajq Jiab* gathered in Oxlajuj Ajpop not only produced the law proposal, but also produced a book about their role in the traditional juridical system (Oxlajuj Ajpop, 2005) as well as a Social-Environmental Agenda for Guatemala based on the indigenous worldviews and the Rights of Mother Earth (Oxlajuj Ajpop, 2009). As part of this work, a dialogue process with the Ministry of Agriculture and Natural Resources took place. Throughout the dialogue process, the need for new constitutional and legal reforms that respect Mother Earth, indigenous territories, biodiversity, and a legally pluralistic state was emphasised. These may be among the reasons why the Ministry of Energy and Mining never expressed interest in a dialogue; a legally pluralistic state would complicate claims on resources, especially those in the subsoil. All of these documents have been based on consultations with representatives of Maya, Garifuna, and Xinca – many of them spiritual leaders – throughout a long and elaborate process. This has been a process of creating common ground by reaching out to many other stakeholders – including various layers of government. According to Ybarra (2013), the work of Oxlajuj Ajpop helps to link a politics of recognition to a politics of distribution and concerns regarding territory. This is important because in Guatemala, territorial rights have not been granted to communities yet.

Clearly, the work on the law proposal for sacred sites has not been undertaken in isolation, it took place as part of a broader movement of Indigenous people in Guatemala claiming their rights and asserting their identity – a larger process of creating common ground. The start of creating this common ground was made when the National Law for Peace Agreements, signed in 1996, came into force. It acknowledged the rights of Indigenous peoples to practise their cultures on a specific territory and sacred (natural) sites as part of that territory. In this context, the Commission for the Definition of Sacred Sites was formed on the initiative of Oxlajuj Ajpop (*Initiative in Support of the Law on Sacred Natural Sites of Indigenous Peoples No. 3835, On the Occasion of the New Mayan Cycle Oxlajuj B'aqtun, Guatemala City*, 2008). An *Ajq'ij* explains:

> In those days, the work we did in the commission was thought to be really valuable and many people were busy working on it. We got a lot of support from civil society and government and started building on ideas of national inventory and legal protection. It was only later when the law proposal started to take shape that the politics changed again and eventually the commission was disbanded by the government.
>
> *(Ajq'ij, senior male person with a leadership role in Oxlajuj Ajpop).*

Negotiations continued until 2013, and after that they have not been continued because the law proposal for sacred sites had not been accepted by all members of the Guatemalan Congress and the government. This is still the case at the time of publishing this chapter. This situation is thought to be caused by the opposition, who fear that the law proposal has the potential to not only lead to a politics of distribution (as Ybarra reasons) but also to a politics of cultural and legal plurality. While this is a politics that many Maya have been fighting for in order to achieve indigenous governance and control over their territories, it faces opposition from those in favour of a neoliberal politics. However, at times there are positive developments, such as some of the first Maya mayors that have been elected into municipal power (Rasch, 2011). While these local wins are not linked to the quest of the law initiative, they can help with changing local politics in favour of the protection of sacred sites.

Starting in 2003, Oxlajuj Ajpop consulted its member organisations, formed its own technical and legal teams, developed a strategic plan, and organised linguistic groups of Mayas, Garífuna, and Xinca to discuss a renewed governmental agreement to support sacred sites. In Santa Cruz del Quiché, Oxlajuj Ajpop implemented various activities on sacred sites. Over 20 communities reflected on the importance of sacred sites, recorded and documented their histories, assessed their current ecological and legal status, and held celebrations to re-sanctify them. In the process, Indigenous communities also became aware of their rights to participate in the administration and governance of sacred sites. Adoption and implementation of the law proposal would enable the Mayan worldview to complement the contemporary Western state-based system currently adopted by the Guatemalan government. In fact, this would mean that a pluri-legal society would be created in which SNS would be recognised as sources of law to Indigenous peoples. In this context, the law proposal serves as an important step towards the development of not only a culturally pluralistic (Nash, 2001), but also a legally pluralistic society in Guatemala.

The Plenary of the Congress of the Republic of Guatemala finally received the law proposal in 2008 and registered it for its study and approval. Technical and legal advisors of different political parties then studied the text, and six articles were revised through a dialogue with Oxlajuj Ajpop. This was a formal process of negotiation in which Oxlajuj Ajpop and the Commission for the Definition of Sacred Sites succeeded in maintaining the essence of the law proposal in the final text (*Initiative in Support of the Law on Sacred Natural Sites of Indigenous Peoples No. 3835, On the Occasion of the New Mayan Cycle Oxlajuj B'aqtun, Guatemala City*, 2008). One could say that this was a written act of building bridges and creating common ground. Surrounding this process, many public meetings took place and various politicians from several parties were contacted and liaised with. Finally, in 2013, in a public meeting, the Commission to Define Sacred Sites and Oxlajuj Ajpop gave a petition to the Congress (channelled through the Peace Commission) to approve the law proposal. The petition never invoked action towards implementation of the law proposal because of counter efforts by interest groups representing the private sector in mining, forestry, and the energy sector. These groups lobbied the Guatemalan parliament arguing that the law proposal affects private property, such as lands with natural resource concessions over them.

## *Ajq Ijab Facing Opposition from the Private Sector*

Within Guatemalan politics, not all parties are supportive of the law proposal or respect historic, spiritual and cultural rights. The processes of privatisation of land meant that many Indigenous communities had to go through difficult periods of reform through which many spiritual leaders decided to leave sacred sites unmarked on the grid of private titles as a means of protecting them from deliberate harm (Ybarra, 2011). Many iconic sacred sites, such as the archaeological monuments in Tikal World Heritage site and National Park have been adopted as symbols of the Guatemalan nation and are regularly co-opted for tourism and marketing purposes (see Figure 12.3).

Many non-indigenous political parties are linked to extractive enterprises and are concerned that the law proposal restricts their ability to exploit natural resources. Interest groups lobbying the Guatemalan parliament on behalf of the private sector have put forward in Parliament that Article 20 affects private property. As a result, the article was modified in 2009 and no longer directly enables indigenous action that may affect ownership, exploitation of, as well as access to private property and natural resources contained therein. This complicates the indigenous custodianship of sacred sites on private land. Custodians

**FIGURE 12.3** The image of Tikal sacred site is used in a bread commercial to appeal to peoples' feeling of nationalism. This traditionally indigenous sacred site represents the Maya culture in which Maize (and not wheat) is traditionally seen as a sacred crop. Photo Credits: Bas Verschuuren.

of sacred sites need access to sacred sites for carrying out their responsibilities and customary use such as performing ceremony and ritual. These activities are vital to keeping a form of spiritual governance in place and in turn instruct indigenous management of SNS and the broader territory, including its natural resources. When owners of the land on which sacred sites are located have made no provisions for such activities, the custodians face problems performing their cultural and spiritual duties (Gomez et al., 2010). In an attempt to build common ground, Oxlajuj Ajpop has been inspired by the ancestral Maya law and ways of managing conflict by transformation and prevention. An *Ajq'ij* describes this traditional governance system as based on the Maya worldview linked closely to the natural world:

> Indigenous people must reclaim the principles of life of our ancestors. They can do this through a process involving sacred natural sites that are located in their communities. The revitalisation and dignification of sacred sites coincide with the conservation of nature and at the same time goes hand in hand with restoring our spiritual governance. It does so through dialogues and by building bridges between the community level up to the national governmental level and in some cases the international level.
>
> *(Ajq Ijab, senior male person who identifies as Maya, and holds a leadership role in Oxlajuj Ajpop).*

The view of the *Ajq Ijab* also alludes to indigenous perspectives on building a common ground, through building of bridges and through dialogue while remaining in connection with the ancestors. This connection has reportedly come up in conversations with the *Ajq Ijab* as key to guiding processes on the interface of Indigenous community interests versus interests from outsiders.

## Conclusions and Discussion: Elements for Creating a Common Ground

Creating a common ground requires different worldviews to come into play in interventions such as mining and agriculture as well as area-based conservation projects. If these worldviews are not taken into account in governance and management decisions, these (neoliberal) interventions typically come at considerable social and environmental costs. Despite discontinuation of this process, empowering traditional authorities and spiritual leaders has been shown to be helpful in terms of recognising and legitimising the spiritual governance systems embedded in worldviews in Guatemala.

Recognition of the importance of myths, divine beings, the ancestors, sacred people, objects, plants, animals, and sites helps in developing a better understanding of the spiritual dimensions of a worldview and opens the way for indigenous ontologies to come into play. We have shown that in Guatemala, the spiritual dimension affects the governance and management of natural resources and land that is regulated through indigenous institutions and community decision-making and that these include a distinct spiritual dimension with a role for the *Ajq Ijab*. Their duties, responsibilities and ceremonial practices are an integral part of ensuring the well-being of the community.

Indigenous communities in Guatemala developed strategies for the protection and conservation of their SNS. Spiritual leaders had played an active role in mobilising the community and helped facilitate efforts to advocate and negotiate their worldviews – as expressed in the development of the law proposal (Gomez et al., 2010). This focus on a rights-based approach also emphasises the need to express and negotiate indigenous realities with actors from outside the community. Changes in worldviews occur due to adjustments to ecological, technological, commercial, political or demographic changes brought about by actors with non-indigenous worldviews. Such changes prompt a response from the community and its spiritual leaders and may be seen as what Foucault terms 'technologies of the self' (Foucault et al., 1988). Dealing with mining companies and government agencies requires a pathway for Indigenous people to enact 'technologies of the self' (Foucault et al., 1988). This can be done by making use of diplomacy and skills that spiritual leaders have been able to attain through collaboration with external NGOs and other supportive actors. At the same time, they maintain their traditional institutions, worldviews and SNS to use these as an ontological starting point for making their voices heard.

Spiritual leaders such as the *Ajq Ijab* have an influential position within and among communities that share a similar worldview, and this attributes agency to SNS. The *Ajq Ijab* have used rights-based approaches to create common ground where elements of traditional law and Maya sciences can be considered as part of indigenous ontologies. We argue that these different ontologies are the basis for developing arguments and positions in the negotiations between Indigenous and non-Indigenous actors. Just as the ontology of Western science qualifies science as a suitable means for decision-making, Indigenous actors should be enabled to include knowledge systems based on their own ontologies (Pascual et al., 2021). Ontological plurality and legal plurality are important, but these do not by themselves guarantee that different worldviews are considered equally. The integration of different knowledge systems is known to be at risk from reproduction of historical patterns of oppression (Tengö et al., 2016). Therefore, we propose that a principle of ontological equity should be established in order to take ontologies and different worldviews seriously in interactions with Indigenous peoples (Verschuuren, 2017, pp. 44, 208).

While many spiritual leaders in Guatemala have shown leadership in governance and management over community-owned natural resources – involving ceremony and ways of mediation with ancestors through sacred sites – their roles and traditional institutions have in recent history been suppressed, disempowered, and rendered invisible. As a result, they are often poorly understood and left out of formal negotiations and legal processes. While the Guatemalan constitution includes provisions for the recognition of traditional governance systems as part of the recognition of rights of Indigenous peoples and traditional communities, these are seldomly recognised or respected by private companies or by the government.

A practice of FPIC involving activities that affect SNS is not standard operation in Guatemala. This enquiry shows that spiritual leaders have been able to improve the positions of communities by assisting in the development of legal positions (Gomez et al., 2010). These legal positions are based on the worldviews and values derived from indigenous ontologies, e.g. other cultural ways of understanding the world around us. We argue that not considering indigenous ontologies on an equal footing with non-indigenous ontologies (such as neoliberal or capitalist ones) is prejudicial. While constitutional rights consider free cultural and spiritual practice and belief, the neoliberal capitalist system provides a governance system based on private property that leaves no space for spirits and ancestors.

Research on neoliberal conservation approaches has seen a shift in focus within capitalism itself, from how nature and natural resources are used, to how nature is being conserved (Büscher et al., 2012). While the culprit identified is often the capitalist system itself – geared at making more profit – critiques of neoliberal conservation approaches rarely contribute any solutions or alternatives to this problem.

Through the analysis of practices and worldviews of *Ajq Ijab* presented in this chapter, we identified forms of cultural and spiritual conservation that existed long before capitalism and neoliberal conservation approaches came into being. Because of the longevity of many spiritual and cultural conservation approaches, we argue that they can form viable alternatives to neoliberal conservation approaches. Such spiritual conservation approaches are, however, easily dismissed or subdued by development trajectories common to neoliberal conservation approaches but may gain support from those who critique neoliberal conservation approaches and identify ideological disjunctures of capitalism and conservation (Buscher & Fletcher, 2020).

Many of the spiritual leaders that we worked with recognise that the current dominant legal system does not allow enough space for their forms of spiritual governance and fear that neoliberal markets will eventually destroy the balance between the spirit world and community well-being because of the threats it poses to SNS. To prevent these problems from getting worse, one *Ajq'ij* explained a vision that Oxlajuj Ajpop should create a common ground and help counteract this issue. According to this *Ajq'ij*, an international body of spiritual leaders should demand that the United Nations develop a convention for keeping states, companies, research centres, and faith groups to a code of conduct that will help respect and protect the sacred sites of Indigenous peoples. The *Ajq Ijab* expressed that an international council of elders would be required to oversee the creation of this international body and the implementation of the code of conduct.

As a consequence of the law initiative not being approved by the government, the conservation of SNS now depends on the willingness of individuals to take responsibility. This is not something that suits most actors currently operating in the resource market in Guatemala, e.g. hydroelectric, forestry, and mining companies. In opposition to activities carried out by such companies, they are often met with violence and force from the police and military. This leaves SNS in serious danger from being destroyed, co-opted and commodified while their guardians' safety is often at risk when they organise opposition. However, the spiritscapes of the Maya remain, but with increasing numbers of significant sacred sites disappearing and/or being dormant, awaiting a re-enactment of their sacred dimensions.

## References

Blaser, M. (2012). Ontology and indigeneity: On the political ontology of heterogeneous assemblages. *Cultural Geographies, 21*(1), 49–58. https://doi.org/10.1177/1474474012462534

Büscher, B., Sullivan, S., Neves, K., Igoe, J., & Brockington, D. (2012). Towards a synthesized critique of neoliberal biodiversity conservation. *Capitalism Nature Socialism, 23*(2), 4–30. https://doi.org/10.1080/10455752.2012.674149

Caillon, S., Sterling, E., Cullman, G., & Verschuuren, B. (2017). A biocultural approach to design indicators: Moving beyond the dichotomy between humans and non-humans. *Ecology and Conservation,* 22(4): 27. https://doi.org/10.5751/ES-09746-220427

Callicott, J. B., Rozzi, R., Delgado, L., Monticino, M., Acevedo, M., & Harcombe, P. (2007). *Biocomplexity and conservation of biodiversity hotspots : Three case studies from the Americas. December 2006* (pp. 321–333). https://doi.org/10.1098/rstb.2006.1989

Conciliation Resources. (1997). *Negotiating rights: The Guatemalan peace process. An international review of peace initiatives accord.* https://rc-services-assets.s3.eu-west-1.amazonaws.com/s3fs-public/02_Guatemala.pdf

Conz, B. W. (2014). Conservation and Maya Autonomy in Guatemala's Western Highlands: The case of Totonicapán. In S. Stevens (Ed.), *Indigenous peoples, National Parks, and protected areas* (p. 380). University of Arizona Press, Tucson.

Costanza, J. N. (2016). Mining conflict and the politics of obtaining a social license: Insight from Guatemala. *World Development, 79*, 97–113.

Delgado, F., & Mariscal, J. C. (2010). *Revalorización y conservación de la diversidad biocultural andina: experiencias y aprendizajes del programa regional BioAndes.* AGRUCO. http://www.agruco.org/bioandes/pdf/1103 Libro MEMORIA BioAndes V2.pdf

Dwiartama, A., & Rosin, C. (2014). Exploring agency beyond humans: The compatibility of Actor-Network Theory (ANT) and resilience thinking. *Ecology and Society, 19*(3). https://doi.org/10.5751/ES-06805-190328

Foucault, M., Martin, L. H., Gutman, H., & Hutton, P. H. (1988). *Technologies of the self: A seminar with Michel Foucault.* Tavistock Publications, London.

Frank, G. (2013). Twenty-first century pragmatism and social justice: Problematic situations and occupational reconstructions in post-civil war Guatemala. In: Cutchin, M. P., and V. A. Dickie, (eds.), *Transactional perspectives on occupation* (pp. 229–243). Springer, Dordrecht.

Gomez, F., & Caal, O. P. (2003). Cosmovision as a basis for development. In B. Haverkort, K. Van het Hoofd, & W. Hiemstra (Eds.), *Ancient roots, new shoots endogenous development in practice. Worldviews and Sciences Series* (pp. 204–208). COMPAS. http://www.compasnet.org/blog/wp-content/uploads/2011/03/ARNS/arns_19.pdf

Gomez, F., Hiemstra, W., & Verschuuren, B. (2010). A law on sacred sites in Guatemala. *Policy Matters, 17*, 119–122.

*Iniciativa de ley de los lugares sagrados de los pueblos indigenas No. 3835, En el marco del Nuevo Ciclo Maya Oxlajuj B'aqtun*, 28 (2008) (testimony of Ministerio de Cultura y Deportes & Conferencia Nacional Oxlajuj Ajpop). http://sacrednaturalsites.org/wp-content/uploads/2012/12/Iniciativa-de-Ley-de-Lugares-Sagrados.pdf

*Initiative in support of the law on Sacred Natural Sites of Indigenous Peoples No. 3835, On the occasion of the new Mayan cycle Oxlajuj B'aqtun, Guatemala City:*, (2008) (testimony of Oxlajuj Ajpop).

Kubota, Y. (2017). Explaining state violence in the Guatemalan civil war: Rebel threat and counterinsurgency. *Latin American Politics and Society, 59*(3), 48–71.

McNiven, I. (2004). Saltwater people: Spiritscapes, maritime rituals and the archaeology of Australian indigenous seascapes. *World Archaeology, 35*(3), 329–349. https://doi.org/10.1080/0043824042000185757

Monterroso, M. I. de, & Bravo, I. A. (2008). *Ciencia y tecnica Maya: con la colaboracion del Consejo Nacional de Ancianos Principles y Guias Espirituales Mayas, Xincas y Garifunas.* Fundación Solar, Guatemala.

Nash, J. C. (2001). *Mayan visions: The quest for autonomy in an age of globalization.* Psychology Press, London.

Oxlajuj Ajpop. (2005). *Aportes del sistema jurídico maya hacia el Estado de Guatemala en materia de prevención y transformación de conflictos.* PNUD/ASDI, Guatemala. Oxlajuj Ajpop. (2008). Oxlajuj Ajpop. 2008b. Organización Para La Recuperación, Conservación, Dignificación Y Administración De Los Lugares Sagrados Mayas. In *2008.*

Oxlajuj Ajpop. (2009). *Agenda Socio Ambiental desde el Pensamiento de los Pueblos Indígenas por los Derechos de la Madre Tierra.* Guatemala: *COMPAS Meso America.*

Pascual, U., Adams, W. M., Díaz, S., Lele, S., Mace, G. M., & Turnhout, E. (2021). Biodiversity and the challenge of pluralism. *Nature Sustainability, 4*(7), 567–572. https://doi.org/10.1038/s41893-021-00694-7

Quang, N. N., Wildemeersch, D., & Masschelein, J. (2013). Community forests as heterotopia. The case of the Mu community forest – Ngoc Son – Ngo Luong Nature Reserve, Vietnam. *International Journal of Environmental Studies, 70*(6), 877–892. https://doi.org/10.1080/00207233.2013.849515

Rasch, E. D. (2011). Quetzaltenango's First Mayan Mayor: Transforming political culture and the politics of belonging? *Journal of Latin American Studies, 43*(01), 121–149. https://doi.org/10.1017/S0022216X10001811

Studley, J. (2010). Uncovering the intangible values of earth care: Using cognition to reveal the eco-spiritual domains and sacred values of the peoples of Eastern Kham. In B. Verschuuren, R. Wild, J. Mcneely, & G. Oviedo (Eds.), *Sacred natural sites: Conserving nature and culture.* Earth Scan, London.

Tengö, M., Hill, R., Malmer, P., Raymond, C., & et al. (2016). Weaving knowledge systems in IPBES, CBD and beyond – Lessons learned for sustainability. *Current Opinions in Environmental Sustainability*, 26(17), 17–25. https://doi.org/10.1016/j.cosust.2016.12.005

Verschuuren, B. (2017). Creating common ground: The role of Indigenous People's sacred natural sites in conservation practice, management and policy. *Wageningen University, Wageningnen, The Netherlands* (Thesis sub). https://doi.org/ 10.18174/420617

Verschuuren, B. (2022). Conceptualising spiritscapes: The Petroglyphic Complexes of the Mongolian Altai World Heritage site and the wider biocultural landscape. In S. Brown & C. Goetcheus (Eds.), *The Routledge handbook of cultural landscapes.* Routledge, London.

Verschuuren, B., Subramanian, S. M., & Hiemstra, W. (2014). *Community wellbeing in biocultural landscapes: Are we living well?* Bourton-on-Dunsmore: Practical Action Publishing.
Verschuuren, B., Wild, R., McNeely, J., & Oviedo, G. (2010). *Sacred natural sites: Conserving nature and culture.* Earth Scan. http://www.amazon.co.uk/Sacred-Natural-Sites-Conserving-Culture/dp/1849711674
Viveiros de Castro, E. (2008). Cosmological Deixis and Amerindian Perspectivism. *The Journal of the Royal Anthropological Institute, 4*(3), 469–488.
Woolgar, S., & Lezaun, J. (2013). The wrong bin bag: A turn to ontology in science and technology studies? *Social Studies of Science, 43*(3), 321–340. https://doi.org/10.1177/0306312713488820
Ybarra, M. (2010). *Living on Scorched Earth: The political ecology of land ownership in Guatemala's Northern Lowlands.* University of California, Berkeley.
Ybarra, M. (2011). Privatizing the Tzuultaq'a? Private property and spiritual reproduction in post-war Guatemala. *Journal of Peasant Studies, 38*(4), 793–810. https://doi.org/10.1080/03066150.2011.607702
Ybarra, M. (2013). "You Cannot Measure a Tzuultaq'a": Cultural politics at the limits of liberal legibility. *Antipode, 45*(3), 584–601. https://doi.org/10.1111/anti.12000

# PART IV

# Broader Perspectives, Applications, and Challenges related to Faith-Based Conservation

# 13

# INTERNATIONAL HUMAN RIGHTS LAW APPROACHES IN SUPPORT OF FAITH-BASED CONSERVATION MOVEMENTS

*Jonathan Liljeblad*

## Introduction

The idea of faith-based approaches to environmental conservation links topics of religion and environment in the formulation and conduct of conservation practices. While the two topics may appear to be distinct, they are not mutually exclusive. There is increasing recognition of the association of specific locations with particular spiritual beliefs along with the significance of the broader environment for spiritual well-being (see, for example, Liljeblad & Verschuuren 2019; Mallarach et al. 2018). The diversity of human cultures and ecologies means different manifestations of the connections between faith and nature, but the persistence of the connection despite the variety in culture and ecology points to a general human practice to associate culture with ecology.

The linkages between religion and environment are largely subordinated in conservation policy discussions revolving around debates between science-driven (alternatively, evidence-based) conservation versus policy-driven conservation. Science-driven approaches determine location-specific conservation requirements from the unique characteristics of a specific ecosystem, whereas policy-driven approaches seek fixed conservation targets to set universal policy (see, for example, Salafsky 2019; Svencara et al. 2005; Sutherland et al. 2004). The content of such debates leaves little room for concerns of spirituality in environmental conservation.

Despite such challenges, there is a basis for faith-based conservation within environmental discourses. In particular, calls for community-based conservation and rights-based approaches (RBA) to conservation involve concerns for human welfare in the course of protecting the environment. Community-based conservation inherently seeks greater sensitivity to the interests of peoples who adjoin or inhabit ecosystems in the management of those ecosystems (see, for example,

DOI: 10.4324/9781003246190-17

Meffe et al. 2009). RBA in conservation promotes the application of human rights to address the concerns of people affected by ecosystem management practices (see, for example, Campese 2009).

This analysis explores legal approaches to enable the presence of religious perspectives in environmental conservation. The law provides a potential tool with the authority of courts to empower the inclusion of faith-based approaches in conservation. The framework for analysis draws upon international human rights law to formulate legal strategies that enable faith-based conservation approaches. The discussion begins with reference to discourses of the United Nations (UN) Special Rapporteur on Human Rights and the Environment (Special Rapporteur), whose work provides a basis in international human rights law to connect rights to religion with environmental conservation. The discussion then addresses how such rights can be accessed via domestic law in terms of the availability of international human rights to religion as legal rights enforceable by domestic courts.

In looking towards the domestic implementation of international human rights law, it should be noted that under international law, a treaty only imposes obligations upon states that become parties to the treaty. Status as a state party means that a state accepts the treaty provisions as legally binding (Shaw 2014). A state becomes a party to a treaty through one of several means: (1) signature and ratification, where a state signs a treaty and then follows domestic procedures to ratify the treaty to indicate consent to its terms; (2) acceptance or approval, which occurs in situations where a state does not have laws regarding ratification and involves a state that expresses an intent to abide by the terms of a treaty; or (3) accession, which arises in cases where a state wants to join a treaty that is already in force, and involves a state announcing that it will adhere to a treaty (UNTC 2021b). Apart from treaties are non-binding instruments such as declarations, whose terms serve as aspirational statements for state signatories (Shaw 2014).

## International Human Rights Law

The topic of faith-based conservation, in associating human spiritual beliefs with treatment of the environment, intrinsically involves a human-nature linkage. The human-nature linkage has been the subject of international policy discourses seeking to bridge the dichotomy into more cohesive frameworks accommodating inter-related cultural and environmental concerns. Of particular relevance with respect to international law has been the discourses within the UN (see, for example, OHCHR 2021a), which has conducted efforts to unify human-nature approaches. The body of such work is addressed by the UN Special Rapporteur on Human Rights and the Environment. The Special Rapporteur holds a mandate under the UN Human Rights Council to examine the human rights obligations relating to the environment, best practices in environmental policymaking for human rights, identifying the issues in promoting the concept of a right to environment, and investigating human rights violations involving the environment (OHCHR 2021a). In 2018, the Special Rapporteur formulated the Framework Principles on Human Rights and the Environment (Framework Principles) that recognize reciprocal interdependencies between both topics (OHCHR 2018a, 2018b). The Framework Principles calls upon states to protect the environment to support the fulfilment of human rights, on the logic that a healthy and sustainable environment provides the conditions necessary for the full enjoyment of human rights (OHCHR 2018a: Arts. 4–5). Conversely, the Framework Principles also expect states to promote human rights to ensure a healthy environment, with the rationale that human rights are necessary to allow the formulation and application of effective measures to protect the environment (OHCHR 2018a: Arts. 4–5).

Religion is not explicitly mentioned within the Framework Principles, but it is encompassed by the use of the term "human rights" in their language. As a product of the UN, the Framework Principles draw upon the corpus of international human rights instruments, including the broad construction of rights to religion presented by the Universal Declaration of Human Rights (UDHR) and International Covenant on Civil and Political Rights (ICCPR) (Human Rights Committee 1993; ICCPR 1966: Art. 18; UDHR 1948: Art. 18). The right to religion involves both individual belief, or *forum internum*, and public manifestations of belief (Parker 2006; Nowak 1993). The right to individual belief is held within the right to freedom of thought, conscience, and religion, and the right to practise religion, individually and communally, is held within the right to public manifestations (ICCPR 1966: Art. 18(1); UDHR 1948: Art. 18; Nowak 1993).

It should be noted that while international treaties pose obligations upon state parties, states must consent to become parties to an individual treaty (OHCHR 2021b). With respect to the international human rights law associated with the discussion of rights of religion vis-à-vis environmental issues, each international human rights treaty has a different slate of state parties. For example, the ICCPR has 173 state parties (OHCHR 2021c). For the faith-based conservation activists within an individual state, the status of their state as a party to a human rights treaty such as the ICCPR means access to the work in international human rights law linking rights to religion and environmental problems. If their state is not a party, then the treaty is not binding and such access is not available. Hence, due diligence for faith-based conservation efforts also includes monitoring the status of individual states as parties to human rights treaties related to rights to religion.

The connection between the Framework Principles and international human rights law assures a place for religion within environmental policymaking. Such a connection opens space for consideration of faith-based conservation approaches. In addition, however, the connection between the Framework Principles and international human rights law also sets minimum expectations for environmental conservation efforts to respect religious concerns, in terms of either explicit or implied rights of religion afforded from existing international human rights law. As a result, the connection identifies the need for environmental policies to accord some measure of attention to faith-based perspectives.

## Domestic Implementation

The preceding section summarized approaches offered by the UN Special Rapporteur for Human Rights and the Environment that address faith-based conservation efforts. The Special Rapporteur links human rights with conservation efforts, allowing human rights protections regarding rights to religion to provide space for religious perspectives in environmental conservation policies. The Special Rapporteur sets Framework Principles that recognize interdependencies between human rights and environment, and which call upon states to promote human rights in environmental conservation. The Framework Principles refer to international human rights law to express human rights as legal rights, such that issues such as rights to religion can be addressed via actions within legal systems.

However, even if an international treaty exists to express a human right as a legal right to be upheld by state parties, there is an additional issue of treaty implementation. In order to be enforceable by the domestic courts of a state, the terms of an international instrument must be implemented by the state. State implementation of international law varies by state in that under the concept of sovereignty, each state holds the power to determine the exercise of law, including domestic exercise of international law, within its jurisdiction. On a global scale, there are two general philosophical trends regarding state implementation of international treaties: Dualist and Monist. Dualist philosophies, held by countries such as the United Kingdom, the United States, or New Zealand, view international and domestic laws as separate, such that international treaty obligations do not become enforceable by domestic courts until they are expressed as domestic law (Shaw 2014). Monist philosophies, such as France, Germany, or Peru, see international and domestic realms as part of the same body of law, so that international treaties can be cited by domestic courts as part of their legal decision-making processes (Shaw 2014). Such issues mean that supporters of faith-based conservation who seek to exercise legal rights to religion offered by international law must engage two steps: (1) identify the status of their country as a state party to international treaties addressing rights to religion, and (2) identify their country's approach in implementing such treaties.

To illustrate the above distinctions between Dualist and Monist approaches, the following sections compare New Zealand and Peru. Comparison of the two countries will help make readers aware of how state parties can differ in their implementation of rights to religion provided in the single human rights treaty of the ICCPR.

### *New Zealand*

The judiciary of New Zealand follows the traditions of common law and constitutionalism inherited from the English legal system (Courts of New Zealand 2021). While it is a sovereign state, it is a monarchy with Queen Elizabeth II serving as the head-of-state through her representative, the Governor-General.

Similar to the United Kingdom, the corpus of New Zealand's constitution is not found within a single document but is rather drawn from multiple sources which comprised court decisions, constitutional practice, the Treaty of Waitangi 1840, historical British laws, and legislative acts (Governor-General of New Zealand 2021).

With respect to the right to religion, New Zealand is a state party to the ICCPR, ratifying the treaty in 1978 (UNTC 2021a). In terms of implementing rights to religion, New Zealand locates it within several different sources: the Bill of Rights Act 1990, which gives protection to *forum internum*, religious practice, and religious minorities; the Human Rights Act 1993, which prohibits discrimination against religious belief; the Education Act 1964, which allows religious education so long as it respects the Bill of Rights Act 1990; and Treaty of Waitangi 1840, which protects Maori religions and beliefs (New Zealand Human Rights Commission 2010). All of the aforementioned sources are legislative acts, and so are not constitutionally protected. The absence of constitutional protection renders rights to religion susceptible to constraints from conflicting laws, such that rights to religion bow to opposing language of conflicting laws (McGregor, Bell, & Wilson 2016). The vulnerability of rights to religion is qualified by New Zealand court practices regarding international treaties, in that New Zealand courts generally seek to find consistency between domestic law and New Zealand's obligations to international law by using the language of international treaties to interpret domestic legislation (McGregor, Bell, & Wilson 2016: 37–39). The implication is that in cases of conflicting laws, the New Zealand courts will use the ICCPR to interpret New Zealand law affecting rights regarding religion.

For faith-based conservation activists, the above parameters point to potential legal strategies available within New Zealand's particular context. To the extent that faith-based initiatives can phrase environmental issues as involving rights to religion, they can access the legal rights granted by domestic New Zealand laws. The rights to religion in New Zealand are limited in situations of conflict with other domestic laws, and so render religious concerns vulnerable to contrary policies or activities. Faith-based activists would have to mitigate such risks by seeking protection through New Zealand courts, whose practices would interpret domestic laws through reference to the ICCPR.

The exercise of the ICCPR, however, is relevant not just in terms of bolstering rights to religion under domestic New Zealand, but also in terms of presenting a legal justification for using rights to religion in association with environmental conservation. The association with international human rights treaties such as the ICCPR means association with the attendant discourse in international human rights law by the Special Rapporteur that justifies the linkage of human rights—including rights to religion—with the environment. Such intersections at the international level, combined with the New Zealand court approaches to international treaties, would bolster domestic claims involving religious perspectives in environmental conservation.

## Peru

Peru's legal system descends from the civil law heritage of the Spanish Civil Code, which itself is a product of Spain's historical exposure to customary law, Catholic canon, Roman law, the Napoleonic Civil Code, and more recent codes. As a sovereign state, Peru is a republic with democratic multi-party system holding a President as the head-of-state; executive, legislative, and judicial branches following separation-of-powers; and central, regional, and local levels of government. Peru's 1993 Constitution, with its amendments, is its sole constitutional source (Legalink 2013; Peru Constitution 1993; Sarmiento & Parras 2009).

In regard to the right to religion, Peru is a state party to the ICCPR and ratified the treaty in 1978 (UNTC 2021a). Domestically, Article 2 of the 1993 Constitution guarantees protection of both *forum internum* and religious practice, with the condition that public expression of faith does not offend morals or disturb public order (Peru Constitution 1993: Art. 2). The constitutional status of religion protects it from conflicting legislation. Such security is strengthened by Peru's status vis-à-vis the ICCPR, in that Peru treats ratified international treaties as holding the same priority as its constitution. As a result, Peru's courts are expected to interpret Peru's obligations to the ICCPR as overriding contrary domestic laws (Venice Commission 2014: 11). In practice, the Peruvian state has been accused of being inconsistent in promoting rights to religion, with critics accusing it of applying religion policies in ways that favour Catholicism (USDOS 2019). The state has undertaken reforms in its religious freedom laws to promote greater accommodation for other religious groups (USDOS 2019; Sarmiento & Porras 2009).

The context of Peru's legal system presents a different potential legal strategy for faith-based conservation initiatives compared to New Zealand. Faith-based conservation activists in Peru enjoy rights to religion that are protected both in terms of having international treaties, given the same status as the constitution but also in terms of being enshrined within the 1993 Constitution itself. As a result, to the extent that activists can phrase environmental issues as involving rights to religion, they enjoy some guarantee of receiving preferential treatment in Peru's legal system against opposing policies or activities.

In such scenarios, connections to international human rights treaties like the ICCPR might not be critical. They would, however, still be relevant in that they would provide additional legal rationales for Peruvian courts that complement the exercise of religious perspectives vis-à-vis environmental conservation. Under Peru's approach to international treaties, faith-based conservation activists would still be able to reference international human rights law and hence continue to be consistent with the Special Rapporteur's work to bridge human rights and environmental issues. Because Peru treats international treaties as holding the same status as the constitution, in Peruvian courts legal arguments through international human rights law linking religion and environment would hold the same status as legal arguments through the constitutional right to religion. Hence, faith-based conservation activists in Peru would enjoy two available strategies to advance the application of religious perspectives in environmental conservation.

The above examples of New Zealand and Peru have the purpose of making readers aware of how different countries can have different approaches to implementing the same international human rights treaty. The purpose of the analysis is to address the utility of international human rights law in enabling faith-based conservation initiatives, and so the above subsections do not address contextual issues of indigenous rights. Such issues are outside the scope of the present analysis and are reserved for future studies.

## Future Directions

Given the space allocated for chapters in the present volume, the above analysis stays within the scope specified in the introduction, which is the utility of international human rights law in enabling faith-based conservation. As a result, the discussion presented the work of the Special Rapporteur to link human rights treaties to environmental conservation. Following the above discussion, there are directions calling for future study with respect to issues of legal personality and legal remedies.

First, the idea of legal personality involves recognition of an entity that is able to hold legal rights. A legal person can be either natural or juridical, with a natural legal person being a human being and a juridical legal person being non-human such as a trust or a corporation (Adriano 2015). There can be different classes of legal persons with varying access to legal rights, with an example being the differences in rights held by citizens and non-citizens. The determination of legal personality is not universal, with each country holding its system to classify legal persons and distribute legal rights. For example, courts in India recognize legal personality for animals, rivers, and religious deities, while courts in the United States do not (Bhattacharya 2019; Srivastav 2019). In addition, there are different systems of legal personality between domestic and international laws, with international law recognizing legal personality with differences such as legal rights held by states, legal rights specified for multi-lateral international institutions in their foundational treaties, or legal rights of non-state entities such as peoples or multi-national corporations (Portman 2010). The availability of legal rights is effectively conditional upon legal personality, such that the issue of legal personality can serve as a bar for faith-based conservation efforts from legal strategies. Without legal personality, it is not possible to exercise legal rights. As a result, faith-based conservation efforts that exist as organic social movements in the sense of having no legal form risk being denied access to legal relief. To seek legal claims in courts, they have to manifest themselves as legal persons. The manner by which faith-based conservation activists gain legal personality is specific to each country's domestic legal system and each venue in international law, and so requires due diligence by activists to meet the requirements for legal personality specific to their respective country.

Second, with respect to legal remedies, the pursuit of legal actions involves in some part a desire for courts to prescribe remedies for the violation of rights. There are different types of legal remedies, with examples including court orders, such as orders that halt or compel behaviour by the defendant; declarations to clarify rights; damages providing monetary amounts representing plaintiff losses; restitution forcing defendants to return benefits taken from plaintiffs; or restoration to return a plaintiff to a state prior to defendant actions. The forms of available remedies vary across legal issues, such as between contracts law or criminal law, and across jurisdictions, particularly diverse legal systems (see, for example, EU 2017a; EU 2018b; Cohen & McKendrick 2005; Herlitz 1966). For activists exercising legal claims in a court of a particular country, the slate of potential remedies is determined by the remedies recognized by the court, domestic legal system, and international law. The court may hold powers to fashion particular remedies for individual cases, the domestic legal system will prescribe its offering of remedies, and international treaties to which the country is a state party may also provide their own remedies. Such issues incur a call for faith-based conservation efforts to consider the types of remedies desired for a specific scenario and to calibrate the attendant legal actions to achieve such remedies in relation to the court, domestic law, and related international law.

Third, in considering how faith-based conservation efforts can pursue domestic legal actions that draw upon international human rights discourses regarding religion and environment, the discussion limited the scope of demonstration to the rights to religion given explicitly by the ICCPR. There are, however, other international human rights treaties such as the International Covenant on Economic, Social, and Cultural Rights that contain provisions that relate implicitly to religion. In focusing discussion on implementation of the ICCPR, the analysis left unaddressed the potential application of additional human rights treaties in supporting rights to religion. As much as the ICCPR served to facilitate to link religion and environment, it does so from its specified framework of civil and political liberties. There is a possibility that the alternative frameworks of other human rights treaties would expand understanding of rights affecting religion and thereby inform the exploration of connections between religion and environment. Hence, there would be value in future studies that sought to enrich the present analysis with investigation of how other human rights treaties beyond the ICCPR can support faith-based conservation efforts.

Fourth, beyond expansion in human rights treaties is the value of more country cases. The previous sections presented a discussion of New Zealand and Peru in delineating the mapping exercise by which faith-based conservation movements can determine the availability of international human rights to religion as legal rights in domestic courts. The choice of New Zealand and Peru provided contrasting cases demonstrating how such a mapping exercise—and the availability of legal rights—would differ across countries. But the choice of both countries should not be construed as representative of all countries, nor should they be viewed as reflecting the full range of variation across legal systems. The limited sample of two countries in the present analysis calls for similar exploration of more countries to better address the gamut of domestic legal systems and their respective approaches towards religion and international human rights law.

Such expansion would provide more insights specific to faith-based conservation activists within the respective contexts of their individual countries.

Last, despite the illustrative use of the ICCPR and country cases of New Zealand and Peru, the preceding analysis was largely theoretical in terms of addressing approaches to construct legal bases that connect rights to religion and environment. To become more relevant for faith-based conservation activists, theoretical discussion needs to be tested by an empirical practice. This involves study of efforts that exercise legal actions related to religion and environment, to determine the issues that challenge their work and the factors that empower their progress. Such insights would help faith-based conservation movements devise more effective legal strategies in support of their goals.

## Conclusion

The preceding sections worked to formulate a legal approach enabling faith-based conservation. The analysis identified existing discourses in international human rights law to connect rights to religion with environmental conservation. The analysis then delineated how such rights can be accessed via domestic law, with cases of New Zealand and Peru used as examples to help readers become aware of how different countries can have different approaches to enforcing international human rights to religion as legal rights in domestic courts. The discussion used the work of the Special Rapporteur on Human Rights and the Environment to identify the connections between the ICCPR's rights to religion and environmental conservation, with the purpose of delineating a legal basis to advance arguments defending faith-based conservation. The preceding sections inform the work of activists, who can tailor the framework presented by the analysis to devise legal strategies specific to the individual contexts of their respective legal systems.

## References

Adriano, E. (2015). The natural person, legal entity, or juridical person and juridical personality. *Pennsylvania State University Journal of Law and International Affairs, 4*(1), 363–380.

Bhattacharya, A. (2019). Birds to holy rivers: A list of everything India considers "Legal Persons." *Quartz India*. Retrieved from https://qz.com/india/1636326/who-apart-from-human-beings-are-legal-persons-in-india/

Campese, J., Sunderland, T., Grieber, T. and Oviedo, G. (Eds.). (2009). *Rights-based approaches: Exploring issues & opportunities for conservation*. Gland: International Union for Conservation of Nature.

Cohen, N., & McKendrick, E. (2005). *Comparative remedies for breach of contract*. London: Hart Publishing.

Committee on Economic, Social, and Cultural Rights (CESCR). (2009). *General Comment No. 21 Right of Everyone to Take Part in Cultural Life (Art. 15, Para.1(a), of the International Covenant on Economic, Social, and Cultural Rights) E/C.12/GC/21*. Retrieved from http://www2.ohchr.org/english/bodies/cescr/docs/gc/E-C-12-GC-21.doc

Courts of New Zealand. (2021). *About the Judiciary*. Retrieved from https://www.courtsofnz.govt.nz/about-the-judiciary/overview/

European Union (EU). (2017a). *Judicial remedies for individuals before the highest jurisdictions, a comparative law perspective: The United Kingdom*. European Union Parliament. Retrieved from https://op.europa.eu/en/publication-detail/-/publication/5d09a9fa-bdee-11e7-a7f8-01aa75ed71a1/language-en

European Union (EU). (2017b). *Judicial remedies for individuals before the highest jurisdictions, a comparative law perspective: The United States*. European Union Parliament. Retrieved from https://op.europa.eu/en/publication-detail/-/publication/3ff23cb1-bdf0-11e7-a7f8-01aa75ed71a1/language-en/format-PDF/source-search

Governor-General of New Zealand. (2021) *New Zealand's Constitution*. Retrieved from https://gg.govt.nz/office-governor-general/roles-and-functions-governor-general/constitutional-role/constitution/constitution

Herlitz, N. (1966). Legal remedies in Nordic Administrative Law. *The American Journal of Comparative Law, 15*(4), 687–708.

Human Rights Committee. (1989). *General comment adopted by the Human Rights Committee Under Article 40, Paragraph 4, of the International Covenant on Civil and Political Rights CCPR/C/21/Rev.1/Add.4, General Comment 18*. Retrieved from https://tbinternet.ohchr.org/_layouts/15/treatybodyexternal/TBSearch.aspx?Lang=en&TreatyID=8&DocTypeID=11

Human Rights Committee. (1993). *General comment adopted by the Human Rights Committee Under Article 40, Paragraph 4, of the International Covenant on Civil and Political Rights CCPR/C/21/Rev.1/Add.4, General Comment 22*. Retrieved from http://tbinternet.ohchr.org/_layouts/treatybodyexternal/TBSearch.aspx?Lang=en&TreatyID=8&DocTypeID=11

Human Rights Council (HRC). (2018). *Report of the Special Rapporteur on the Issue of Human Rights Obligations Relating to the Enjoyment of a Safe, Clean, Healthy, & Sustainable Environment A/HRC/37/59*. Retrieved from https://undocs.org/A/HRC/37/59

International Covenant on Civil and Political Rights (ICCPR). (1966). *International covenant on civil and political rights*. Retrieved from http://www.ohchr.org/EN/ProfessionalInterest/Pages/CCPR.aspx

Legalink. (2013). *Peru anticorruption laws*. Retrieved from https://www.legalink.ch/xms/files/CROSS_BORDER_QUESTIONNAIRES/CORRUPTION/Peru_Anticorruption_Laws_LEGALINK2013_GRAU_ABOGADOS.pdf

Liljeblad, J., & Verschuuren, B. (Eds.). (2019). *Indigenous perspectives on sacred natural sites*. Oxfordshire: Routledge.

Mallarach, J.-M., et al. (2018). Implications of the diversity of concepts and values of nature in the management and governance of protected and conserved areas. In B. Verschuuren & S. Brown (Eds.), *Cultural and spiritual significance of nature in protected areas governance, management, and policy* (pp. 14–30). Oxfordshire: Routledge.

McGregor, J., Bell, S., & Wilson, M. (2016). *Fault lines: Human rights in New Zealand*. Retrieved from https://www.waikato.ac.nz/__data/assets/pdf_file/0004/248782/NO-watermark-Fault-lines-Human-rights-in-New-Zealand.pdf

Meffe, G., et al. (2009). *Ecosystem management: Adaptive, community-based conservation*. Oxfordshire: Routledge.

New Zealand Human Rights Commission. (2010). *Freedom of religion and belief*. Retrieved from https://www.hrc.co.nz/files/4614/2388/0493/HRNZ_10_Freedom_of_Religion.pdf

Nowak, M. (1993). *U.N. covenant on civil and political rights: CCPR commentary*. Kehl am Rhein: NP Engel Publishers.

Office of the High Commissioner for Human Rights (OHCHR). (2018a). *Report of the Special Rapporteur on the Issue of Human Rights Obligations Relating to the Enjoyment of a Safe, Clean, Healthy, and Sustainable Environment, A/HRC/37/59*. Available at: https://undocs.org/en/A/HRC/37/59

Office of the High Commissioner for Human Rights (OHCHR). (2018b). *Framework Principles on Human Rights & the Environment*. Available at: https://www.ohchr.org/Documents/Issues/Environment/SREnvironment/FrameworkPrinciplesUserFriendlyVersion.pdf

Office of the High Commissioner for Human Rights (OHCHR). (2021a). *Special Rapporteur on Human Rights & the Environment*. Retrieved from https://www.ohchr.org/en/Issues/environment/SRenvironment/Pages/SRenvironmentIndex.aspx

Office of the High Commissioner for Human Rights (OHCHR). (2021b). *International Human Rights Law*. Retrieved from https://www.ohchr.org/en/professionalinterest/pages/internationallaw.aspx

Office of the High Commissioner for Human Rights (OHCHR). (2021c). *Status of ratification*. Retrieved from https://indicators.ohchr.org/

Parker, M. T. (2006). The freedom to manifest religious belief: An analysis of the necessity clauses of the ICCPR and the ECHR. *Duke Journal of Comparative and International Law, 17*, 91–129.

Peru Constitution 1993. (1993). *Peru's Constitution of 1993 With Amendments Through 2009*. Constitute Project. Retrieved from https://www.constituteproject.org/constitution/Peru_2009.pdf?lang=en

Portman, R. (2010). *Legal personality in International Law*. Cambridge: Cambridge University Press.

Salafsky, N., et al. (2019). Defining and using evidence in conservation practice. *Conservation Science and Practice, 1*(5). Retrieved from https://conbio.onlinelibrary.wiley.com/doi/full/10.1111/csp2.27

Sarmiento, G., & Porras, D. (2009). The long road to religious freedom in Peru. *Brigham Young Law Review, 3*, 597–650.

Shaw, M. (2014). *International Law* (7th edn.). Cambridge: Cambridge University Press.

Srivastav, P. (2019). Legal personality of Ganga and ecocentrism: A critical review. *Cambridge Law Review, 4*(1), 151–168.

Sutherland, W., et al. (2004). The need for evidence-based conservation. *Trends in Ecology and Evolution, 19*(6), 305–308.

Svencara, L., et al. (2005). Policy-driven versus evidence-based conservation: A review of political targets and biological needs. *BioScience, 55*(11), 989–995.

United Nations Treaty Collection (UNTC). (2021a). *International covenant on civil and political rights*. Retrieved from https://treaties.un.org/Pages/ViewDetails.aspx?src=TREATY&mtdsg_no=IV-4&chapter=4&clang=_en

United Nations Treaty Collection (UNTC). (2021b). *Glossary of terms relating to treaty actions*. Retrieved from https://treaties.un.org/pages/overview.aspx?path=overview/glossary/page1_en.xml

United States Department of State (USDOS). (2019). *Peru International Religious Freedom Report 2019*. Retrieved from https://www.state.gov/wp-content/uploads/2020/05/PERU-2019-INTERNATIONAL-RELIGIOUS-FREEDOM-REPORT.pdf

Universal Declaration of Human Rights (UDHR). (1948). *Universal declaration of human rights*. Retrieved from http://www.ohchr.org/EN/UDHR/Pages/UDHRIndex.aspx

Venice Commission. (2014). *Report on the implementation of International Human Rights Treaties in domestic law and the role of courts*. Retrieved from https://www.venice.coe.int/webforms/documents/default.aspx?pdffile=CDL-AD(2014)036-e

# 14

# PRAYING IN BRAZILIAN PROTECTED AREAS

## From Conflicts to Potential Alliances with Society for Nature Conservation

*Érika Fernandes-Pinto*

### Introduction

The role of religious organizations in contributing to nature conservation (Palmer & Finlay, 2003; Bhagwat *et al.*, 2011; Awoyemi *et al.*, 2012; McLeod & Palmer, 2015) and the importance of the sacred dimension of protected areas (PAs) (Dudley *et al.*, 2005; Verschuuren & Brown, 2019; Verschuuren *et al.*, 2021) have been increasingly recognized in international publications and in global forums on environmental discussions. However, these perspectives still represent a novelty in Brazil as managers of PAs are accustomed to looking at nature through the technical-scientific lenses of the natural sciences (Fernandes-Pinto, 2017).

I have previously carried out an exploratory study on sacred natural sites (SNSs) in Brazil (Fernandes-Pinto, 2017) which showed that several natural formations spread across all regions of the country have religious and spiritual meanings recognized by different cultural groups.[1] Despite the social importance of many of these areas, few national public policies recognize sacred sites and the intangible values of nature. Moreover, many SNSs are threatened by large infrastructure projects (such as hydroelectric plants and roads), land speculation and other pressures (Fernandes-Pinto & Irving, 2018).

Some of these previously identified SNSs listed in Fernandes-Pinto (2017) are located within the limits of legally established PAs. In spite of the fact that this helps to contain external threats to these sites, the cultural aspects of PAs are still little considered in Brazilian nature conservation strategies, which tend to focus on biodiversity.

This scenario is not unique to Brazil. Dudley (2008), for example, reported that existing policies and legal frameworks, in most countries, generally did not adequately support the protection of sacred sites and that they were not effectively

DOI: 10.4324/9781003246190-18

reflected in PA designations and their management plans. Also, according to Wild and McLeod (2008), SNSs all over the world were inadvertently integrated into PAs without recognition of their meanings for local communities. The adoption of measures restricting access and the use of sacred places without dialogue with their traditional guardians has resulted in animosities and the lack of social support for the management of many PAs.

Some recommendations to change this situation have been proposed in recent decades at world congresses of the International Union for the Conservation of Nature (IUCN) and in publications of guidelines for good practices in PA management – such as Wild and McLeod (2008) and Verschuuren *et al.* (2021). They reinforce the importance of recognizing and integrating the cultural and spiritual meanings of nature into PA management strategies. It is imperative, however, that these initiatives become better known among PA managers of different countries, so that they can be tested and adjusted to each national legal system.

As will be shown in this chapter, initiatives to discuss these issues, on a national scale in Brazil, are relatively recent. Until 2018, religious use was practically not mentioned in official documents related to a potential visitor's interest in the country's PAs. Religious practices in nature are still a polemic topic that raises controversies (Fernandes-Pinto, 2017).

Given this context, this chapter seeks to demonstrate the religious and spiritual relevance of Brazilian PAs for different social groups; analyse how the subject has been treated in management bodies and explore the main challenges that must be overcome so that these values can be integrated into the governance of these areas.

Research methodology included literature searches and consultations of government documents. In-depth interviews were carried out with managers, researchers and representatives of social movements during my doctoral studies – this forms the basis for the information presented in this chapter (Fernandes-Pinto, 2017). I also bring my professional experience working in the governmental agency for the management of federal PAs in Brazil, as well as my personal experiences in sacred places in some countries.

The chapter is divided into four sections: (1) A general contextualization of the situation of PAs in Brazil and dilemmas related to the management of human interactions in the most restrictive PA categories. (2) A short portrayal of the diversity of spiritual and religious interests associated with the country's PAs. (3) An analysis of how religious use has been addressed in the management of PAs and its main problems. (4) Some of the initiatives that have contributed to awakening discussion on this theme in the country.

I look at this reality in an interdisciplinary and critical manner. I also reinforce my ethical engagement with actions that aim at overcoming patterns of coloniality and structural racism in the management of PAs.

## Protected Areas in Brazil and the Dilemmas of the Relationship between Society and Nature

With its extensive territory and variety of terrestrial and aquatic environments, Brazil is a country with great biological wealth (Lewinsohn & Prado, 2005). Allied to its natural exuberance, it is home to a social plurality – represented by approximately 250 Indigenous peoples, remaining *quilombo* communities (African descendants who, in the process of resistance to slavery, gave rise to groups that share cultural characteristics and occupy territories of common use) and other traditional groups (a diverse set of social groups of plural origins, with ways of life adapted to certain environments). In addition, Brazil received immigrants from different regions of the world, resulting in the formation of a multicultural nation that also stands out for its rich religious diversity (Fernandes-Pinto & Irving, 2018).

PAs are an important pillar of policies for the protection of natural heritage in Brazil, where they are called *unidades de conservação*.[2] There are more than 2,500 PAs in the country, covering about 18% of its land (CNUC/MMA, 2021). The National System of Conservation Units – SNUC (Federal Law No. 9.985/2000) defined 12 management categories, arranged into two main groups: (1) integral protection and (2) sustainable use. The first (represented mainly by national parks and biological reserves) is equivalent to IUCN's PAs categories I, II and III and does not allow direct use of natural resources. The second group (represented by Extractive Reserves and National Forests, among others) is equivalent to their categories IV, V and VI. It allows human settlements and sustainable management of part of the natural resources.

However, not only sustainable-use PAs are inhabited or used by local communities. It is estimated that at least 70% of Brazilian integral protection PAs overlap with areas of Indigenous peoples, *quilombola* communities or traditional groups (Madeira *et al.*, 2015). This situation was considered a "collision of rights" and the subject of a strong dispute over which right should prevail – whether social interests or the need to protect nature, as if they were exclusionary goals.

This scenario began to change only in 2009, with a decision of the Federal Supreme Court that determined a dual allocation legal regime between the *Monte Roraima* National Park and the *Raposa Serra do Sol* Indigenous Land (MPF, 2014).

In this debate, it is important to contextualize that during the most part of Brazilian PAs' history, the *social* was not part of what was understood as *environmental*. Although PA categories of sustainable use have been recognized by the national system established in 2000, the conception of most nature protection policies followed a clearly preservationist ideology (Diegues, 2008). From this perspective, the use of PAs by society is seen primarily with a focus on the risks posed to the degradation of nature. And long-term conservation of biodiversity is conceived as incompatible with the existence of human groups.

In this context, many types of cultural relationships that people form with nature, which do not necessarily involve the direct use of natural resources, were neglected in the management of PAs. After all, if nature cannot have people, it cannot have cultural value either. Thus, it can be said that a dichotomous view of nature versus culture contributed to obscuring religious uses associated with nature, even though they occur in many Brazilian PAs, as will be discussed next.

## Diversity of Religious Use in Conserved Areas

A preliminary study (Fernandes-Pinto, 2017) identified 115 Brazilian PAs with occurrence of SNSs and/or religious use in natural areas, including the most restrictive categories. Spiritual and religious places found in PAs embrace various types of natural formations, such as mountains, caves, rivers, lakes, waterfalls, springs, forests, trees, dunes, islands, beaches and others. Some areas are kept in their natural state, while others may be consecrated with human artefacts and structures.

The religious and spiritual values of Brazilian PAs are related to different social groups – Indigenous and traditional peoples, followers of institutionalized religions and of different spiritualist lineages. There are also places associated with legends, mysteries, reports of inexplicable phenomena and the presence of mythological beings that permeate the imaginary (used here in the sense of a system that structures thought) of local communities – these may not be linked to religious practices.

Multiple religious uses can be seen in natural areas – from individual practices of contemplation, meditation and prayer to large collective celebrations. Each tradition, however, conceives the sacredness of nature in their own way and this is reflected in different forms of interaction with PAs, as will be illustrated next.[3]

### *Catholicism*

In the context of Catholicism (largest belief system in Brazil), nature commonly provides a background for religious ceremonies, such as baptisms, masses and processions. The areas of interest for this tradition are usually marked with human elements – from small sacred images, altars and crosses (as in *Serra do Caparaó* National Park) to large statues, Stations of the Cross and churches (as in *Serra da Piedade* Natural Monument). In several PAs, there are religious monuments of historical importance, listed as national cultural heritage, like the statue of *Cristo Redentor* (Christ the Redeemer), an internationally recognized symbol of Brazil (Figure 14.1).

Some special celebrations, such as Holy Week, bring thousands of people to some natural places. In addition, several PAs are part of sacred paths and pilgrimage routes, such as the *Terra Ronca* State Park, destination of an annual religious cavalcade in honour of *Bom Jesus da Lapa*.

Given the importance of Catholicism in the country, in many PAs, the interests of these groups are "accommodated" in PA regulations, even in the most restrictive categories. In some cases, there are allowances for infrastructure to meet religious needs (Cavalcante, 2012). As will be discussed next, the same may not occur when it comes to other faiths.

FIGURE 14.1 (Left) Nossa Senhora da Piedade Catholic Sanctuary in the Natural Monument of Minas Gerais State. Photo credits: Miguel Andrade/SNS Brazil Collection; (Top Right) Statue of Cristo Redentor in Morro do Corcovado, Tijuca National Park. Photo credits: ICMBio Collection. (Bottom Right) Cross on top of Pico da Bandeira, in Serra do Caparaó National Park. Photo credits: ICMBio Collection.

### Pentecostalism

A relatively recent and expanding phenomenon in Brazil is the so-called *montes de oração* (prayer hills), linked to Pentecostal traditions. Following biblical precepts, climbing certain natural heights has the connotation of sacrifice in search of purification and spiritual ascent (Santos & Matiolli, 2011).

Prayer vigil practices of these traditions involve a growing number of PAs, especially those located close to urban centres, causing conflicts with their management (Maciel & Gonçalves, 2017). This is the case with many PAs in Rio de Janeiro State, like the *Tinguá* Biological Reserve, where more than a hundred infraction notices were issued against people who were praying in the area, with the confiscation of sacred and personal objects.

### Afro-Brazilian Traditions

Many practices of Afro-Brazilian traditions, which include various religions, such as Candomblé and Umbanda, are carried out in native forests, streams and waterfalls. Their deities, called *Orixás*, are representations of the forces of nature. Some PAs are used for individual or collective rites of these traditions. These may involve singing and dancing ceremonies, purification baths, the harvesting of sacred plants and offerings (of food, drinks, flowers or objects) in special places in nature (Sobreira, 2011).

These traditions' practices are often considered threats to PAs, mainly because of the leftovers resulting from offerings and the fire risk associated with the use of candles (Santos-Júnior *et al.*, 2021). Although these are concrete problems, the lack of dialogue with these groups and the criminalization of their practices are noteworthy. According to authors such as Moutinho-Da-Costa (2008), these traditions are the most affected by behaviours of environmental racism and religious intolerance, often practised by state officials themselves.

There are also PAs that keep emblematic SNS for the history of Afro-descendant populations in Brazil. This is the case of *Pedra de Xangô*, in the Municipal Environmental Protection Area of *Vale do Assis Valente/Bahia*, which marks an escape route used during the slavery period (Silva, 2017). The protection of this place took place, thanks to a strong mobilization of the town's Afro-religious people. This also helped to preserve a wider area of forests and rivers (Figure 14.2).

### Alternative Spiritual Traditions

In the broad and eclectic universe of alternative spiritual traditions in Brazil, elements of different classical religious traditions are articulated. Many seek the awakening of a new consciousness that would reestablish the vital unity between humans and nature (Siqueira, 2002).

In this context, the SNSs are often considered *Earth chakras* – points of high energy concentration interconnected in a planetary circulation grid (known as *ley lines*). *Chapada dos Veadeiros* National Park is at the centre of one of the

**FIGURE 14.2** Pedra de Xangô in the Municipal Environmental Protection Area of Vale do Assis Valente. Photo credits: Maria Alice Silva/SNS Brazil Collection.

**FIGURE 14.3** (Left) Jardim de Maytrea in Chapada dos Veadeiros National Park. Photo credits: ICMBio Collection and (Right) Serra Azul State Park in Serra do Roncador. Photo credits: Edson Rodrigues/ Secom-MT.

most significant regions for this movement, attracting national and international mystical groups. Many PAs are also associated with the occurrence of phenomena related to UFOs. Foreseeing a potential touristic interest, a *discoport* (airport for flying saucers) was built in the *Serra Azul* State Park in the area where the image of an "extraterrestrial" welcomes visitors (Figure 14.3).

There are also PAs that are centres of origin and dispersion of traditions that have spread worldwide, such as the Purus National Forest in Acre, the birthplace of the *Santo Daime* doctrine (Brandão, 2005). This religion was born in the interior of the Amazon rainforest. It is characterized by a deep devotion to nature and the ceremonial use of an entheogenic drink produced with native plants (also known as Ayahuasca).

## *Indigenous and Traditional Peoples*

Usually, in Indigenous and traditional cultures, there is not a separation between nature and spirituality. In the worldview of these groups, SNSs are commonly associated with dwellings of ancestral spirits and places where mythological events occur, such as the emergence of people on the surface of the Earth. This makes some places reference centres for their cultural identity and essential for the shaping of their territories (Scolfaro *et al.*, 2013).

This is the case of the *Monte Roraima* National Park, on the triple border between Brazil, Guyana and Venezuela. This mountain holds sacred places for the *Ingarikó* Indigenous people in Brazil and the *Pemon* in Venezuela. Another such example is the *Pico da Neblina* National Park. There, the highest point of Brazil – called *Yaripo* by the *Yanomami* People – is considered the home of the spirits that protect the world from the *falling of the sky* (Figure 14.4).[4]

Indigenous sacred places, in contrast to those recognized by institutionalized religions, tend to be maintained without direct human intervention. They can be visited on special occasions or left reserved for spiritual beings. These values can conflict with touristic incentive policies in PAs, as has happened in the two PAs previously mentioned.

Sacred values can be relevant even in areas where native peoples have been exterminated or de-territorialized decades ago. As occurred in the national park of the famous Iguazu Falls (one of the five great waterfalls in the world, on the border between Brazil and Argentina), an area that was part of the *Guarani Tekoa Iguassu* ancestral territory (Figure 14.5).

**FIGURE 14.4** (Left) Monte Roraima and (Right) Pico da Neblina – sacred mountains for Indigenous people. Photo credits: Marcos Amend/SNS Brazil Collection.

**FIGURE 14.5** The Cataratas do Iguaçu in Iguaçu National Park. Photo credits: Marcos Amend/ SNS Brazil Collection.

It is worth mentioning that some PAs may have a multi-religious character, such as the Tijuca National Park, where different traditions appear (Moutinho-Da-Costa, 2008). This coexistence of beliefs in the same area, however, is not always peaceful and religious intolerance becomes another topic to be considered in the management of PAs.

## Religious Interests and PAs Management Tools

Despite the variety of SNSs found in Brazilian PAs and the recurrence of religious use in these areas throughout the country, these aspects are still poorly recognized and valued in nature management strategies, in both categories: sustainable use and full protection. Until a few years ago, religious and spiritual interests were not usually considered among the possibilities of public use of PAs. This started to happen, albeit timidly, in 2018.[5] More than that, some PA management plans contain a general rule expressing the prohibition of religious manifestations within those areas.

As far as it was possible to investigate with the interviews carried out, this rule originated via a peculiar interpretation of the *state's secularity*. As PAs are public spaces, it was considered that religious symbols and practices should not be allowed in those areas. It seems that there was a misunderstanding that confused a *secular state* with an *atheist state* – the latter characterized by opposing religious practices in public places.

The Brazilian Federal Constitution (1988) guarantees the rights to freedom of belief and the protection of places of worship (Article 5). Still, there are conflicts involving the repressive action of policing within PAs – notably with indigenous, African-Brazilian and Pentecostals traditions (Moutinho-Da-Costa, 2008; Léo Neto, 2015; Maciel & Gonçalves, 2017).

In a country marked by deep social asymmetries, the strategies of *command and control* of nature management can disproportionately affect certain social groups. While some religious interests of majority religions are "accommodated" in the management of many PAs, the interests of others are not considered. Thus, PA managers can reproduce patterns of social injustice and environmental racism – for example, when environmental norms or policies fall disproportionately on vulnerable ethnicities, from conduct that doesn't necessarily have a racist intention, but implies a racial impact (Moutinho-Da-Costa, 2008).

Without an adequate contextualization of the cultural dimension, religious interest in PAs is commonly addressed in management plans as a threat to be fought against. Religious uses are often equated to illegal hunting and vandalism, for example, extolling the potential for damage to biodiversity and disregarding the symbolic universes of religious traditions (Santos-Júnior *et al.*, 2021). In this context, religious groups are not seen as legitimate interlocutors for dialogues on PA management and their participation in councils of the PAs is not usual.

Over the years, however, the result of repressive actions doesn't seem to be diminishing socio-cultural interest in PAs, as religious practices continue to occur regularly. In some cases, this repressive government conduct doesn't facilitate positive change, for example, organizing activities to mitigate environmental damage. Instead, it contributes to feeding resentment and reinforces animosity between the parties involved. Moreover, there is also the ignored dimension of care and feelings of affection and belonging that many traditions can express to areas that are considered *natural sanctuaries*.

This is not to say that religious and spiritual practices cannot have negative impacts on biodiversity. In many cases, it is really necessary to consider adjustments or restrictions in the use of natural environments. Erosion of trails because of a high number of visitors, litter accumulation, risks of forest fires, damage to native fauna and flora and areas of greater environmental sensitivity (such as mountain tops or river sources) are pointed out by PA managers as common negative consequences of these practices (Malta, 2016; Santos-Júnior *et al.*, 2021).

The potential impacts listed above, however, accompany almost all human activities in nature – regardless of whether the motivation is recreational, educational or religious. Some areas of special touristic interest in Brazilian PAs – such as the *Cataratas do Iguaçu* and *Morro do Corcovado* – receive thousands of visitors daily. These PAs have public use programmes that aim to reconcile the public interest with conservation objectives. They include the installation of adequate infrastructure for their level of visitation and management models with public-private partnerships. I argue that the same logic of "conciliation of objectives" should be applied to areas of religious interest.

In fact, there are some initiatives being implemented in PAs that seek to equate the needs of religious groups that perform rituals in nature with good environmental practices and impact mitigation strategies. In the *Tijuca* National Park/RJ, for example, the dialogue with Afro-Brazilian traditions representatives helped to identify specific places for offerings and establish joint responsibilities regarding waste management (Boniolo, 2018). Also, in accordance with religious movements, planning measures were adopted by the government to minimize the impacts of pilgrimage in the caves of the *Terra Ronca* State Park.

Examples from international academic literature also illustrate how the adoption of good practice protocols and adaptive measures with respect to religious traditions and their symbologies can be implemented aiming at better PA management, benefiting all (Awoyemi *et al.*, 2012; Verschuuren & Brown, 2019; Verschuuren *et al.*, 2021). Publications on these topics, developed and endorsed by the international PA community, can play the role of guiding documents to help bring about changes at the national level and link the Brazilian debate to international policies and trends. However, the highly complex realities of Brazilian PAs, added to the lack of financial and human resources in the managing agencies of these areas, make this task challenging.

I argue that part of the problems related to religious use in Brazilian PAs could be solved by the proper application of management tools already provided by national legislation. To this end, the dimension of cultural and spiritual values of nature must be integrated into existing protocols. Religious movements that already work on ethical issues for a more harmonious relationship between people and nature could help to think about these methods, without romanticizing the debate or "closing their eyes" to the problems.

It is worth highlighting some initiatives developed in the last decade that have paved the way towards a new conservation paradigm and are helping to recognize the importance of cultural and spiritual values of PAs in Brazil.

## Towards Recognition of the Sacred Dimension in Brazilian Protected Areas

The *Sítios Naturais Sagrados do Brasil* initiative, coordinated by the author since 2013, has been involved in the dissemination of relevant themes through lectures, interviews, scientific and non-scientific publications, in the articulation of networks of interested parties and in the construction of bridges with the international movement for the recognition of spiritual and culture values in nature management. Some actions undertaken by the author which are worth mentioning are as follows:

- host of the *1st SNS Brasil Network Meeting* and the *1st Colloquium on Spirituality and Nature Conservation*, at the VIII *Brazilian Seminar on Protected Areas and Social Inclusion* – SAPIS, in 2017 (in Niterói/Rio de Janeiro);
- participation in the *International Expert Workshop* on the *Cultural and Spiritual Significance of Nature in the Governance and Management of Protected and Conserved Areas* (Isle of Vilm/Germany), in 2017, to build IUCN guidelines on the subject (Verschuuren *et al.*, 2021);
- promotion of thematic events at the *III Congreso de Áreas Protegidas de Latinoamérica y el Caribe*, in 2019 (in Lima/Peru), when a Latin American articulation of those interested in these themes began.[6]

Within the scope of nature management, it is worth highlighting initiatives promoted by ICMBio, the managing body of the federal PAs in Brazil:

- the *1st Seminar on Cultural Values of Nature: new challenges for public conservation policies*, in 2019 (Brasília/DF);
- the training course on *Integration of Cultural Values in the Management of Conservation Units*, for PA managers and collaborators, with 40 hours of content on a virtual learning platform (released in 2021).

These initiatives had hundreds of participants, helping to disseminate/spread the seeds of these new perspectives on the integration of society and nature, to a broad public directly or indirectly involved in the management of PAs. Also, they contributed to reducing the knowledge gap in the country about the international debate related to these themes.

Via ICMBio, some priority PAs were selected to undergo more detailed analyses of their historical and cultural values – notably the *Aparados da Serra* National

Park and the *Iguaçu* National Park. These studies have helped to reveal historical invisibilities about the relationship of traditional peoples and communities with the PAs, which were disregarded throughout their administration. They also help to exemplify the wide range of cultural meanings that can be associated with nature, even in the most restrictive PA categories. I hope that these studies on emblematic PAs will inspire similar initiatives in other areas.

As for management tools, a new participatory methodology for preparing management plans at the federal level allows cultural aspects to be also listed as fundamental values of the PAs. In the *Pico da Neblina* National Park plan, for example, the significance of the PA in protecting sacred places of Indigenous peoples was highlighted.

In 2020, a new regulation regarding visits with educational purpose in full protection PAs contemplates a broader vision of public use, including possibilities of recreational, sports, sensorial, therapeutic and religious activities (ICMBio, 2020).[7]

These initiatives add to an effort to establish a programme aimed at the integration of cultural values in the management of PAs and the construction of institutional competences to deal with these themes. Little by little, more inclusive perspectives involving the recognition of the multiple meanings of nature have permeated different areas of the federal agency responsible for the management of PAs.

## Final Considerations

As discussed in this chapter, the social importance of the religious use of natural areas evidenced in several PAs contrasts with the invisibility of the topic in PA management in Brazil. The gap in the recognition of cultural and spiritual meanings in these areas has contributed to the view that historical relationships of local communities with nature are "threats to conservation", their traditional practices are "environmental crimes" and their very presence in their ancestral territories are "invasions".

Despite the advances and promising initiatives of global debates on a more inclusive view of society in nature conservation strategies, this continues to be an unresolved issue in Brazil. The dichotomy between culture and nature observed in national PAs, however, reflects historical social problems that permeate the roots of the structure of Brazilian society as a whole, still marked by remnants of a colonialist inheritance, with prejudices and inequalities mirrored in strategies of conservation.

Repressive practices for PA protection can reproduce patterns of social injustice and environmental racism in relation to certain segments of the population. The mere prohibition and criminalization of religious practices have not proved to be an efficient strategy to tackle resulting problems, while the dialogue and management strategies implemented in some cases have generated positive results.

It is essential to understand the phenomenon of religious use in PA from an integrated view of nature and culture and the ties that connect people to specific places. Therefore, it is not just about the evaluation of the impacts of practices, but also about the understanding of their knowledge, symbologies and meanings. And to ponder if they may not be helping to protect natural areas in different ways from those recognized so far by PA managers.

Given the urgent need for social support for the maintenance of PAs in Brazil, the alignment of religious movements with the goals of nature conservation is not just an interesting possibility – this may be imperative.

The initiatives developed in Brazil in the last decade that stimulated the debate on cultural and spiritual values of nature and the religious use of PAs, although carried out in a fragmented way, marked the beginning of a more consistent debate on these topics at the national level. It has been helping to weave a network system that has highlighted these previously almost unknown themes. Whether in PA management or in academia, it can be said that these seeds have germinated, and various relevant initiatives have begun to sprout across the country. They demand, however, continuity and depth so that they can effectively change the foundation of the management of Brazil's PAs.

As someone deeply involved in this process of changing the conservation paradigm, I dream of a time when Brazil's PAs will cease to be mirrors of society's illnesses and rather become lighthouses illuminating forms of harmonious connection with nature, with even more beauty and justice.

The journey, after all, continues…

## Acknowledgement

Thanks to Bas Verschuuren, Silvana Meira and Maria Tereza Fernandes Abrahão for their careful reading and helpful comments on this manuscript.

## Notes

1 About 480 SNSs were listed, in areas subject to different property regimes. Examples and explanations of the methodology used can be found in Fernandes-Pinto and Irving (2018). This survey continues with the initiative *Sítios Naturais Sagrados do Brasil*, expanding the records to more than a thousand SNSs in 2021 (more information: sitiosnaturaissagrados.org; Instagram @snsbrasil).

2 In Brazilian legislation, there is a distinction between the terms *protected areas* and *conservation units* – the first being broader than the second. However, I am using them interchangeably, considering only areas legally established by the government for the purpose of nature protection.

3 You can find more about the examples mentioned in this chapter on the blog snsbrasil.blogspot.com.

4 Title of a book that recounts the wisdom of a *Yanomami* shaman about the forest and the struggle of Indigenous peoples (A Queda do Céu: palavras de um xamã Yanomami by Albert & Kopenawa, 2015).

5 According to Santos-Júnior (2021), the first reference related to religious interest, in legal norms on public use in Brazilian Pas, is ICMBio's norm No 5 of 2018 (available at: https://www.icmbio.gov.br/portal/images/stories/portarias/intrucao_normativa_5_2018.pdf).

6 With the formation of the Study Group on Sacred Natural Sites of Latin America on Facebook (https://www.facebook.com/groups/snslatinoamerica/), with 74 members in 2021.
7 ICMBio Normative No. 12 de 2020 (Available in: https://www.in.gov.br/web/dou/-/instrucao-normativa-n-12/gabin/icmbio-de-21-de-outubro-de-2020-286696270).

## References


Albert, B., & Kopenawa, D. (2015). *A Queda do Céu: palavras de um xamã Yanomami.* Brasil: Companhia das Letras.

Awoyemi, S., Gambrill, A., Ormsby, A., & Vyas, D. (2012). Global efforts to bridge religion and conservation: Are they really working? In T. Povilitis (Ed.), *Topics in conservation biology.* https://doi.org/10.5772/35587

Bhagwat, S. A., Ormsby, A., & Rutte, C. (2011). The role of religion in linking conservation and development: Challenges and opportunities. *Journal for the Study of Religion Nature and Culture, 5*(1), 39–60. https://doi.org/10.1558/jsrnc.v5i1.39

Boniolo, R. M. (2018). *"Como ordenar o que não pode ser ordenado?": criação de regras de uso do Espaço Sagrado da Curva do S* (Parque Nacional da Tijuca/Rio de Janeiro). Cadernos de Campo, São Paulo, vol. 27, n.1. https://www.revistas.usp.br/cadernosdecampo/article/view/112449

Brandão, P. C. (2005). *Diagnóstico geoambiental e planejamento do uso do espaço na FLONA do Purus, Amazônia Ocidental: um subsídio ao plano de manejo.* Tese de Doutorado. Universidade Federal de Viçosa. http://www.neip.info/downloads/pedro_cristo/tese_pedro.pdf

Cavalcante, M. B. (2012). Políticas de Turismo em Áreas Protegidas: o caso do Parque Estadual da Pedra da Boca (Araruna/PB). *Revista de Geografia, 29*(2). http://www.revista.ufpe.br/revistageografia/index.php/revista/article/viewFile/426/43

CNUC/MMA. (2021). *Painel Unidades de Conservação Brasileiras.* Ministério do Meio Ambiente. https://dados.gov.br/dataset/unidadesdeconservacao

Diegues, A. C. (2008). *O mito moderno da natureza intocada.* 6 Ed. São Paulo: NUPAUB- HUCITEC.

Dudley, N. (Ed.). (2008). *Guidelines for applying protected area management categories.* Gland: IUCN. https://portals.iucn.org/library/sites/library/files/documents/pag-021.pdf

Dudley, N., Higgings-Zogib, L., & Mansourian, S. (2005). *Beyond belief: Linking faiths and protected areas to support biodiversity conservation.* World Wildlife Fund for Nature, Equilibrium and Alliance of Religions and Conservation.

Fernandes-Pinto, E. (2017). *Sítios Naturais Sagrados do Brasil: Inspirações Para O Reencantamento Das Áreas Protegidas.* Tese de Doutorado. Programa EICOS/UFRJ. https://goo.gl/ZNCE11

Fernandes-Pinto, E., & Irving, M. A. (2018). Among Saints, Enchanted and Orixás: A journey through the diversity of sacred natural sites in Brazil. Desenvolv. *Meio Ambiente, 46*, 37–60. https://doi.org/10.5380/dma.v46i0.57281

Léo Neto, N. A. (2015). *"Nós somos os donos": conflitos socioambientais entre os índios Pipipã de Kambixuru e o ICMBIO no sertão de Pernambuco.* In: REA. https://evento.ufal.br/anaisreaabanne/gts_download/Nivaldo%20Aureliano%20Leo%20Neto%20-%201020291%20-%203795%20-%20corrigido.pdf

Lewinsohn, T. M., & Prado, P. I. (2005). Quantas espécies há no Brasil? *Megadiversidade, 1*(1), 36–42. https://www.researchgate.net/publication/271644747_Quantas_especies_ha_no_Brasil/stats#fullTextFileContent

Maciel, G. G., & Gonçalves, R. S. (2017). Educação ambiental e mediação de conflitos neopentecostais no Parque Nacional da Tijuca. *Argum, Vitória, 9*(1), 134–149. https://doi.org/10.18315/argum..v9i1.13666

Madeira, J. A., Abirached, C. F. A., Francis, P. A., Castro, D. M. P., Barbanti, O., Cavallini, M. M., & Melo, M. M. (2015). Interfaces e Sobreposições entre Unidades de Conservação e territórios de Povos e Comunidades Tradicionais: dimensionando o desafio. In N. Hanazaki, et al. (Orgs.). *Culturas e Biodiversidade.* Anais do VII SAPIS. P. 617–625. http://sapiselapis2015.paginas.ufsc.br/anais

Malta, R. R. (2016). *A significância religiosa do Parque Nacional da Tijuca*: as paisagens valorizadas pelos usuários religiosos de uma unidade de conservação. Tese de Doutorado. UERJ, Rio de Janeiro. https://www.bdtd.uerj.br:8443/handle/1/13239

McLeod, E., & Palmer, M. (2015). Why conservation needs religion. *Coastal Management, 43*, 238–252. https://doi.org/10.1080/08920753.2015.1030297

Moutinho-Da-Costa, L. (2008). *A Floresta sagrada da Tijuca*: estudo de caso de conflito envolvendo uso público religioso de parque nacional. Dissertação Mestrado, UFRJ, Rio de Janeiro. http://pos.eicos.psicologia.ufrj.br/wp-content/uploads/2008_MEST_Lara_Moutinho_da_Costa.pdf

Ministério Público Federal – MPF (2014). *Territórios de Povos e Comunidades Tradicionais e as Unidades de Conservação de Proteção Integral*: alternativas para o asseguramento de direitos socioambientais. Série Manual de Atuação. Brasília, 121p.

Palmer, M., & Finlay, V. (2003). *Faith in conservation: New approaches to religions and the environment.* World Bank. http://web.worldbank.org/archive/website00530/WEB/PDF/FAITHINC.PDF

Santos, C. C., & Matiolli, T. (2011). *Os Montes de Oração*: etnografia de uma experiência religiosa. *Mneme – Revista De Humanidades, 11*(29). http://www.periodicos.ufrn.br/ojs/index.php/mneme

Santos-Júnior, V. J., Machado, C. J. S., & Vilani, R. M. (2021). *Turismo e práticas afror-religiosas em unidades de conservação*: uma relação conflituosa? Revista Iberoamericana de Turismo- RITUR, Penedo, v. 11, Dossiê Especial, p. 297–32. https://doi.org/10.2436/20.8070.01.215

Scolfaro, A. et al. (Orgs.). (2013). *Cartografia dos sítios sagrados: iniciativa binacional Brasil-Colômbia.* São Paulo: Instituto Socioambiental; Brasília: IPHAN/Ministério da Cultura do Brasil; Bogotá: Ministerio de Cultura de Colombia. https://www.socioambiental.org/sites/blog.socioambiental.org/files/publicacoes/informe_binacional_cartografia_web.pdf

Silva, M. A. P. (2017). *Pedra de Xangô: um lugar sagrado afro-brasileiro na cidade de Salvador.* Dissertação de Mestrado. https://repositorio.ufba.br/handle/ri/24875

Siqueira, D. (2002). Novas religiosidades na capital do país. *Tempo Social, 14*(1), 177–197. São Paulo. http://www.scielo.br/pdf/ts/v14n1/v14n01a09.pdf

Sobreira, R. F. F. (2011). *Práticas religiosas afro-brasileiras, marco regulatório e uso do meio ambiente.* Tese de Doutorado em Meio Ambiente. UERJ, Rio de Janeiro. https://www.bdtd.uerj.br:8443/handle/1/14257

Verschuuren, B., & Brown, S. (2019). *Cultural and spiritual significance of nature in protected areas.* Routledge. https://www.routledge.com/Cultural-and-Spiritual-Significance-of-Nature-in-Protected-Areas-Governance/Verschuuren-Brown/p/book/9781138091191

Verschuuren, B. et al. (2021). *Cultural and spiritual significance of nature.* Guidance for protected and conserved area governance and management. Best Practice Protected Area Guidelines Series No. 32, Gland, Switzerland: IUCN. https://portals.iucn.org/library/node/49268

Wild, R., & Mcleod, C. (2008). *Sacred natural sites: Guidelines for protected area managers.* Gland: IUCN; Paris: UNESCO, Best Practice Protected Area Guidelines Series, 16. http://sacrednaturalsites.org/items/sacred-natural-sites-guidelines-for-protected-area-managers/

# 15

# FAITH IN THE WEST? ENGAGING CHRISTIAN GROUPS IN CONSERVATION IN 'SECULAR WESTERN SOCIETIES'

*Brittany Ederer, Rachel Mander, Dave Bookless, Robert Sluka and Jeremy Lindsell*

## Introduction

The historical relationship between Christian faith communities and wildlife conservation has been complex and often contested. Lynn White's famous allegation that 'Christianity is the most anthropocentric religion the world has seen' (White, 1967, p. 1205) has been enormously influential (Lodge, 2006), despite being widely critiqued on both historical and theological grounds (Bauckham, 2011; Harrison, 1999; Pepper, 1984). Yet, the modern conservation movements in both Europe and North America trace their roots to individuals motivated to study the natural world by their Christian faith. Both John Ray in the 17th century (Armstrong, 2000), who was the first to define something close to a biological species concept (Mayr, 1982), and proto-ecologist Gilbert White in the 18th century (Barton, 2002; Oriel College, n.d.), were ordained Christian ministers whose belief in a benevolent Creator and an ordered, investigable creation inspired their studies. Other European pioneers such as the Swedish taxonomist Carl Linnaeus were similarly driven towards the scientific study of nature by their Christian faith (Graves, 1996, pp. 80–83).

Indeed, as the modern conservation movement emerged in North America, the debate between the 'preservationist' approach which argued for National Parks and nature's pristine integrity protected from human interference, and the 'conservationist' approach which sought sustainable management of 'natural resources' effectively became a disagreement about biblical interpretation. Preservationist John Muir advocated nature's divinely sanctioned integrity (Muir, 2017), while scholar and naturalist George Perkins Marsh (1864) among others such as Gifford Pinchot (1910) reiterated traditional Christian thinking on humanity's responsibility for careful stewardship of the earth's resources (Santmire, 1985).

DOI: 10.4324/9781003246190-19

Despite these Christian roots, the late 19th and 20th centuries saw a growing divergence between Christianity and the conservation movement. The reasons were complex but included greater secularisation in Western society, increasing professionalism of conservation science, and the rise of fundamentalism in North America resulting in suspicion of science and a focus on 'spiritual' concerns (Moberg, 1977). During this period, few Christian voices advocated for environment or wildlife conservation. However, following the popular environmentalism of the 1960s, there was a gradual reawakening of concern among Christians.

Today, while much media attention is given to the close relationship between US Evangelicalism and ultra-conservative anti-environmentalist politics, the vast majority of mainstream Christian denominations have made significant progress with their environmental positions both theologically and practically (Bartholomew & Bartholomew, 2012; Francis, 2015), including from the Evangelical tradition (Bell, 2016).

In turn, major global conservation organisations and academic researchers have begun to seek partnerships with faith groups, including Christian ones, recognising that religions and the values they espouse can be the strongest possible motivating factor in changing behaviour towards nature (Alliance of Religions and Conservation, 2007; Bhagwat et al., 2011; O'Connor, 2011; Weidensaul, 2018). What is striking, however, is that while the global leadership of most major Christian bodies remains based in Europe and North America, and Christian ecological awakening is advancing rapidly in the West,[1] nearly all the partnerships initiated by conservation organisations have been in the Global South.

Such initiatives stem from a broad realisation that faith groups are key partners in achieving the aims of wildlife conservation (Bhagwat & Palmer, 2009; Bhagwat et al., 2011). Yet, it is hard to find examples of such partnerships in Europe and North America. This may reflect conservation priorities in terms of the location of global biodiversity hotspots (Bhagwat et al., 2011). There may also be an unconscious premise that the Global South is more open to 'religious messaging' than the increasingly 'secular' West where there has been a steady decline in membership of religious institutions and lessening of religious influence in the public sphere (Pew Research Center, 2018; Smith et al., 2015). South American, African, and Southeast Asian nations continue to hold strong religious affiliations (Hackett et al., 2012); future global projections anticipate that these regions will continue to be strongly religious (Stonawski et al., 2015).

This chapter presents case studies of faith-based conservation work within four 'secular' Western countries. Each project is linked to A Rocha, a faith-based organisation within the Evangelical Christian tradition. Our intention is to demonstrate that Evangelical Christian religious faith is a meaningful motivation for conservation in Western societies that are often perceived as predominantly 'secular' and that such motivation can result in action that leads to a positive conservation impact.

## Methods

In order to explore the contribution of faith-based motivation for conservation in secularised societies, we selected case studies from the A Rocha family of organisations in Europe and North America. Ten of the 22 countries (45%) where A Rocha has an institutional presence appear in the top 25 most secularised countries in the world, defined here as the percentage of the population claiming 'no religious affiliation' according to Hackett et al. (2012). We selected cases that illustrate a variety of modes of operation, diverse religious and cultural contexts, and varied approaches to conservation. As each of the cases pertains to the work of A Rocha organisations that share a 'basis of faith' (i.e. agreement with a brief theological statement of key beliefs) and core values that give prominence to cooperation, community, and cross-cultural work (A Rocha International, n.d.), we have assumed that the faith basis that motivates the described actions is the same in each case study. A Rocha's core motivation is an understanding that God imparts an intrinsic value to ecosystems and biodiversity, that the universe exists for God's purposes, and that human beings have a vocation to 'serve and preserve' their fellow species (Bookless, 2014).

For each case study, we examined reports provided by the respective local A Rocha organisations and conducted a semi-structured interview with a key member of staff. In one of the cases (USA), those staff members are also co-authors here. We defined the conservation value and impact of the work undertaken and explored the local cultural context, the prevailing religious adherence, and the range of actors engaged in the work. Three of the four cases (the UK, the USA, and the Netherlands) were in strongly secularised countries, while a fourth (Portugal) was less strongly so. Portugal was also distinguished by being a predominantly Roman Catholic societal context in contrast to the other three cases that were predominantly Protestant.

## Case Studies

### *Partners in Action, A Rocha UK*

The UK has a strong Christian heritage, although the proportion of practising Christians is low (Saghal, 2018) (Figure 15.1). The largest denomination in the UK remains the Anglican Church, the Church of England. Their records reflect the wider trend of religious decline, with most key measures of attendance falling by between 15% and 20% from 2009 to 2019 (Church of England, 2020). Nevertheless, regular worshippers in the Church of England comprise over one million people (Church of England, 2020).

A Rocha UK (ARUK) began in 2001, and now manages a range of conservation programmes working with churches and individual Christians. Their Partners in Action programme forms a collaborative network of Christian institutional land managers demonstrating active care for the natural world. There

**FIGURE 15.1** Children planting potatoes in a raised bed at Hazelnut Community Farm, a member of the Partners in Action programme. Photo Credits: Hazelnut Community Farm.

are 16 existing partners accounting for *c 600* ha of land, varying from substantial areas within National Parks and Sites of Special Scientific Interest, to smaller community-managed urban spaces, and with over 30 potential partners seeking to join. Partners are varied by geography and setting; they include youth outdoor centres, community groups, and residential centres. Each of these partners selects a conservation focus from ARUK's 'Target 25' list comprising a selection of species and habitats drawn from the UK list of threatened birds (Eaton et al., 2015), the UK Biodiversity Action Plan list of priority habitats and species (BRIG, 2007), and the Natural Environment and Rural Communities Act (Parliament, 2006). Survey data are fed into the UK's National Biodiversity Network (National Biodiversity Network Trust, 2021).

One of ARUK's most recent Partners in Action is Hazelnut Community Farm, which joined in September 2020. Hazelnut Community Farm is based on disused land in Lockleaze, Bristol. It began in May 2019 as a result of local churches across Bristol reviewing how they could incorporate a theological understanding of caring for the environment into their church community. Hazelnut Community Farm is sited on land owned by the local council and land privately owned by a local shop, which has ensured that their work has been in partnership with the wider community from the outset. They continue to engage the community around them through urban agriculture, food production, and waste management, gifting half of their produce to those outside the community. Their current conservation focus is promoting native wildflowers and their associated pollinators (J. White pers. comm., 21 February 2021).

## Local Volunteer Action Groups, A Rocha Netherlands

The Netherlands is one of the most secularised countries in the world, with *c.* 42% of the population religiously unaffiliated (Hackett et al., 2012). A Rocha Netherlands (ARNL) began in 2002 out of a desire to resource existing conservation organisations that were struggling with a lack of volunteers. ARNL has four main activities: practical conservation, environmental education, inspiring churches, and theological teaching. This is grounded in place-based action accessible to community members that want to volunteer.

ARNL started by attracting volunteers from churches to support the implementation of management plans by organisations or government entities with management authority over a particular area, and now has 14 projects around the country. In some cases, ARNL has been asked to take on the active role of management. Typically, one local church takes on leadership at a site but participation is open to other churches and to those with no church association. It is primarily Protestant churches that approach ARNL with an interest in starting a local project.

One of the fourteen projects is in the city of Zwolle, located near the IJssel River. The Scheller- and Oldeneler Buitenwaarden is a government-owned 63 ha site jointly managed by the ARNL local group and an organic farming cooperative, Uw Stadsboer. It is part of a site designated under the EU Birds and Habitats Directives Natura 2000 European Ecological Network. One of

**FIGURE 15.2** Black Tern parent and chick on a nest raft. Photo Credits: Margreet van Middelkoop.

the conservation objectives for the area is to develop it into herb and fauna-rich grassland, through a combination of extensive grazing and, in stages, mowing and removal. This includes working towards Natura 2000 management goals for the Corncrake *Crex crex* by restoring and protecting habitat. Nest rafts were utilised to provide breeding habitat for the Black Tern *Chlidonias niger*[2] (Figure 15.2). An Otter *Lutra lutra* shelter was built to support the local population of this Near Threatened species. The ARNL group began by offering themselves to the governing agency for voluntary work, and, as experience and trust grew, were given greater responsibility (E. Messelink pers comm., 18 January 2021).

Many of the core group members come from one local church congregation, but they cooperate with other groups such as the Scouts, as well as local residents (E. Messelink pers comm., 18 January 2021). The project is marked by the local group's enthusiasm for enjoying the beauty of the IJssel River area, including observing and reporting species such as otters and beavers and organising activities such as art, photo exhibitions, educational walks, regular partner/local citizen meetings, and species conservation projects.

The ARNL team see their faith as motivating all that they do. However, they also see their role as challenging the Dutch church to regard a commitment to conservation as a Christian imperative. The practical and theological works of ARNL are at the forefront of inspiring this change. They also exercise a role within the wider conservation community to encourage hopefulness (E. Messelink pers comm., 18 January 2021).

## Field Study Centre, A Rocha Portugal

Roman Catholicism has been the main religion in Portugal for centuries, but adherence in recent decades shows a shift towards secularism. Nonetheless, Portugal remains the most religiously active Catholic country in Western Europe, with roughly one in four attending church weekly (Saghal, 2017).

Within this context, A Rocha Portugal (ARP) was founded in 1983, and a Christian field study centre for conservation in the Algarve region was established near the Alvor Estuary on the outskirts of Portimão in 1987. Although the initial activities focussed on bird conservation, this expanded to include plants, butterflies, moths, microplastics monitoring, and water quality monitoring. Today, the centre actively monitors *c.1455* ha of land and estuary and hosts conservation professionals, volunteers, and students from around the world, many of whom view their participation in terms of their faith convictions to care for the environment.

The Alvor Estuary is an internationally important wetland site, listed under the International Convention on Wetlands (Ramsar Convention), and as a Special Area of Conservation under the EU Habitats Directive Natura 2000 European Ecological Network. The site is listed for its range of coastal habitats, importance for migratory birds, and significant shellfishery (Mateus et al., 2016). It is also a site of intense anthropogenic pressure, principally agriculture, building development, land reclamation, coastal engineering, aquaculture, and shellfishing (Mateus et al., 2016).

**FIGURE 15.3** Quinta da Rocha, the conservation site under threat of development. Photo Credits: A Rocha International.

Through regular monitoring of proposed land development plans in the Alvor, ARP detected when a new landowner and property development company attempted to change the protected status of coastal lagoon habitat in Quinta da Rocha (Figure 15.3) in order to avoid costly development mitigation requirements. ARP's long-term fieldwork showed that this area held rare plants protected under the European law, including the Portuguese endemic toadflax *Linaria algarvinia* and the Camphor thyme *Thymus camphoratus*. ARP filed an administrative court case against the landowner in 2007 that it won in 2018 after multiple appeals by the owner. In 2009, the Attorney General also filed criminal court cases against the property development company and the company managers that concluded in favour of the Attorney General in 2020 and 2017, respectively. As a result, all the coastal lagoon habitats of the Alvor Estuary are more robustly protected amidst heavy development pressure. By using both the administrative and the criminal courts, the ARP and the Attorney General have begun to establish case law that would prevent companies from reducing conservation requirements merely by altering the mapping of key habitats (M. Felguieras pers. comm., 25 January 2021).

## Field Study Centre, A Rocha USA

The USA is predominantly Christian: 68% of citizens claim adherence to Evangelical Protestantism, Mainline Protestantism[3] (Lantzer, 2012), or Catholicism; about 23% describe themselves as atheist, agnostic, or 'nothing in particular'; and only 6% identify with another named religion (Smith et al., 2015). Though some

**FIGURE 15.4** A Rocha USA staff and volunteers searching for nurdles within Canaveral National Seashore. Photo Credits: A Rocha USA.

Evangelical individuals, nonprofits, church congregations, and denominations have embraced the notions of creation care and/or environmental justice (Stuart et al., 2005), many more Evangelicals in the USA do not recognise nor prioritise environmental conservation or action (Zaleha and Szasz, 2015). Within this milieu, A Rocha USA (ARUSA) occupies a distinct niche conserving biodiversity through national programmes and local projects using both theological and scientific lenses to highlight species, habitats, and threats. ARUSA engages mainly Protestants, both mainline and Evangelical. We focus on the marine conservation work for this case study.

ARUSA's marine conservation work (A Rocha USA, n.d.) seeks to integrate four major themes shared by A Rocha's global Marine Conservation Programme (A Rocha, n.d.): Biodiversity and Beauty, An Ocean of Hope, Livelihoods and Wellbeing, and Climate Change. The Florida Project began in 2017 with an emphasis on plastic pollution, examining microplastics and studying the spatial and temporal distribution of plastic pellets, called nurdles in Canaveral National Seashore, a US Federal Protected Area (Figure 15.4).[4] The methods utilised were chosen to be amenable to community scientists, involving individuals, schools, and churches in collecting data that have been contributed to global databases and reported to the US National Park Service on microplastic pollution. A collaboration with Arkansas-based Williams Baptist University, an institution affiliated with the Southern Baptist Convention, trained university students on faith and conservation issues, culminating in an STEM research experience field trip to the Florida Project study sites.

These experiential programmes for locals and visitors to the Florida Project have sought to integrate faith and conservation science in order to increase participation in conservation activities. They have been successful in recruiting local high school students of varied religious backgrounds to participate, aided by state-based funding initiatives to incentivise volunteerism. Several local churches and a Christian school have participated in plastic pollution monitoring and beach cleanups, although widespread interest from local churches still remains low (R. Sluka pers. comm., 20 January 2021).

The Florida Project has also provided volunteers donating hundreds of person-hours to support activities of other local conservation organisations and universities, including a horseshoe crab conservation project and mangrove and oyster restoration in the Indian River Lagoon. This cooperation has sought to enhance the work of others and to provide an opportunity for Christians to volunteer, exploring faith and conservation through service and action (R. Sluka pers. comm., 20 January 2021).

## Discussion

'The West' is not a secular monolith, but rather a patchwork of societies with varying religious and environmental histories that influence the present approach to conservation. A Rocha organisations, though united in a shared Christian identity, nonetheless move towards a conservation impact in diverse ways. The Christian conviction of ARP towards long-term commitment to protect and restore an exceptional site provided the grounds for them to lead a science-informed legal challenge against commercial development tactics. This has resulted in hope for establishing legal precedent for habitat protection that will extend far beyond this one site. By contrast, ARNL's Christian motivation to serve, rather than compete, in the conservation space led them to cooperate and resource the habitat management work of others with volunteer workers. In the USA, the hope for experiential programmes to offer personal reflection and integration of faith and conservation convictions manifests in community science opportunities within existing monitoring programmes. In the UK, the Partners in Action programme of ARUK is not just seen as a means to an end, but as a tool for social innovation, economic inclusion, and political and spiritual transformation.

Although the conservation impact within some of our cases may seem modest, they are genuine and coherent. The much discussed potential for faith communities to bring something new and significant to conservation is being realised even if not always at a great scale. Despite early misgivings about a faith-based organisation engaging in scientific research and suspicions about data quality, it was ARP's monitoring data that proved central to the legal defence of an internationally important wetland under the European law. In the Florida

Project, community science, microplastics monitoring and research, and habitat restoration have happened because of ARUSA's Christian convictions. In the Netherlands, ARNL's conservation of species and habitats has come out of their Christian hope, manifested in a sense of wonder, dedication to volunteerism, and clear connection to theological underpinnings, distinguishing them from the secular milieu. And *c.* 600 ha of land in the UK has been managed for biodiversity conservation because ARUK forged long-lasting relationships with Christian landowners and provided a framework for faith-based partnership and community.

## Adaptation to the Local Context

In each case study, the local context influenced the particular project, resulting in tailored approaches to address environmental needs. While we have shown that the mode of engagement of faith-motivated conservation work can be varied, and that there can be many common characteristics to the ways such work proceeds, it is also apparent that the local socio-cultural context does influence the opportunities present. The prospects for working with Christian land managers are quite different in the UK compared to Portugal, for example. In the Netherlands, where the conservation space is more crowded, the concern to avoid competition and detracting from the work of others led ARNL to a model of service. In the USA, the Florida Project met the needs and desires of Christian university students from around the country to complete STEM research in a faith and conservation context. More than the other cases, ARP has struggled to connect meaningfully in their local context. Local churches, NGOs, universities, and government representatives were initially wary of an organisation with a faith basis conducting scientific work, and the centre's data were questioned by some because it originated from a Christian NGO. Meanwhile, local churches and other faith-based groups were sceptical of their strong science focus. Engagement by the local church in A Rocha's work in Portugal remains modest but partnerships with research institutions, local schools, and churches have grown.

## The Importance of Community

Faith has been intentionally integrated into A Rocha's conservation work, drawing in people who might otherwise not be engaged. For the field study centre model used in Portugal and the USA, this is worked out in a holistic experience of community living, where all are welcomed and there is an expectation that faith motivations for the conservation work will be explored and shared. In the Netherlands and the UK, the community concerned has substantial definition beyond the project work itself, often as a local Christian church community, which seeks to act together on a common task. For example, Partners in Action in the UK embeds conservation work into a community context, understanding that environmental outcomes cannot be achieved in isolation from other aspects of life, including expressions of faith.

## The Influence of Partnership

In all our cases, partnerships provided opportunities for cooperation with people and organisations of varying beliefs and religious adherence, from environmental NGOs to university faculty and students to governments and farm cooperatives. As a result, A Rocha organisations added value both to their own initiatives and to local actors striving to achieve similar conservation outcomes while minimising competition. Long-lasting connections forged by existing communities within their local area were evident especially in three of the cases, namely the Netherlands, the UK, and the USA. Partnership has allowed ARUK to create a network of Christian landowners that chose and conserved species and habitats identified as in need of monitoring and protection; by working in a partnership model they achieved ecological action at a broader scale and for a wider community than would have been possible through their own activities alone. In the Netherlands, co-management of the Scheller- and Oldeneler Buitenwaarden site was shared between ARNL and an organic farming cooperative, enabled by partnering with local churches to supply volunteers for conservation activities. In the USA, the Marine Conservation programme supported the monitoring and restoration of the Indian River Lagoon in partnership with a diversity of local universities, government entities, and schools.

## Conclusion

Our case studies clearly challenge assumptions that faith is irrelevant to conservation in Western secular societies. Since its foundation in Portugal in 1983, A Rocha has grown in size and influence as Christians have become convinced of the importance of biodiversity conservation. Significant potential for scaling up these efforts exists, but given the dispersed, diverse, and local nature of Christian congregations, it is likely that this will mostly go unattributed and be of a form difficult to generalise.

Western Christians, and particularly young people, are often passionate about environmental issues. A recent survey revealed that 90% of Christian teenagers in the UK regard tackling climate change as a very high priority; yet, less than 10% believe that their church is doing enough about it (Paveley, 2021). Although this survey was about climate, our experience suggests that many young people who care about climate change care equally about biodiversity loss, plastic waste, and animal ethics. Our cases also stand as a challenge to a prevailing narrative that Evangelical Christians have a pietistic, other-worldly faith (Orr, 2005), or that they inevitably hold negative environmental attitudes. We thus end with a note of hope, that seeds of conservation planted now will germinate into a rich garden of faith-inspired biodiversity conservation.

## Notes

1 The Eco Church programme in England and Wales, launched in 2016, with over 3,200 churches participating by early 2021 is a good example: https://ecochurch.arocha.org.uk/
2 This short film documents work beginning on the Black Tern project. Watch A Rocha members installing nesting rafts in the Schellerwade – a pond near the IJssel River: https://vimeo.com/303517970
3 Mainline or Mainstream Protestant denominations include the Congregational Church, the Episcopal Church, the Evangelical Lutheran Church, the Presbyterian Church, the United Methodist Church, the American Baptist Convention, and the Disciples of Christ.
4 See the education section here: www.arocha.org/plastics-toolbox

## References

A Rocha. (n.d.) *Marine conservation*. Retrieved 9 March 2021 from https://www.arocha.org/en/work/scientific-research/marine-coastal/

A Rocha International. (n.d.) *Our values*. Retrieved 10 March 2021 from https://www.arocha.org/en/values/

A Rocha USA. (n.d.) *Marine Conservation Program*. Retrieved 9 March 2021 from https://arocha.us/marine

Alliance of Religions and Conservation. (2007). *Whales, turtles and timber: News from Indonesia* [archived report] University of Bath Library, Bath, England.

Armstrong, P. (2000). *The English parson-naturalist: A companionship between science and religion*. Herefordshire, England: Gracewing Publishing.

Bartholomew, I., & Bartholomew, E. P. (2012). *On earth as in heaven: Ecological vision and initiatives of Ecumenical Patriarch Bartholomew*. New York: Fordham University Press.

Barton, G. A. (2002). *Empire forestry and the origins of environmentalism* (Vol. 34). Cambridge, United Kingdom.

Bauckham, R. (2011). *Living with other creatures: Green exegesis and theology*. Melton Keynes: Authentic Media Inc.

Bell, C. R. (Ed.). (2016). *Creation care and the gospel: Reconsidering the mission of the church*. Hendrickson Publishers.

Bhagwat, S., & Palmer, M. (2009). Conservation: The world's religions can help. *Nature, 461*(7260), 37–37.

Bhagwat, S. A., Dudley, N., & Harrop, S. R. (2011). Religious following in biodiversity hotspots: Challenges and opportunities for conservation and development. *Conservation Letters, 4*(3), 234–240.

Bookless, D. (2014). Let everything that has breath Praise the Lord: The Bible and Biodiversity. *Cambridge Papers*. Cambridge, Jubilee Centre.

BRIG – Biodiversity Reporting and Information Group. (2007). *Report on the Species and Habitat Review to the UK Standing Committee*. Peterborough: JNCC.

Church of England. (2020). *Statistics for mission 2019*. London: Church of England Research and Statistics.

Eaton, M. A., Aebischer, N. J., Brown, A. F., Hearn, R. D., Lock, L., Musgrove, A. J., … Gregory, R. D. (2015). Birds of conservation concern 4: The population status of birds in the United Kingdom, Channel Islands and Isle of Man. *British Birds, 108*, 708–746.

Francis, P. (2015). *Laudato si*. Vatican City: Vatican Press, May, 24, w2.

Graves, D. (1996). *Scientists of faith: Forty-Eight biographies of historic scientists and their Christian faith* (pp. 80–83). Grand Rapids, MI: Kregel Publications.

Hackett, C., Grim, B., Stonawski, M., Skirbekk, V., Potančoková, M., & Abel, G. (2012). *The global religious landscape.* Washington, DC: Pew Research Center.

Harrison, P. (1999). Subduing the earth: Genesis 1, early modern science, and the exploitation of nature. *The Journal of Religion, 79*(1), 86–109.

Lantzer, J. (2012). Mainline Christianity: The past and future of America's majority faith. New York; London: NYU Press.

Lodge, D. M. (2006). *Religion and the new ecology: Environmental responsibility in a world in flux.* Notre Dame, IN: University of Notre Dame Press.

Marsh, G. P. (1864). Man and Nature; or. *Physical Geography as Modified by Human Action, 43,* 149–56.

Mateus, M., Almeida, D., Simonson, W., Felgueiras, M., Banza, P., & Batty, L. (2016). Conflictive uses of coastal areas: A case study in a southern European coastal lagoon (Ria de Alvor, Portugal). *Ocean & Coastal Management, 132,* 90–100.

Mayr, E. (1982). *The growth of biological thought: Diversity, evolution, and inheritance.* Cambridge, MA: Harvard University Press.

Moberg, D. O. (1977). *The great reversal: Evangelism and social concern.* Eugene, OR: Wipf and Stock Publishers.

Muir, J. (2017). *A thousand-mile walk to the gulf.* Houghton, MI: Mifflin Harcourt.National Biodiversity Network Trust. (2021, January 19). Retrieved 15 March 2021 from http://www.nbnatlas.org/

*Natural Environment and Rural Communities Act 2006,* c. 3. https://www.legislation.gov.uk/ukpga/2006/16/contents

O'Connor, E. (2011). Gau Island Feral Pig Workshop and follow-up report. *Suva, Fiji: Nature Fiji–Marqueti Viti.*

Oriel College. (n.d.) *Gilbert White (1720–1793).* Retrieved 10 March 2021 from https://www.oriel.ox.ac.uk/gilbert-white-1720-1793

Orr, D. W. (2005). Armageddon versus extinction. *Conservation Biology, 19*(2), 290–292.

Paveley, R. (2021, February 12). Church inaction on climate could cause young people to leave church, says report. *Church Times.* Retrieved from https://www.churchtimes.co.uk/articles/2021/12-february/news/uk/church-inaction-on-climate-could-cause-young-people-to-leave-church-says-report

Pepper, D. (1984). *The roots of modern environmentalism* (Vol. 6). Oxfordshire: Routledge.

Pew Research Center. (2018, October 29). *Eastern and Western Europeans differ on importance of religion, views of minorities, and key social issues.* Washington, DC: Pew Research Center.

Pinchot, G. (1910). *The fight for conservation.* Doubleday, 95–6.

Pew Research Center (2017). *Five centuries after reformation, Catholic-Protestant divide in Western Europe has faded.* Washington, DC: Pew Research Center.

Saghal, N. (2018, May 29). 10 *Key findings about religion in Western Europe.* Washington, DC: Pew Research Center.

Santmire, H. P. (1985). *The travail of nature: The ambiguous ecological promise of Christian theology.* Minneapolis, MN: Fortress Press.

Smith, G., Cooperman, A., Mohamed, B., Martinez, J., Alper, B., Sciupac, E., & Ochoa, J. (2015). *America's changing religious landscape.* Washington, DC: Pew Research Center.

Stonawski, M., Skirbekk, V., Hackett, C., Potančoková, M., Connor, P., & Grim, B. (2015). Global population projections by religion: 2010–2050. In *Yearbook of International Religious Demography 2015* (pp. 99–116). Leiden: Brill.

Stuart, S. N., Archibald, G. W., Ball, J., Berry, R. J., Emmerich, S. D., Evans, D. M., ... Whitten, T. (2005). Conservation theology for conservation biologists: A reply to David Orr. *Conservation Biology, 19*, 1689–1692.

Weidensaul, S. (2018, June 13). A Galaxy of Falcons: Witnessing the Amur Falcon's Massive Migration Flocks. *Living Bird magazine, 37*(3).

White, L. (1967). The historical roots of our ecologic crisis. *Science, 155*(3767), 1203–1207.

Zaleha, B. D., & Szasz, A. (2015). Why conservative Christians don't believe in climate change. *Bulletin of the Atomic Scientists, 71*(5), 19–30.

# 16

# *HIMA* AS A PROTECTED AREA – OPPORTUNITIES AND CHALLENGES IN THE 21ST CENTURY

*Rianne C. ten Veen*

## Introduction

The Islamic view on the environment and, in particular, the interaction between people and nature in achieving environmental conservation and sustainable use of natural resources, is clear, though not getting as much attention as it merits. For example, a '*Hima*' or Shariah-based conservation zone guarantees an area's continued protection. The system of *Hima* is the most widespread and longstanding indigenous/traditional conservation institution in the Middle East, and perhaps on Earth (Kilani et al., 2007). *Hima* (environmental protection), an Arabic word meaning 'inviolate zone' (boundary), is a system of environmental protection in Islam. *Himas* are one of the earliest recorded examples of successful land management. This idea of holding land in a reserved status appeared as a social practice and antedates Islam. It was later adopted by Islam and sanctioned by Shariah law to enforce application.

During the 20th century, political and socio-economic changes led to the deterioration of the *Hima* system (Gari, 2006). And more widely, sustainable systems of land use declined and so did the diversity of habitats, contributing to biodiversity loss (which even suffered a 'dangerous decline' according to the 2019 IPBES Global Assessment Report on Biodiversity and Ecosystem Services). It was the same case with Indigenous peoples' ways of life. Biodiversity loss is dangerous because biodiversity forms the basis for our existence, providing us with food, clean drinking water, climate adaptation and a buffer against disease. According to the IPBES report, and reports since, it is still possible to turn the tide of biodiversity loss, if solutions are considered and implemented. It could be hard, but, to quote Nelson Mandela: 'It always seems impossible until it is done'.

This chapter summarises the Islamic view on environmental conservation and sustainable use of natural resources and then reflects on the opportunities, but

DOI: 10.4324/9781003246190-20

also the challenges, in further implementing this concept of Islamic conservation in the 21st century.

## Islamic View on Environmental Conservation and Sustainable Use of Natural Resources

Before going into the opportunities and challenges of the *Hima* conservation system, and the Islamic view on environmental conservation and sustainable use of natural resources, it is relevant to share the objectives (*maqasid*) of Islamic *Shariah*, and of Islam more generally: the protection of life, the protection of dignity, the protection of property, the protection of health and the protection of religion (Auda, 2008). The sustainable protection of these logically requires a habitable Earth as a foundational element to be able to achieve these *Shariah* objectives. With increased climate change and higher prevalence and severity of, for example, heatwaves, droughts and floods (IPCC, 2021), the protection of life, health, property and dignity is clearly at risk.

According to Islamic teachings, humans have been appointed as caretakers of Creation (Quran 6:165): 'It is God Who has made you God's agents, inheritors of the earth: God has raised you in ranks, some above others: that God may try you in the gifts God has given you: for your God is quick in punishment: yet God is indeed Oft-forgiving, Most Merciful'. This different ranking is not in terms of difference in human worthiness, but in terms of different levels of skills, abilities and power to affect change. And this caretaking refers to all beings, not just humans, as shared in the verses: 'There is not an animal (that lives) on the earth, nor a being that flies on its wings, but (forms part of) communities like you' (Quran 6:38) and 'Indeed the creation of the heavens and the earth is greater than the creation of mankind; but most of mankind do not know' (Quran 40:57). That humanity might affect the Earth negatively had already been anticipated by the Quran: 'Corruption has appeared on land and sea because of (the evil) which human's hands have done, that God may make them taste a part of that which they have done, in order that they may return (to the straight path)' (Quran, 30:41). The idea here is that this 'corruption' is a blessing for humans – that they do not get confronted with the consequences of their deeds on the Day of Judgment (when it would be too late to go back and alter their deeds), but 'come back to the straight path' while still in earthly life. And since Creation was for everyone, and there's one humanity (only), humans are guided not to see disasters happening in other parts of the world as happening to 'them', but to 'us', as a test of how humans react (Aksa, 2020).

Ever since the earliest Arabic writings on medicine, treatises were dedicated to widespread illnesses or crowd diseases, including epidemics. Some of those works discuss the causes (including environmental pollution) and treatment; others deal with the treatment only. They cover subjects like air and water contamination, solid waste mishandling and environmental assessments of certain localities. Some of these written treatises from the 13th century (to indicate how far back this environmental interest goes) are by Al-Kindi, Qusta b. Luqa, Al-Razi, Ibn al-Jazzar, Al-Tamimi, Abu Sahl al-Masihi, Ibn Sina, Ali b. Ridwan, Ibn Jumay, Yaqub al-Israili, Abdullatif al-Baghdadi, Ibn al-Quff and Ibn al-Nafis. Some of these authors were merely copying the Greek theory of humours and miasma; others made original contributions to the field (Gari, 2002).

The term *Hima*, in Arabic, literally means a protected place or protected area. For rural communities living in the Arab world, the term holds connotations that appeal to their collective memory and hence when evoked, the term is not only readily recognised as familiar but also valued and triggers an air of acceptance and ease among communities living anywhere in the Arabian Peninsula or Arab Middle East. The tradition of *Hima* involves humans and nature coexisting harmoniously through the sustainable use and extraction of natural resources, and it promotes food, energy and water security.

Culturally, there are parallels to this linguistically, in terms referring to conservation institutions and prohibitions rooted in Islamic legislation. The word *haram* means wife and at the same time a sanctified inviolable zone such as *haram* Makkah and *haram* Madina. In the same vein, *haram* or *moharram* means prohibited and *harim* zones are greenbelts and easements – whatever is near developed land and pertains to its well-being, such as its pathways and watercourses, its rubbish dump, its square (Llewellyn, 2003:20). Few established systems of protected areas are known that have a history comparable with traditional *Himas* (Llewellyn, 2003).

Gari (2006) demonstrates how, in the late 1960s, Omar Draz, a Syrian Food and Agricultural Organization (FAO) advisor, was able to combine rangeland development, conservation and management by building upon small-scale protected zones (*Himas*) as the conceptual basis for a successful grazing cooperative system in Syria. The results were remarkable: the episodic disastrous impact of drought on livestock numbers was drastically reduced. The system was all based on convincing the Bedouin that all one was talking about was reverting to a traditional concept led by chieftains and *arafah* (dispute settlers) of the tribes in order to protect and preserve tribal rights and land capabilities (Manyena, 2016).

## Opportunities

It is increasingly recognised that environmental conservation is not possible without the active involvement of local communities (Reed, 2008). Conventional models of conservation tend to focus on centralised planning and decision-making processes, relying heavily on governments for conservation. While there has been some success in saving threatened species and protecting their habitats, this success is increasingly undermined by growing conflicts over natural resources and local communities' increasing opposition to government-led conservation efforts due to violations of their land rights. This has also been recognised by the International Union for Conservation of Nature (IUCN 2012). The cultural and spiritual significance of nature has been defined as the spiritual, cultural, inspirational, aesthetic, historic and social meanings, values, feelings, ideas and associations that natural features and nature in general have for past, present and future generations of people – both individuals and groups. IUCN guidelines on the cultural and spiritual significance of nature respond to a growing need to make conservation more inclusive, effective and socially just, by accommodating multiple worldviews – by treating natural and cultural heritage as interlinked and by suggesting ways for engaging and empowering all relevant groups and stakeholders in protected area design, governance and management. The guidelines also assist with creating common ground, resolving conflicts and implementing rights-based approaches that recognise human rights and legal pluralism (Verschuuren et al., 2021).

*Hima* is recognised as a Community-Based Natural Resources Management (CBNRM) System, which promotes sustainable livelihood, resource conservation and environmental protection for all humans (Walid and Hashemi, 2011). The subsequent part of this chapter discusses how the concept of *Hima* has been applied in different contexts.

### Case Study 1 – Indonesia

In Indonesia, a formal conservation approach has not been enough in terms of a response to today's environmental challenges. Since 20% of global emissions are caused by forest degradation from tropical rainforests (Pearson et al., 2014), an initiative to protect any natural forest would be a need. Considering the local religious prevalence, the Islamic *Hima* and Harim System were introduced for a new approach to nature conservation. The project initiated community awareness initiatives through Islamic teachings to contribute to conservation around Mt. Gede Pangrango National Park and Mt. Halimun and Salak National Park boundaries. The two parks constitute strongholds of biological diversity and contain endangered species from the Java region such as the Javan gibbon, Javan hawk-eagle and clouded leopard (Mangunjaya, 2011).

### Case Study 2 – Lebanon

The Society for the Protection of Nature in Lebanon (SPNL), established in 1984 as a national environmental NGO, works to raise awareness of environmental issues and protect natural areas in Lebanon for the purpose of protecting birds, nature and biodiversity in collaboration with municipalities and local people. SPNL recognises the importance of involving local communities in the conservation of their natural resources. It has succeeded in reviving the *Hima* approach. Upon the establishment of its

*Hima* sites, SPNL has promoted local conservation groups and Homat Al Hima (Hima Youth Nature Guardians), and empowered and supported them to protect and maintain nature, species and resources in the *Himas*. Thus, each *Hima* has a group of enthusiastic young protectors who aim to prevent illegal hunting and ultimately to conserve the ecosystem of the area. For more on the work of SPNL in relation to reviving the *Hima* approach, please see https://www.spnl.org/hima/

## Case Study 3 – Mizan/Pemba: Misali Marine *Hima*

Misali Island is a small (one-hectare) coral islet located off the west coast of Pemba Island, north of two main islands that make up the Zanzibar archipelago, off Tanzania. During the nineties, a foreign investor wanted to establish a tourist resort in Misali, but the local community successfully fought this off, because in addition to its benefit as a food source, Misali had religious connotations for local people. With the encouragement of the Islamic Foundation for Ecology and Environmental Sciences (IFEES), Misali was designated a marine *Hima*, the first of its kind (Kilani et al., 2007).

A nearby project is the *Hima* REDD+ project.[1] The forests of Zanzibar form part of the East Africa Coastal Forests Eco-region, one of the world's 200 biodiversity hotspots. Additionally, the forests harbour a wide range of other endemic plants, mammals, reptiles and birds, some only found in Pemba. These species have significant ecological value, and at least 35 of them are listed as threatened. The project focuses on delivering climate benefits that fall into two main categories: mitigation of carbon emissions through the preservation of forests and mangroves, and helping communities adapt to new, more sustainable land use practices to adapt to a changing climate. The project has helped facilitate the establishment of forest governance structures, demarcation of Community Forest Management Areas (COFMAs) and community patrolling, providing the resources necessary for local communities to sustainably manage their own forests. The project is abbreviated as HIMA from 'Hifadhi ya Misitu ya Asili ya jamii'.

## Case Study 4 – Iraq

The biodiversity and cultural integrity of the Tigris-Euphrates River Basin is jeopardised by water scarcity, inequitable allocation of water rights and a high risk of desertification. Dams and upstream water diversions in Turkey, Syria and Iran have reduced mean annual flows, resulting in water scarcity and impaired water quality throughout the watershed. The Mesopotamian Marshes, or *al Ahwar* in Arabic, are a culturalised landscape, consisting of a reciprocal relationship formed over thousands of years between Marsh Arab cultures (or *Ma'dan*) and the marshes, through traditional resource management. The Marsh Arabs are now becoming environmental refugees without land tenure (Young, 2011). 'Hima Mesopotamia' is an initiative aimed at protecting the Mesopotamian marshes of Iraq, and committed to nurturing the eco-cultural heritage of the Tigris Euphrates watershed through the following: outreach, coordination and capacity building; synthesis of scientific information, traditional and local knowledge systems; and to provide a forum for cultural and environmental information exchange (Stevens, 2013).

### Case Study 5 – Saudi Arabia

The Arabian Peninsula is the birthplace of the *Hima* and the region where the *Hima* has been most widespread. While some *Himas* were established and managed directly by central governments for their cavalry or for other purposes, most of them were established and managed by local communities but recognised by central governments. Most functioned as grazing reserves for restricted use by a village community, clan or tribe, which were set aside to allow regeneration as part of a grazing management strategy. In the 1960s, it was estimated that there were about 3,000 *Himas* in Saudi Arabia (Llewellyn, 2013). Nearly every village in the southwestern mountains of the country was associated with one or more *Himas*. Saudi Arabia's national strategy for the conservation of biodiversity, which was submitted in 2005 to the Convention on Biological Diversity, includes 10 mentions of *Hima*, and it advocates for their implementation (Abuzinada et al., 2005). According to Kilani et al. (2007), strengthening legal and policy frameworks is needed to enable existing *Himas* that are viable to gain legal recognition, to enable *Himas* that are not functioning well to be revived, and to establish new *Himas*.

## Challenges

Challenges in relation to strengthening the *Hima* system can be argued to be as follows:– Challenges in prioritisation of Islam-inspired nature conservation concepts

The challenges in prioritisation of religion-inspired nature conservation concepts among Muslims are often due to ignorance about this and/or also fear of being deemed extremist (Allen, 2016). Where there is understanding and interest, fundraising for Islam-inspired projects is challenging.– Challenges from mainstream environmental protection concepts

Mainstream environmental protection concepts are secular and although there is now inclusion of concepts such as traditional ecological knowledge (TEK), there is still significant decolonialisation needed.– Wider challenges in the prevalent economic paradigms

The development-oriented economic model puts huge pressures on local resources (Yang and Meyer, 2015). This, for example, becomes clear from the use of terms such as Least Developed Countries (LDCs) or 'developing country' where low levels of international trade are seen as core identifiers of these, with no use of the term 'overdeveloped country', which seems factually appropriate for countries consistently living above the Earth's carrying capacity. However, even the Sustainable Development Goals (SDGs) focus on economic growth (see specifically SDG8), even when in 2021 'Earth Overshoot Day' (EOD) was the 29th of July. EOD is the date when humanity's demand for ecological resources and services in a given year exceeds what Earth can regenerate in that year.[2] Also, globally, growth has not been decoupled from resource consumption and environmental pressures and is not likely to become so (Parrique et al., 2019; Hickel and Kallis, 2020; Wiedmann et al., 2020).

## Conclusion

A *Hima*, a sustainable, holistically and justly managed protected area, is not a new concept by any means, nor outdated. With the environmental challenges that the world faces, it should not be ignored – alongside a focus on the SDGs, there is plenty of scope for their implementation.

To complement secular approaches, and similar concepts as they may exist in other faith traditions or religions, *Himas* can play a significant role in protecting and restoring land. To be a *Hima*, it should be constituted by a legitimate authoritative body and be established in the Way of God, for purposes pertaining to public welfare. It should not cause undue hardship to local people and not deprive them of resources that are indispensable to their subsistence. Finally, it should realise greater actual benefits to society than detriments.

The opportunities are many, especially when all can see that ideas and concepts to protect areas are complementary, not intended to compete or be contradictory nor exclusionary. The fear of proselytising or promoting any particular faith would not be relevant where the prevalent existing religion is Islam (or other faith, in the case of applying a faith-inspired inclusive response pertinent to that particular area). At the same time, it is important to be realistic about pertinent challenges in a time of complex geopolitics.

## Notes

1 *Hima* REDD+ project in Zanzibar: https://www.terraglobalcapital.com/hima-hifadhi-ya-misitu-ya-asili-ya-jamii-redd-program-zanzibar-tanzania
2 https://www.overshootday.org/

## References

Abuzinada, A. H., Al-Wetaid, Y. I., & Al-Basyouni, S. Z. M. (2005). The National Strategy for Conservation of Biodiversity in the Kingdom of Saudi Arabia. Prepared and issued by: The National Commission for Wildlife Conservation and Development. Conservation of Biological Diversity, Riyadh, Saudi Arabia.

Agreement, P. (2015, December). Paris agreement. In *Report of the Conference of the Parties to the United Nations Framework Convention on Climate Change (21st Session, 2015: Paris).* Retrieved December (Vol. 4, p. 2017).

Aksa, F. I. (2020). Islamic perspectives in disaster: An alternative to changing fatalistic attitudes. *Jàmbá: Journal of Disaster Risk Studies, 12*(1), 2072–845X.

Al-Ajlani, D. (2006). Towards a twenty-first century Hima System: Natural resource and wildlife conservation policy in Saudi Arabia. *Master's thesis.* Beirut, Lebanon: American University of Beirut.

Allen, C. (2016). *Islamophobia.* London: Routledge.

Abuzinada, A. H., Al-Wetaid, Y. I., & Al-Basyouni, S. Z. M. (2005). The National Strategy for Conservation of Biodiversity in the Kingdom of Saudi Arabia. Prepared and issued by: The National Commission for Wildlife Conservation and Development. Conservation of Biological Diversity, Riyadh, Saudi Arabia.

Auda, J. (2008). *Maqasid al-Shariah as philosophy of Islamic law: A systems approach.* Herndon, VA: International Institute of Islamic Thought (IIIT).

Gari, L. (2002). Arabic treatises on environmental pollution up to the end of the thirteenth century. *Environment and History* (pp. 475–488). Cambridgeshire: White Horse Press.

Gari, L. (2006). A history of the Hima conservation system. *Environment and History, 12* (2) (May 2006): 213D28. Retrieved from http://www.environmentandsociety.org/node/3267

Hickel, J., & Kallis, G. (2020). Is green growth possible? *New political economy, 25*(4), 469–486.

IPCC. (2021). Climate change 2021: The physical science basis. *Contribution of Working Group I to the Sixth Assessment Report of the Intergovernmental Panel on Climate Change.* Cambridge University Press.

IUCN. (2012). Motion 122. *Promoting and supporting community resource management and conservation as a foundation for sustainable development.* Adopted at the 5th IUCN Congress, Jeju, Republic of Korea, Sept. 2012.

Kilani, H., Serhal, A., & Llewellyn, O. (2007). Al-Hima: A way of life. *IUCN West Asia Regional.*

Mangunjaya, F. M., & McKay, J. E. (2012). Reviving an Islamic approach for environmental conservation in Indonesia. *Worldviews: Global Religions, Culture, and Ecology, 16*(3), 286–305.

Mangunjaya, F. M., & Praharawati, G. (2019). Fatwas on boosting environmental conservation in Indonesia. *Religions, 10*(10), 570. https://doi.org/10.3390/rel10100570

Mangunjaya, F. M., & McKay, J. E. (2012). Reviving an Islamic approach for environmental conservation in Indonesia. *Worldviews: Global Religions, Culture, and Ecology, 16*(3), 286–305.

Mangunjaya, F. M. (2011). Introducing the Islamic Hima and Harim system as a new approach to nature conservation in Indonesia Phase II. Retrieved from https://ruffordorg.s3.amazonaws.com/media/project_reports/5750-2%20Progress%20Report.pdf

Manyena, B. (2016). After Sendai: Is Africa bouncing back or bouncing forward from disasters? *International Journal of Disaster Risk Science, 7*(1), 41–53.

Narayanan, Y. (2013). Religion and sustainable development: Analysing the connections. *Sustainable Development, 21*(2), 131–139.

Parrique, T., Barth, J., Briens, F., Kerschner, C., Kraus-Polk, A., Kuokkanen, A., & Spangenberg, J. H. (2019). Decoupling debunked. *Evidence and arguments against green growth as a sole strategy for sustainability. A study edited by the European Environment Bureau EEB.*

Parvaiz, A. (2009). Reorienting Islamic theology towards environmental consciousness & sustainability–A model for incorporating environmental education in Islamic Education. *Religion and Sustainable Development: Opportunities and Challenges for Higher Education, 46*, 151.

Pearson, T. R., Brown, S., & Casarim, F. M. (2014). Carbon emissions from tropical forest degradation caused by logging. *Environmental Research Letters, 9*(3), 034017.

Plieninger, T., Bieling, C., Fagerholm, N., Byg, A., Hartel, T., Hurley, P., … Huntsinger, L. (2015). The role of cultural ecosystem services in landscape management and planning. *Current Opinion in Environmental Sustainability, 14*, 28–33.

Praharawati, G., Mangunjaya, F. M., Saragih, H. M., Firdaus, A. Y., Mulyana, T. M., Ilmi, F., … Sinaga, E. (2021). A model of religious moral approach for Peatland ecosystem restoration in Indonesia. *Jurnal Manajemen Hutan Tropika, 27*(2), 132.

Ramadan-Jaradi, G. (2016) *Impact of conservation efforts on the Syrian Serin at Hima Anjar.* Report. SPNL. Beirut-Lebanon.

Reed, M. S. (2008). Stakeholder participation for environmental management: a literature review. *Biological Conservation, 141*(10), 2417–2431.

Sayem, M. A. (2021). Islam and environmental ethics. *Islamic Studies, 60*(2), 157–172.

Serhal, A. (2016). *Homat Al Hima Guideline Manual.* Report-publication. SPNL. Beirut-Lebanon. Retrieved from http://www.spnl.org/homat-al-Hima-guideline-manual/

Serhal, A. (May 2019). Hima for peace: Actions speak louder than birds. *Parks*, 10.2305/IUCN.CH.2019.PARKS-25-1AS.en; Retrieved from https://parksjournal.com/wp-content/uploads/2019/05/PARKS-25.1-Serhal-10.2305-IUCN.CH_.2019.PARKS-25-1AS.en_.pdf

Shaikh, S. A. A., & Ismail, A. G. (2015, October). Islamic approach to environmental sustainability: Review of worldview, philosophy & teachings. In *IAIS Conference Proceedings Islam, Science & Sustainable Development*; MPRA Paper 68747, University Library of Munich, Germany.

Stein, S. A. B. R. I. N. A., & Walch, C. (2017). *The Sendai framework for disaster risk reduction as a tool for conflict prevention.* New York City: The Social Science Research Council.

Stevens, M. L. (2013). Hima Mesopotamia: Community generated conservation in the Tigris Euphrates Watershed. In *International Workshop: Towards an implementation strategy for the human integrated management approach governance system*; undated (p. 220).

Suleiman, M. M. K., Saleh, W., Hashemi, M., & Bhat, N. R. (2013). *International workshop: Towards an implementation strategy for the human integrated management approach governance system.* Kuwait: Kuwait Institute of Scientific Research.

Tomalin, E., Haustein, J., & Kidy, S. (2019). Religion and the sustainable development goals. *The Review of Faith & International Affairs, 17*(2), 102–118.

Verschuuren, B. (2006, October). An overview of cultural and spiritual values in ecosystem management and conservation strategies. In *International conference on endogenous development and bio-cultural diversity* (pp. 299–325), Series on Worldviews and Sciences (6), WUR, Leusden (NL).

Verschuuren, B., Mallarach, J. M., Bernbaum, E., Spoon, J., Brown, S., Borde, R., … Lee, E. (2021). Cultural and spiritual significance of nature. *Guidance for protected and conserved area governance and management. Best Practice Protected Area Guidelines Series* (32).

Walid, S., & Hashemi, M. (2011). Principles of Hima Governance: A Community-Based Natural Resources Management (BNRM) System. In W. Saleh & M. Hashemi (Eds.), *The Governance of Hima: A Community-Based Natural Resources Management (BNRM) System*. Policy Proposal presented at the World Justice Forum 111, Palau De Congresos De Catalunya, Barcelona, Spain, June 2–23.

Wiedmann, T., Lenzen, M., Keyßer, L. T., & Steinberger, J. K. (2020). Scientists' warning on affluence. *Nature Communications, 11*(1), 1–10.

Yang, W., & Meyer, K. E. (2015). Competitive dynamics in an emerging economy: Competitive pressures, resources, and the speed of action. *Journal of Business Research, 68*(6), 1176–1185.

Young, G. (2011) *Return to the marshes: Life with the Marsh Arabs of Iraq*. London: Faber & Faber.

# 17

# BIRDS AS BRIDGES BETWEEN RELIGIONS AND PEOPLE

*Mark Coreth, Yossi Leshem and Alexandre Roulin*

## Introduction

Through their ability to fly, their beauty and songs, birds are symbols of freedom for people across the world providing inspiration for all religions. Their wondrous migrations beyond political borders and continents have similarly inspired people throughout human history.

Since the founding of the Abrahamic or monotheistic religions, Christianity, Islam and Judaism, the Middle East has been a contested territory. The region's geopolitical location, at the intersection between the three continents of Asia, Europe and Africa, alongside its unusual concentration of holy places, including the Holy City of Jerusalem and the Temple Mount, has brought to the region countless wars, crusades and occupations that have passed deep-seated resentments from generation to generation up to the present day. Yet, the Middle East's location in this tri-continent junction has turned it into a bird migration "bottleneck", with global significance, as 500 million birds pass over it twice a year. For some species, such as The Lesser Spotted Eagle *Clanga pomarina*, Levant Sparrowhawk *Accipiter brevipes* and The Great White Pelican *Pelecanus onocrotalus*, the entire global population migrate through the region (Leshem & Yom-Tov, 1996).

Despite its history of conflict, the Middle East's focus for bird migration makes it a place where shared delight in, and concern for, birds might also act to mediate between the three Abrahamic religions. Here then is an opportunity for cooperation between nature conservationists, educators, scientists, farmers and faiths.

The potential for peacebuilding through a shared love of birds has long been recognised. In 1994, in his first visit to the Middle East, the Dalai Lama spoke before about 7,000 bird and nature lovers, emphasising the connection between nature, bird migration and the vital importance of connecting people and religions. In 2019, an eminent group spanning the principal religious traditions of

DOI: 10.4324/9781003246190-21

the Middle East met with Pope Francis in the Vatican to share a vision to unite all people on earth through a shared concern for the well-being of nature in general, and birds in particular. Recognising the Pope's message of universal human fraternity, the group included Professor Yossi Leshem, General Mansour Abu Rashid, Professor Alex Roulin and a Palestinian partner.

The story we wish to share emphasises the extraordinary work of uniting people through nature already undertaken by so many people reinforced by the world's great religious leaders. It is a story of collaboration between people from all the Abrahamic faiths, and one which could never have taken place without the close teamwork between the authors, General Mansour Abu Rashid, the faith communities and nature itself. We present our story in three parts.

## The Tree of Hope – Mark's Story

As a wildlife sculptor, I was asked by the CEO of the St John of Jerusalem Eye Hospital Group to consider creating a piece of sculpture to go in the hospital's garden in Muristan. The garden is situated in the very centre of the Old City of Jerusalem where the four Quarters meet, and beside which the Order of St John was planning to open an eye clinic that would serve all people of all nations and all religions within the city.

The Biblical symbolism of gaining sight at the sacred centre of three of the world's great faith traditions was not lost on me (Jeremiah 5:21; Matthew 13:13). However, the commission was not a simple one as the work needed to tell the history of Jerusalem from its beginning, through to the present day and indeed into the future. It had to tell of the passage of people, from those that occupied the city, to the many pilgrims from all corners of the earth. It had to be all-inclusive; it had in effect to fly over and above the historic issues of Jerusalem itself and so doing, like the promise of sight itself, spread a message of hope for the region and the world.

The City of Jerusalem is *the* Holy City, the centre of the world to the Abrahamic faiths. As Simon Sebag Montefiore aptly describes in his book *Jerusalem*, this is a city that exists twice – in heaven and on earth, a city that is as important to Jews as it is to Christians and Muslims. From the time of King David to the present, the city has seen beauty and turmoil (Montefiore, 2011). The walls have been built only to be destroyed and then rebuilt by its numerous invaders and defenders.

So how could I symbolise this turmoil and beauty, the bustle of people trading, living and visiting, in an eye-catching way? It was this dilemma that made the sculpture so difficult to conceive, but in equal measure such a fantastic challenge, and a challenge, when complete, to be placed in the centre of a uniquely dramatic city.

I visited Jerusalem for the first time in February 2016. I was taken on a tour of the Holy Land. We drove down to Jericho, saw the Dead Sea, the Sea of Galilee, Nazareth, Caesarea and down via Bethlehem back to Jerusalem. By the end of the day, I knew that I had the core element of the sculpture in mind. Wherever you go in the Holy Land, you see how the olive tree (*Olea europaea*) is central in the lives of all people. It is mentioned in all the great texts, where the olive branch is the symbol of peace. But when you look upon an old olive tree you see a tired, battle-worn trunk; age is evident in its form. The trunk is dead in places and yet as a whole is full of life, new leaves and fruit! It was, in essence, the perfect description of the city walls and an extraordinary symbol of the history of Jerusalem. But did it speak fully of the dynamic movement of the people?

Should you visit the Holy Land between March and June, you will witness the migration of the Common Swift (*Apus apus*, hereon swift). The swift flies from Sub-Saharan Africa through the Holy Land and onto both Europe and Asia. They have done so since before religious and political borders were ever formed. Swifts nest and breed in the Western Wall of the Temple and have done so since the days of the first Temple as well as all the Holy sites, and so I decided to make the canopy of my olive tree not out of leaves but out of swifts. The sculpture was now to be a tree full of life, which is itself the source of hope (Figure 17.1).

Circumstance took me to a garden belonging to the Order of Malta just outside of Bethlehem called Tantur. There I found a tree that gave me the balance of age and beauty. I approached a foundry in Netanya, who agreed to mould and cast a copy of this tree in bronze. I then made and cast one hundred and fifty swifts, which I placed on 'flight lines' and formed those into the canopy of the tree. The trunk speaks symbolically of the history of Jerusalem, the swifts of the

**FIGURE 17.1** Mark Coreth's Sisim Tree, the Tree of Hope, in the garden of the St John of Jerusalem Eye Hospital in Muristan, in the centre of the Old City of Jerusalem. Photo Credits: Baruch Gian.

movement of peoples over millennia. The process of making it was also symbolic, for Palestinian Muslims and Christians worked harmoniously with Jews, and with me, a Catholic, on the sculpture, which was sponsored by a Canadian Jew for a Christian Order that works for all people, though their patients are predominantly Palestinian.

Extending the symbolism, I have placed around the walls of Muristan and through the St John Eye Hospital in East Jerusalem several flights of swifts, in groupings of three: a prayer flag that Abrahamic faiths may fly together in unity and mutual respect. These groups are all part of the sculpture in Muristan. I hope to spread them much further afield in the future. Every swift that migrates to the far reaches of the world will stretch this message of hope for the Holy Land, understanding, tolerance and mutual respect between the Abrahamic faiths and peoples everywhere. Furthermore, we have placed nest boxes around the Muristan garden, enough for 15 pairs of swifts. The birds themselves will become a visible part of the sculpture and help spread that vital message of peace.

## The Barn Owl as the Dove of Peace – Yossi and Alex's Story

Beyond the migrations of birds, the biodiversity of Israel is also unusual. In a small country with a total area of 23,500 km2, about the size of New Jersey, you can observe 550 species of birds – both residents and/or wintering or summering migrants (Shirihai, 1996). This holds great significance for the Biblical obligation (Genesis 2:15), respected by the three Abrahamic faiths, to stewardship of the earth, and thus to conserve its rich biodiversity. It also has economic significance through ecotourism.

In the 1970s, we discovered that all the regions' countries had been using pesticides extensively against rodents. Together with hunting, these pesticides caused the deaths of thousands of migrating birds each year.

The intensive use of agricultural pesticides throughout Israel and across the region has posed a grave threat to the environment and its biological diversity, as well as to public health for over half a century. Many species of raptors went extinct in the Middle East as breeding species due to secondary poisoning resulting from rodenticides used in agricultural fields. The project we describe here had two objectives, therefore. First, a long-term significant reduction in the use of pesticides and changing farming habits to become more environmentally sustainable. Second, with the help of Barn Owls (*Tyto alba*) as biological pest control agents, to build a bridge between hundreds of Muslims, Jews and Christians. The strategy also rested on the fact that, as with migratory birds, Barn Owls know no political or religious boundaries (Roulin *et al.*, 2017).

Barn Owls are very common in the Middle East, especially in agricultural fields. Each pair of Barn Owls feeds on between 2,000 and 6,000 rodents per year, making them an efficient alternative to pesticides for the farmers (Kan *et al.*, 2014). As this species makes its nests in cavities, either natural or manmade, they readily accept specially designed nest boxes erected for the purpose. These considerations led to the initiation of a project some 40 years ago promoted by the Society for the Protection of Nature in Israel (SPNI), which by 2008 had evolved into a national project involving three government ministries.

To have an impact on pesticide use for migratory birds, the project had to cover the entire region, as improving the situation in only one country through which birds migrated would be of only partial value. In this spirit, and under the heading of "Barn Owls Know No Boundaries", we began to interest our neighbours in Jordan and the Palestinian Authority in the benefits of the project. Working together, the whole agricultural area was soon crisscrossed with nest boxes, enabling the Barn Owl population to expand and establish itself. Through this essential collaboration, the project constituted a platform for bridging between Jordanians, Palestinians and Israelis (Glausiusz, 2018; see Figure 17.2).

The project started in 1983 at Kibbutz Sde Eliyahu, in the Beit She'an Valley, on the Jordanian border. From 14 nest boxes initially, by 2008, it had involved the Israeli Ministries of Agriculture, of Regional Cooperation and of Environmental Protection. There are currently 5,000 nest boxes for Barn Owls (Bloch & Leshem, 2021). In 2002, the Jordanians, led by General Mansour Abu Rashid, the chairperson of the Amman Center for Peace and Development (ACPD), and the Palestinians joined the regional project that continues to this day despite the ongoing political tensions in the region. In 2015, Greece and Cyprus (where there are already 500 nest boxes) also joined the project. The Cypriot Minister

**FIGURE 17.2** Jordanian and Israeli farmers taking part in a joint seminar on Barn Owls. The picture also shows one of the 5,000 nest boxes erected across the entire region taking part in the project. Photo Credits: Hagai Aharon.

of Agriculture, Prof. Costas Kadis, has led the project in Cyprus, establishing a parallel system replicating that developed in Israel. In 2018, a seminar was held in Jordan, with the Maghreb countries, and Morocco decided to lead a similar project. The project is a unique model bridging cultures and people through cooperation between religions, farmers, academics, conservationists, government and public organisations.

In 2010, the ornithologist Prof. Alexandre Roulin joined the project as a representative of a neutral country. Switzerland supports the joint activity between the three religions. As a representative of a mainly Christian country, Prof. Roulin promoted, through an extensive activity with his Jewish and Muslim friends, meetings at the Geneva Centre for Security Policy, meetings with key government figures, and joined seminars in the Middle East. The seminars we held included participation by Muslim farmers from Jordan and the Palestinian Authority, Jewish farmers from Israel, scientists and nature conservationists from all countries. At one such meeting on June 28th, 2018, the Swiss president Alain Berset held a meeting with 150 ambassadors in Switzerland, on a farm with Barn Owl nest boxes. Both Muslim and Jewish representatives participated in presenting their mutual activities (Alon *et al.*, 2018). A further meeting took place, this time with Pope Francis in the Vatican, on May 11th, 2019, with two Muslim representatives from Jordan and the Palestinian Authority, respectively, a Jewish representative from Israel and a Catholic representative from Switzerland (Bloch & Leshem, 2021). Together, they presented the project to the Pope, displaying the cooperation between the three religions and the promotion of a healthier environment for both people and wildlife (see Figure 17.3).

The Pope showed great interest in the project, echoing the voice and sentiment of another spiritual world leader some 25 years earlier. In 1994, while Yossi was its CEO, the SPNI celebrated its 40th anniversary. During the festivities, the society hosted the Dalai Lama for a week. For 20 years, the Dalai Lama had tried to reach Israel as the first Dalai Lama, and perhaps the first Tibetan Buddhist, to make a pilgrimage to the Holy Land (bridging Eastern and Western spiritual traditions). While there, he gave a talk in the mountains of Eilat emphasising the connection between all religions, peace and nature conservation. Together with 'earth music' composed and performed by Paul Winter, the message was very strong and heard by an audience of 7,000 who came to the desert to experience the event. During his visit, the Dalai Lama travelled to nature sites and sacred places for Jews, Christians and Muslims in the Holy Land (Sagi & Leshem, 1994; see Figure 17.4).

Bringing the history up to date, Prof. Imad Cherkaoui, a representative from Morocco, joined the project in 2021, with plans to realise the project in Morocco in 2022, and further hopes to promote the project in Tunisia and Algeria (two countries with no political connections with Israel), all the way to Egypt. In October 2021, the same team who visited the Pope, joined by Prof. Cherkaoui, attracted much attention when they presented the project in the Dubai EXPO. The UAE Minister of Climate Change and Environment, H.E. Dr. Mariam

**FIGURE 17.3** May 11th, 2019, meeting with Pope Francis in the Vatican. Left: General Mansour Abu-Rashid (Jordan) Prof. Yossi Leshem (Israel) Pope Francis and Prof. Alex Roulin (Switzerland). Photo Credits: Vatican Media.

**FIGURE 17.4** March 1994, the Dalai Lama gave a lecture to 7,000 guests on top of Eilat Mountains, on religions, peace and the environment. Photo Credits: Amihud Naor.

bint Mohammed Saeed Hareb Almheiri, has decided to join the project. In the Muslim culture, the Barn Owl is a symbol for bad luck. An old Muslim proverb says: "Follow in the path of the Barn Owl and you will be met with destruction". Through this joint-religions initiative, Imams in Mosques and Rabbis in Synagogues have lectured on the subject, emphasising the Barn Owls' contribution to nature conservation and potential as a bridge to peace.

## Flight2Hope – Our Combined Stories

In 2017, Mark Coreth was asked to speak at the International Swifts Conference in Tel Aviv about the *Tree of Hope*. Inspired by the journey of the Swifts and still furthering the theme of the sculpture, it was here that he and Yossi decided that they should 'migrate' into Jerusalem alongside the Swifts. The conference attendees included General Baruch Spiegel from Israel and General Mansour Abu Rashid from Jordan, both retired and instrumental in the 1994 Peace Accord between Israel and Jordan. *Flight2Hope* was born. Our aim was to fly with nature over religious and political borders, alongside the Swifts and all migratory birds. Birds have migrated enormous distances in the delicate balance of life for thousands upon thousands of years, and long before religious and political borders ever existed. Crucially, however, we wanted to do this flight with crews from across the Abrahamic faiths.

Mark managed to enthuse seven fellow aviators, all friends with the same vision, to bring their aeroplanes and crews on a 2,500-mile journey from the UK. This journey was to be no easy feat as apart from the distance flown, much over long stretches of water, as he says, "we had to fly into a militarily sensitive part of the Middle East as our destination. Bureaucracy alongside security was understandably very tight".

Every aeroplane and every crew member has their own story to tell about their flights out and back. All were caught in foul weather at some point or another and yet at other times had the most glorious passages through the snowy Alps to Split and Dubrovnik, flying over the Greek Islands, the Corinth Canal and most important of all, the length of the Holy Land. Some aeroplanes were equipped to fly on instruments, whereas others, including Mark's, were not. Equally, some pilots were far more experienced than others. Two pilots had circumnavigated the world, while others had flown in Africa, the North Atlantic and Pakistan. Yet, *Flight2Hope* asked all to tackle some very challenging weather, with deep low-pressure systems causing high winds and weather fronts that grounded even the most capable pilots and aeroplanes. One of the pilots, Bill Hall, said:

> The best words to describe flying great distances in a light aircraft might be: unpredictable, unexpected, beautiful. Plans are laid, routes are plotted … and then the weather gods decide you do something completely different. Though, of course, this is part of the charm, one ends up going to places one would not have planned to go to. One thing is for certain, you will see beautiful skies and breathtaking views of the Earth.

How akin to the migration of the Swift.

Mark describes his experience:

> my aeroplane, a Jodel D1051, was made from wood and canvas in 1964. She is a basic stick and rudder aeroplane and not in any way equipped for inclement weather. I have fuel enough for 6 hours flying at 100 kts or 110 MPH. I can carry one passenger in the tight cockpit. Over the course of Flight2Hope I flew 58 hours and almost half of which was over water, the longest leg being 5.25 hours from Crete to Corfu battling strong headwinds. For my aeroplane and experience I had to ensure that I kept in sight of the ground or sea which at times meant long passages at low level and sometimes in marginal visibility. Three days were lost to weather and re-employed to much needed rest and recuperation as well as contingency planning.
>
> These challenges were magnified by travel through Greek airspace which is not used to catering for light aircraft and therefore often only has jet fuel but no 100LL petrol and they always have sky high landing and service charges! Often airports are closed to light aircraft on certain days or hours, but we managed to navigate all these hurdles both outward and homeward bound.
>
> That might paint a challenging picture, but we were flying through some extraordinarily beautiful country all the way from the United Kingdom, over the Massif Central to Avignon, Corsica, through the Apennines, the spine of Italy, to the olive groves in Italy's heel. Oh, the pleasure of approaching Corfu at dusk knowing that soon we would be outside with that cold beer, delicious squid dish and, as it turned out, the overhead scream of Swifts at nightfall! Early departures, long flights over Greek Islands one after another to Rhodes, an Island so full of history and the one-time home of the Knights of Malta, before flying on to Larnaka in Cyprus. Corfu to Rhodes took four hours, Rhodes to Larnaka a further three and a half, almost all over water. Others took different routes with equally beautiful, but sometimes challenging flying.
>
> At this stage I, along with one other of my aeroplanes, was running happily ahead of the storm which we knew was following and which we knew would cause a problem to those crews following behind. Even those aeroplanes that could be flown on instruments, had autopilots and whose crews were instrument rated were on some critical occasions stuck on the ground with 50+ mph surface winds and pouring rain. Weather and in particular the threat of icing conditions has to be respected.

Although the team had challenging flights with inclement weather systems hounding them most of the way, everyone made it on time to Haifa, Israel's most northerly port and subsequently to Ein Yahav in the south of the Jordan Valley, north of Eilat. We were joined by three Israeli guide pilots led by Eli Peretz from Ayit Aviation.

Apart from the regulatory need to have guide pilots with us, their reassurance and local knowledge were invaluable as we flew through what amounts to very tightly controlled military airspace. But just as importantly, they made for a wonderful expansion of the *Flight2Hope* team, as did the addition of a further Israeli aeroplane flown by Roy Ritter, which brought our flight to a total of nine aircraft.

*Flight2Hope* took place on April 2nd, 2019. So much had militated against success, from all the political and military issues in the Middle East to unusual weather patterns. But determination from the whole team (and Mark says, especially from Yossi and Eli Perez) ensured that we hit the start line on time and in great shape. Somewhat against the odds, the poor weather that threatened the mission broke as if by a miracle enabling us to fly up the Jordan Valley and through the Holy Land. We landed at Masada, 1,280 feet below sea level, before flying north around the Dead Sea, crossing into Jordan and on towards Amman before retracing our steps and achieving the ultimate goal of circling Jerusalem.

Seeing the golden dome of the Rock and all of Jerusalem below our wingtips was an emotional moment beyond most others. "What is more", says Mark, "as we circled the city, I had numerous sightings of flocks of Swifts flitting past my aeroplane as if to welcome us there". The flight alongside nature represented the link of nature to people, as well as highlighted the stories of people beneath and the cooperation between them. We noted, for example, below our wings the Palestinian, Israeli and Jordanian farmers working together on various agricultural projects, including the Barn Owls project headed by two crew members of *Flight2Hope*, General Mansour Abu Rashid and Professor Yossi Leshem.

The flight ended, symbolically and actually, at Muristan back with the *Tree of Hope*. We had flown with the Swifts, a symbol of people from across the ages, and most crucially with members of the Abrahamic faiths: Jews, Christians and Muslims. We had returned to the beating heart of the Old City of Jerusalem and of those faiths, the city that exists twice, in Heaven as it does on earth.

There was a palpable feeling of relief and gratitude on the evening of April 2nd. Gratitude that the skies had cleared and that all plans had come to be a reality. We had flown the Abrahamic faiths, crews, male and female, from motorbike mechanics to high-ranking generals, academics and even a NASA astronaut, Ricky Arnold. As they each spread their own unique stories, they will become ambassadors for the Swifts and the *Flight2Hope* message, sculptors helping to carve that message on high.

The *Flight2Hope* programme continued over the next days with a series of talks arranged by Yossi calling for and demonstrating unity through nature, people, mutual respect and cooperation. The talks were followed by a visit to Muristan and the *Tree of Hope*, before going to the Western Wall to witness the extraordinary sight of the Swifts' screaming flights at dusk. Team members went on to see two outstanding hospitals, both reaching out over the divides, the St John of Jerusalem Eye Hospital in East Jerusalem, and the Holy Family Hospital in Bethlehem, which is run by the ancient Order of Malta. We also saw Swift nest boxes above the school in Bethlehem, another show of cooperation through nature. Then came the time to fly home. For crews from the UK, this involved a further 2,500 miles, and again through more challenging Mediterranean weather. We were relieved on landing back to learn that all the aircraft had returned safely from this enthralling and unique journey.

## Conclusion

*Flight2Hope* was only made possible through the cooperation of some extraordinary partners to whom we would like to pay tribute, including General Mansour Abu Rashid from Jordan, Eli Peretz CEO of Ayit Aviation, Yossi's secretary Adi and many more working behind the scenes. Yossi was supported by the SPNI, the Hoopoe Foundation and the Amman Centre for Peace & Development (ACPD). Key to so much also was the St John of Jerusalem Eye Hospital Group whose Chairman and Hospitaler flew as crew alongside many mentioned above. They hosted the flight back in Muristan and the original commissioners of the *Tree of Hope*. Ricky Arnold and his wife Eloise from NASA joined *Flight2Hope* and are, like all, ambassadors to the energy henceforth.

The *Flight2Hope* message rings with urgency and importance in this world of mistrust and pandemic; there is a need for mutual respect and understanding between faiths, people and nature across the world. The difficulties encountered in organising and actioning *Flight2Hope* form a perfect metaphor for the challenges that face the world, but against all adversity it was a huge success, which we trust will bring hope to others. We believe that the relationship between nature and faithful people everywhere constitutes a strong bridge to mediation and mutual understanding, and for a healthier and more environment-friendly earth (Leshem & Ezer, 2013). In this spirit, we end with a quote from a remarkable peacemaker, friend and partner and one of the crews in our *Flight2Hope*, General (Ret) Mansour Abu Rashid, Chairman of ACPD:

> Today, the Middle East remains the greatest threat to international peace and security and despite the chronic failure to resolve the central conflict in the Middle East, we believe that it is possible to settle it. Flight2Hope, resembles that; people's genuine efforts to make it possible to construct a better future, to amplify the voices of those who are not afraid to build bridges and pave the way for peace, those who are courageous enough to reach out and reconnect the bonds of our common humanity.

## References

Alon, D., Aviel, S., Bloch, I., Charter, M., Haviv, R., Leitner, G., … Shimshon, I. (2018). *Barn Owls and Kestrels as pest control agents in agriculture – Summary report 2008–2017* (In Hebrew available at: http://bit.ly/2Cx1kl9).

Bloch, I., & Leshem, Y. (2021). *Barn Owls as pest control agents in agriculture – Summary report 2020* (In Hebrew available at: https://bit.ly/2TDnyQR).

Glausiusz, J. (2018). Owls for peace: How conservation science is reaching across borders in the Middle East. *Nature, 554*(7690), 22–23.

Kan, I., Motro, Y., Horvitz, N., Kimhi, A., Leshem, Y., Yom-Tov, Y., & Nathan, R. (2014). Agricultural Rodent control using Barn Owls: Is it profitable? *American Journal of Agricultural Economics, 96*, 733–752.

Leshem, Y., & Ezer, A. (2013). *Buma The Barn Owl, The Farmer's Friend*, The Hoopoe Foundation Library (69 pp.) SPNI and Tel Aviv University (in Hebrew, Arabic and English).

Leshem, Y., & Yom-Tov, Y. (1996). The magnitude and timing of migration by soaring raptors, pelicans and storks over Israel. *Israel Journal of Zoology, 138*, 188–203.

Montefiore, S. S. (2011). *Jerusalem, the biography*. London: Weidenfeld & Nicholson.

Roulin, A., Abu Rashid, M., Spiegel, B., Charter, M., Dreiss, A.N., & Leshem, Y. (2017). 'Nature knows no boundaries': The role of nature conservation in peacebuilding. *Trends in Ecology & Evolution, 32*, 305–310.

Sagi, Y., & Leshem, Y. (1994). *Four decades of environmental Stewardship – Conference Report with the Dalai Lama* (44 pp.), SPNI, Tel Aviv.

Shirihai, H. (1996). *The birds of Israel*. London: Academic Press Limited.

# 18

# GREEN SABBATHS

## Putting the Anthropocene Era to Rest (Once a Week)

*Jonathan Schorsch*

### Introduction

I founded the Green Sabbath Project campaign in 2019 to adapt the ancient biblical concept of *shabbat* as a remedy for our collective environmental crises. The idea is to create a ritualised weekly vehicle for ecological consciousness-raising and action that stems from a practice that has been indigenous to Western culture for well over a millennia. As a creative, updated practice organically woven into daily life, a weekly green sabbath and earth day is grounded on the two pillars of the Jewish and Christian sabbath: (a) rest, withdrawal, doing "nothing"; and (b) regeneration, renewal and repair.

#### *Multidimensional Problems Require Multidimensional Remedies*

The Coronavirus pandemic has taught us once again, the hard way, how interconnected we all are and how interconnected are all aspects of our planetary existence. This is also a lesson of our worsening, intertwined environmental crises, and the results of multidimensional human shortcomings that are technological, economic, political, cultural, psychological, emotional and spiritual. The all-too successful Facebook motto, attributed to founder Mark Zuckerberg—"Move fast and break things"—crystallised and amplified the already existing systemic imperatives of capitalism. The system of corporate-captured governments and polities, along with its seemingly opposite, but actually similarly materialist, extraction-oriented forms of industrialism, the communist economies, all represent the culmination of several centuries of Western *pathocracy* (a term coined by Polish psychiatrist Andrew Łobaczewski), governance by sociopaths.

The multidimensional nature of our environmental crises demands holistic and multiple responses that are technological, political, economic, ethical,

DOI: 10.4324/9781003246190-22

cultural and behavioural. Many in the environmental movement now understand that technocratic solutions, while necessary, are not sufficient. The prominent U.S. environmentalist James Gustave Speth famously acknowledged (Rahman, Mackenzie and Falcon 2017):

> I used to think that the top global environmental problems were biodiversity loss, ecosystem collapse, and climate change. I thought that with 30 years of good science we could address these problems. I was wrong. The top environmental problems are selfishness, greed, and apathy, and to deal with these we need a spiritual and cultural transformation.

The question of work makes an excellent example. The shutdowns forced on the world by the Coronavirus pandemic focused attention even more urgently on questions related to work: economic inequities, exploitation of workers, work's meaningfulness, the quality of the work environment and work's ecological ramifications. Creative and radical technocratic analysis and solutions are being offered, for instance, by researcher Philipp Frey, the Zentrum Emanzipatorische Technikforschung (Center for Emancipatory Technology) and the Autonomy think tank in England (Frey 2019; Frey and Schneider 2019; Center for Emancipatory Technology 2020). Noting the link between overly long working hours, worker burnout, overproduction and environmental harm, Frey and his colleagues propose work-free Fridays, called Freeday for Future, which are modelled on the Fridays for Future environmental youth movement. Though attuned to cultural factors, they were unaware of or chose not to reference an age-old, similar spiritual technology, the idea and practice of a weekly sabbath. This is surprising, given that the idea of the sabbath explicitly opposes servitude to work—one of the two central motivations for the weekly day of rest is the liberation of the Israelite slaves from Egypt (Deut. 5:15)—while the modern concept of the weekend itself comes directly from the sabbath.

In the call for environmental solutions, new and alternative kinds of stories, myths, relevant meaningful practices and rituals have arisen with increasing frequency. The reasons for this are many. Secular, rationalist, scientistic modernism is often understood to be one of the main causes of today's planetary environmental collapse. In some ways, postmodernism has (re)turned to and reassessed the cognitive and cultural aspects that have been ex(or)cised by modernism. Within the environmental movement, a growing number of people yearn for procedures, discussion and approaches that take into account and address the emotional, psychological and even spiritual/metaphysical layers of ecological work. In 2016, Stephen Cave and Sarah Darwin wrote a provocative recognition of ritual's potency for the ecologically minded (Cave and Darwin 2016; see also Grimes 2003). To mourn species that have gone extinct, some people came up with Lost Species Day, now an annual commemoration each November 30th (Remembrance Day for Lost Species; Read 2018). Marcus Coates produced a thought-provoking video about a 2017 Irish project to issue a public apology to the Great Auk, which had just been declared extinct (Coates 2017). Others have turned to indigenous and non-Western practices as models. Scientific research seems to confirm that negative individual human behaviour can be modified, affecting the whole, as religious, spiritual and wisdom traditions have long understood (Corner 2018).

Religion remains one of the world's most powerful socio-political forces. Like all complex phenomena, it has many faces. It is particularly capable of keeping groups of people devoted to long-term collective goals and ideals. Ritual is one means by which this is accomplished (Sosis 2007; Alcorta and Sosis 2005). Given the urgency of our ecological crises and the fact that a large majority of people in the world continue to live and think cosmologically or spiritually, environmentalists—a term I use in the widest possible sense—and religious-spiritual practitioners ought to use, (re)interpret, revise and (re)create practices from traditions in a manner that is self-conscious, thoughtful, playful, careful and bold. This is how the history of religions works: change masks itself as tradition's original intent. When it comes to mitigating human harm to our host ecosystems, some religious leaders have proven themselves creative and intrepid in putting their traditions to use. Buddhist monks in Thailand, Cambodia and Sri Lanka ordain trees, wrapping them with coloured sashes, to remind people of their sacredness and prevent such trees from being cut down (Darlington 2012; Morrow 2012; Tannenbaum 2000). Cláudio Carvalhaes develops and uses creative, unorthodox, profound and profoundly thought-provoking new rituals, such as confessing to plants or meditating on bee corpses held in the hand (Carvalhaes 2019a, 2019b).

## *From Shabbat to Green Sabbaths*

I propose a hybrid of an ancient Jewish practice that has had widespread attractiveness over millennia for billions of people around the world, and the more recent "green virtues" promulgated by environmentalists based on such ancient spiritual technologies and traditions long lived by members of most religious cultures: cooperativeness, mindfulness, simplicity, temperance and respect for nature (see, for instance, Jamieson 2014). The word technology derives from the Greek word *technê*, meaning craft, skill or art. The idea and practice of *shabbat*, of the sabbath day, is a technology and one obvious cure among the many we should have already begun implementing.

The biblical and rabbinic idea and practice of a weekly day of rest emerged from a holistic spiritual ecology. As described (and idealised) in the Bible and rabbinic literature, early Israelite and Judaean rabbinic societies are now understood to have been Indigenous societies (regardless of actual geographic origins) grounded in an ecotheology devoted to the intertwined well-being of land and people, a quasi-animist community opposed to internal socioeconomic inequality and the empires of the region. The rabbis of the talmudic era forged from the biblical sabbath a ritualised "temple in time" (Abraham Joshua Heschel) whose main theme was (and remains) avoiding the manipulation of nature for one day a week. For religious Jews, shabbat has remained central, meaningful and beloved, even as they have modified the way they observe it. The day is welcomed and appreciated as a celebration, refuge and opportunity for decolonisation of the self from society's economic and administrative demands.

In the biblical texts, we are told that the commandment to cease from working is not only a metaphysical, spiritual matter. Sabbath rest is not just because God wants it. It is not merely beneficial to humans. *The world needs its rest.* The trans-species community of living beings—work animals, for instance—cannot be continuously worked; the social hierarchy, whose differentiation between those at the top and those at the bottom results from human behaviour, cannot survive endless demands (Ex. 23:12; Deut. 5:14). The Bible connects the weekly sabbath to the seven-year cycle of the sabbatical year (Ex. 23:10–12), which comes, we are told explicitly, because the land itself desires to rest (Lev. 26:34–35) to be healed. In certain respects, this cultural system acknowledges that the natural world has a voice and that we are not free to ignore it.

The spiritual ecology implied by the biblical shabbat was expanded by the rabbis. Basing themselves on the biblical text and oral tradition, the ancient rabbis saw in the sabbath day not merely a symbolic reflection of God's resting after having created existence as we know it, nor as a mere metaphorical form of *imitatio dei*. From the biblical description of the construction of the mobile desert Tabernacle, they derived 39 types of labour, all of which they prohibited on the sabbath under the general biblical commandment to refrain from working on that day (Ex. 20:10–11, 23:12, 31:14–15; Deut. 5:14). Thus, the Mishnah, Judaism's earliest extant legal code, edited (ca. 200 CE) by Rabbi Judah Hanasi, forbade

> sowing, ploughing, reaping, binding sheaves, threshing, winnowing, sorting grain, grinding, sifting, kneading, baking, shearing wool, cleaning it, beating it, dyeing it, spinning, weaving, making two loops, weaving two threads, separating two threads, tying, untying, sewing two stitches, tearing in order to sew two stitches; hunting a deer, slaughtering it, skinning it, salting it, curing its hide, scraping it, cutting it up, writing down two letters, erasing in order to write two letters; building, taking down; extinguishing a fire, kindling a fire, striking with a hammer, carrying [something] from one domain to another
>
> *(M. Shabbat 7:2).*

For the rabbis, the Tabernacle stood as a human-built microcosm of the natural macrocosm that is God's creation.[1] Therefore, any kind of labour that contributed to the erection of this human mini-cosmos was to be avoided on shabbat, a means of emulating God's refraining from the work of creating nature on the seventh day. Rest means cessation from work, leaving the world as it is without human intervention, since work means transforming what is given, what exists. The repeated emphasis in this Mishnah on the number two shows that for the rabbis, the leap from one to two reflected a leap from nature as unity to the multiplicity of culture. On the sabbath day, like God resting, Israelites/Jews are to stop changing the world around them.[2] Shabbat was and remains a temporary return to Eden, to the wilderness through which the ancient Israelites wandered after being freed from Egyptian slavery according to tradition. Shabbat was a recurring, temporary antidote to the society of accumulation and exploitation represented by Pharaonic Egypt. The counter-cultural impulse of shabbat is visible even in its trans-species reach: not only are work/domestic animals supposed to rest, but according to some sources, they are to be allowed "to graze unhindered, so that they, too, might experience the pleasure of Sabbath" (Winkler 2003, p. 165).[3]

The spiritual virtues of sabbath have been developed for millennia. The literature on the sabbath in Judaism and Christianity (I am not as familiar with its significance in Islam), its meanings and historical transformations are well known. In Hellenistic fashion, Philo thought that the sabbath was a day to avoid physical labour and concentrate on the "higher" activities of the mind and soul (Philo, *De Specialibus Legibus* 2.61; *De Decalago* 20; *De Vita Mosis* 2.215–216; *De Vita Contemplativa* 3–4). Medieval Jewish mystics in Egypt (influenced by Sufism) saw in the sabbath an opportunity "to curb worldly preoccupations in order to give oneself over to worship and seclusion (*ḥalwa*), 'to delight in God's recollection (*ḏikr*) and to fill one's thoughts with him'" ('Obadyah Maimonides 1981, p. 69). So essential are the spiritual values of observing sabbath, especially in our era of commercialism, technological hubris and distraction, that a host of contemporary Jews have (re)turned to the sabbath with an eagerness that perhaps reflects desperation (Lieberman and Klinghoffer 2011; Shulevitz 2011; Elkins 1998; Kaplan 1993; Peli 1991; Fromm 1972; Heschel 1951). The same is true among Christians (Amstutz 2018; Brueggemann 2013; Wirzba 2006; Dawn 1989).

The various qualities of the sabbath are discussed at length by many: its sacred nature, communing with the divine, prioritising what is truly important, training in self-restraint, curtailing of desire, revival of family togetherness, cultivation of self-sufficiency, restoration of personal energy, psychological cushioning in the face of taxing work-life, temporary decrease in environmental harm, the intentional interruption of material accumulation and so on. There is no need here to add to these insightful and urgently relevant understandings of the beauties and benefits of the sabbath.

## *Doing "Nothing"*

I argue that from the perspective of our current ecological crises, we can no longer afford "minimal" or symbolic observance of the sabbath. In a more "maximal" form, along the lines envisioned by the ancient rabbis, sabbath observers on their sabbath day would not build, operate or work in factories, do business, farm, produce clothes at home, drive cars, fly, use engines of any kind, spend money, hunt, etc.

Along with many others, I have come to see a weekly day of rest, a green sabbath, as an environmental remedy with unique potential. Earth day, I concluded, comes 51 times too infrequently each year. A green sabbath or weekly earth day restores shabbat to its original intention of commemorating the creation of the world—*zekher lema'aseh bereishit* (Gen. 2:1–3; Exod. 20:7–10). Rabbi Micha Odenheimer notes that the only concrete biblical sabbath prohibition is against the burning of fires (Exod. 35:3), which he reads half-playfully as a warning against emitting carbon dioxide and greenhouse gases on shabbat (personal communication, 2019). While some seek eco-salvation in expensive and untried new technological fixes such as biofuels, carbon capture or geoengineering, shabbat offers a nearly cost-free remedy that acknowledges and addresses the root cause of our problems: the human psyche.

I am by no means the first to issue such a plea or challenge around sabbaths. Among others, some 30 years ago, German Evangelical theologian Jürgen Moltmann called on Christians to observe the sabbath "in the original sense of abstaining for one day a week from all commercial, productive, and industrial activities" (Nasr 1996, p. 218; and see Moltmann 1985). Like Moltmann and many others, I believe that for religious believers and secular people alike, our sabbath days must become a time of active avoidance of environmental vandalism. But more, they must become a time for programmatic individual and collective reflection on how we are undoing creation.

I am drawn to the sabbath's imperative—counter-intuitive, provocative, never timelier yet still often suppressed even in environmental discourse—to "do nothing." "Doing nothing" one day weekly is not meant to replace political, economic or technological environmental solutions, but to help reinforce them, to unify them in a meaningful armature of culture and intention. Even the most secular individuals increasingly appreciate these insightful and urgently relevant understandings of the beauties and benefits of the sabbath. We see more and more calls for "sabbaths" from our modern technologies, such as filmmaker Tiffany Shlain's Technology Shabbats (Technology Shabbats), cities implementing car-free days, as is done in Bogotá, Colombia, to great popular acclaim (Cross 2019; Guillermoprieto 2019), or The Center for Emancipatory Technology's Freeday for Future. Hinting at the enormous impact of human behaviour, the temporary dip in carbon emissions due to the COVID pandemic has been dubbed an "anthropause" (Rutz, Loretto, Bates, et al. 2020). The pandemic and its lockdowns have become known to many as "the Great Pause." The lesson of taking a break, making do with less, of finding genuine contentment within, the sabbath ethos is meant to infuse behaviour for the rest of the week. This ancient spiritual technology, creatively updated and harnessed for the anthropocene era, offers a holistic vehicle for addressing the intersecting political, economic, cultural, spiritual and psychic causes of human environmental degradation.

Imagine if most of the world's monotheists—I am beginning with them—and the so-called post-religious citizens chose to eliminate their own harm to the environment for one day out of seven on a consistent basis. This might prove to be the one of the cheapest solutions at humanity's disposal, though it would not necessarily be easier than implementing complex policies. In theory, as others have noted, more shabbat observance could produce a one-seventh or 14.3% reduction in carbon emissions without additional spending, new technologies or unintended environmental consequences.

The green sabbath, or a weekly earth day, appeals to me for many reasons. The cosmic and trans-species design of shabbat (it is for the entire creation) challenges comfortable and dangerous anthropocentrism. Shabbat has comprised a recurring inoculation against human dominionism, a reminder of cosmic balance, serving as a weekly "little death" in line with Buddhist emptiness teachings.[4] Without darkness, we challenge sleep; without silence, we cannot hear well and without stillness, we die of exhaustion (Figure 18.1).

Drawing on the radical anti-econometric premise of shabbat—a world not ruled by servitude to work and profit-making—green sabbaths cut across the misguided binarism that pits personal change against systemic structural transformation by recognising that our nearly global political-economic system prospers because of our consent; green sabbaths foreground the intention to undo this consent. Asks Robin Wall Kimmerer, why "have we permitted the dominance of economic systems that commoditize everything? That create scarcity instead of abundance, that promote accumulation rather than sharing? We've surrendered our values to an economic system that actively harms what we love" (Kimmerer 2020) (Figure 18.2).

**FIGURE 18.1** Attaining balance is harder than it looks. Rock sculpture, Martha's Vineyard, Massachusetts, 2021. Photo Credits: Jonathan Schorsch.

**FIGURE 18.2** Sabbath anchors a system of seven-year cycles whose purpose is to prevent dangerous socioeconomic disequilibrium. Sign in Beacon, New York. Photo Credits: Jonathan Schorsch.

The idea and practice of a weekly day of rest is Indigenous to Judaism, Christianity and, in a more ambiguous way, Islam. Given that the West has been primarily responsible for the destruction of our planetary ecosystems, urging westerners to curtail their environmental harm through a familiar practice from their own cultures makes particular sense and promises to have a particularly significant impact. Abstract calls to transition to new, sustainable forms of society have not proven effective enough. Today's (post)monotheistic societies cannot just take up the ways of Indigenous peoples. What is needed are practices that embody mindfulness and sustainability, drawn from local and regional cultures and organically interwoven into the fabric of daily personal and collective life.

### *Doing Something*

While "doing nothing" comprises one main element of green sabbaths, the other is the doing of acts that embody and nurture ecological consciousness. With the distractions of modern life eliminated, green sabbaths can serve as a day for taking a walk, meditating, prayers, playing with your children, reading on your own or reading out loud, conversing with friends or singing, or a host of other activities. As an idea and practice adapted and updated from its traditional roots in *shabbat*, green sabbaths can serve as a weekly vehicle for a wide range of creative practices and activities—religious, spiritual, pantheistic, animistic, agnostic or materialist—particularly on the level of the organisation, neighbourhood, congregation or municipality. While striving to keep environmental impacts to a minimum, groups might organise on their weekly earth day a cleanup or invasive species removal at a local site that needs some love and care; take a group walk outdoors; make a temporary sculpture or mandala out of found natural materials (twigs, leaves, stones, etc.); read nature poetry out loud; invite elders from within or beyond your community to share stories about the local landscape, ecosystems and history; hold prayer services or a meditation circle outdoors, in a local garden or park; play games that teach environmental lessons, including group simulation games to mitigate climate catastrophe. In addition to such activities, congregations or meditation centres might get members together to write their own prayers for the earth, the rain, their favourite trees or recently lost species and incorporate them into their liturgy or practice; devote sermons or dharma talks to environmental themes; organise a choral performance or community sing-along of selections from *The Lost Words*, Joanne Shenandoah's "Dancing on Mother Earth," Malcolm Dalglish's "Hymnody of Earth" or his setting of Wendell Berry's poem, "Great Trees," Miriam Makeba's "Meet Me at the River," "A Place in the Choir" by Bill Staines, or J. David Moore's setting of Jack Manno's "Earth Blessing." The possibilities are as vast as the imagination.

In the interest of nurturing community in the midst of a global pandemic, the Green Sabbath Project (www.greensabbathproject.net) has launched a series of twice-monthly online Friday Gatherings. Meant to serve as a bridge from the work week to the weekend, day of rest or weekly earth day, each loose-format, hour-long get-together is an opportunity for reflection, meditation, readings, melodies, short videos, prayer, conversation, showing some visual art, etc.—in religious, spiritual or secular modalities—led by a diverse group of spiritual leaders, environmentalists, activists, writers, artists and the like. As far as possible given the lockdowns, and certainly when conditions permit, the Green Sabbath Project plans to co-sponsor creative, local, and partially or fully ritualised outdoor walks, site visits and cleanup activities already led by organisations such as Radical Joy for Hard Times, SanghaSeva and the Center for Spirituality in Nature. The Green Sabbath Project has also established a global network of institutional members: organisations, congregations and communities dedicated to wielding a weekly day of rest for ecological reflection, connection and action.

Local green sabbath activities of all kinds, organised by network members and others, are listed on the online Green Sabbath Project event calendar. Organised for the weekend following earth day, the Green Sabbath Project is sponsoring a Global Green Sabbath Weekend, a ritual event through which groups around the world can commemorate a green sabbath and foster the idea of conducting the practice more frequently than once a year.

## Conclusion

A weekly green sabbath or earth day properly practised—as radical intervention and not mere lip-service—balances self-restraint and voluntary withdrawal with joyous celebration of contentment, delight in pure being. It offers a weekly interruption of the suicidal economic fantasy of infinite growth, a weekly divestment from fossil fuels, a weekly investment in family and local community, a weekly bit of rewilding, a respite for both humans and other-than-humans and a ritualised forum for meditating on how we want to live. As Greta Thunberg has reminded us repeatedly, we already know what the solutions are for our environmental crises. Green sabbaths can provide a recurring weekly time for incubating the required collective consciousness and willpower—the ultimate renewable energies—to make the solutions reality. Green sabbaths could constitute both a model and a foretaste of an ecologically sane world to come.

In order to foster this vision, I founded the Green Sabbath Project in 2019. As described by Donna Haraway (2016, p. 51), in their book, *La sorcellerie capitaliste*, Philippe Pignarre and Isabelle Stengers "affirm on-the-ground collectives capable of inventing new practices of imagination, resistance, revolt, repair, and mourning, and of living and dying well."[5] Given the abundance of wise and skilful practices within monotheistic traditional lifeways, and given the troublesome consequences of chasing after the always new, we (post)monotheists might not want to fetishise novelty in ecological repair. Practices such as green sabbaths promise to fruitfully graft the old with the new. Come, let's do nothing together.

## Notes

1 On the linguistic and symbolic correspondences between the biblical accounts of the building of the Tabernacle (and hence Temple) and the creation, see Levenson (1985), pp. 142–145, 178–184. On rabbinic development of sabbath, see Green (1980). Attempting to replace the now-destroyed Jerusalem Temple with the sabbath as a central unifying symbol, the rabbis renew the biblical text: "By *doing* all these labours in the particular prescribed configuration, one creates sacred space. By *refraining* from these same acts, in the context of the Sabbath, one creates sacred time" (294).

2 On labour and shabbat, see Auld (1986) and Roth (1982). More general analyses are offered in Neusner (2000); Kraemer (1997); Hidary (2015) and Noam and Qimron 2009.

3 Winkler is citing Midrash Mechilta and the medieval commentator Rashi on Ex. 23:12.

4 Marilyn Paul suggested this understanding to me (personal communication, 2020). See also Paul (2017).

5 Haraway is citing Pignarre and Stengers (2005).

## References


Alcorta, C. S., & Sosis, R. (2005). Ritual, emotion, and sacred symbols: The evolution of religion as an adaptive complex. *Human Nature, 16*(4) (Winter 2005), 323–359.

Amstutz, A. (2018). *Soul tending: A journey into the heart of Sabbath*. Nashville, TN: Skylight Paths Publishing.

Auld, A. G. (1986). Sabbath, work and creation: 'Melaha' reconsidered. *Henoch: Studi storicofilologici sull'ebraismo (Torino), 8*(3), 273–280.

Brueggemann, W. (2013). *Sabbath as resistance: Saying no to the culture of now*. Westminster John Knox Press.

Carvalhaes, C. (2019a). Why I created a Chapel Service where people Confess to plants. *Sojourners*, 26 September 2019. Retrieved from https://sojo.net/articles/why-i-created-chapel-service-where-people-confess-plants.

Carvalhaes, C. (2019b). *Ritual – Ice melting*. Retrieved from https://www.claudiocarvalhaes.com/liturgies-2/ritual-ice-melting.

Cave, S., & Darwin, S. (2016). It's not easy being green. *Aeon, 15* (March 2016). Retrieved from https://aeon.co/essays/rituals-and-guilt-can-these-save-environmentalism.

Coates, M. (2017). *Apology to the Great Auk, 2017*. Retrieved from https://www.workplacegallery.co.uk/artists/9-marcus-coates/works/10309.

Corner, A. (2018). Tiny individual decisions really could help avert climate chaos. *New Scientist*, 12 January 2018. Retrieved from https://www.newscientist.com/article/2158432-tiny-individual-decisions-really-could-help-avert-climate-chaos.

Cross, L. (April 15, 2019). Car-free Sundays are the norm in Colombia's capital city, Bogotá. *Inhabitat*. https://inhabitat.com/car-free-sundays-are-the-norm-in-colombias-capital-city-bogota.

Darlington, S. M. (2012). *The ordination of a tree: The Thai Buddhist environmental movement*. Albany, NY: SUNY University Press.

Dawn, M. J. (1989). *Keeping the Sabbath wholly: Ceasing, resting, embracing, feasting*. Gran Rapids, Mich.: William Eerdmans Publishing Co.

Elkins, D. P. (1998). *A Shabbat Reader: Universe of Cosmic Joy* (Dov Peretz Elkins, Ed.) New York: UAHC Press.

Frey, P. (2019). *The ecological limits of work: On carbon emissions, carbon budgets and working time*. Autonomy Research. Retrieved from http:/autonomy.work/wp-content/uploads/2019/05/The-Ecological-Limits-of-Work-final.pdf.

Frey, P., & Schneider, C. (2019). *FreeDay for future: Klimaschutz durch Arbeitszeitverkürzungen*. Zentrum für emanzipatorische Technikforschung. Retrieved from https://content.emancipatory.technology/wp-content/uploads/2020/08/free-day-for-future.pdf.

Fromm, E. (1972). Der Sabbath. In Y. Spiegel (Ed.), *Psychoanalytische Interpretationen biblischer Texte* (pp. 174–184). Munich: Kaiser.

Green, A. (1980). Sabbath as temple: Some thoughts on space and time in Judaism. In S. Z. Fishman & R. Jospe (Ed.), *Go and study: Essays and studies in honor of Alfred Jospe* (pp. 287–305). Washington, DC: B'nai Brith Hillel Foundation.

Grimes, R. (2003). Ritual theory and the environment. *The Sociological Review, 51*(2), 31–45.

Guillermoprieto, A. (2019). This City Bans Cars every sunday — and people love it: Spurred by environmental concerns, an experiment in Bogotá, Colombia, is spreading worldwide. *National Geographic*, 27 March 2019. Retrieved from https://www.nationalgeographic.com/environment/2019/03/bogota-colombia-ciclovia-bans-cars-on-roads-each-sunday.

Haraway, D. J. (2016). *Staying with the trouble: Making Kin in the Chthulucene*. Durham, NC: Duke University Press.

Heschel, A. J. (1951). *The Sabbath: Its meaning for modern man.* New York: Farrar, Straus and Giroux.
Hidary, R. (2015). Revisiting the Sabbath Laws in 4Q264a and their contribution to early Halakha. *Dead Sea Discoveries, 22*(1), 68–92.
Jamieson, D. (2014). *Reason in a dark time: Why the struggle against climate change failed – And what it means for our future.* New York: Oxford University Press.
Kaplan, A. (1993). *Sabbath: Day of eternity.* OU/NCSY.
Kimmerer, R. W. (2020). The serviceberry: An economy of abundance. *Emergence Magazine*, December 2020. Retrieved from https://emergencemagazine.org/story/the-serviceberry.
Kraemer, D. C. (1997). The spirit of the Rabbinic Sabbath. *Conservative Judaism, 49*(4), 42–49.
Levenson, J. D. (1985). *Sinai and Zion: An entry into the Jewish Bible.* Binghamton, NY: Winston Press.
Lieberman, J. I., & Klinghoffer, D. (2011). *The gift of rest: Rediscovering the beauty of the Sabbath.* New York: Howard Press.
Moltmann, J. (1985). *God in creation: An ecological Doctrine of creation.* Harper Row.
Morrow, A. (2012). Tree ordination as invented tradition. *ASIANetwork Exchange: A Journal for Asian Studies in the Liberal Arts, 19*(1). Retrieved from https://asianetwork-exchange.org/articles/10.16995/ane.11/galley/6/download.
Nasr, S. H. (1996). The rediscovery of nature: Religion and the environmental crisis. In *Religion and the Order of Nature, the 1994 Cadbury Lectures at the University of Birmingham.* New York: Oxford University Press.
Neusner, J. (2000). The Sabbath Halakhah in the context of Rabbinic Judaism's theological system. In R. A. Argall, B. A. Bow, & R. A. Werline (Ed.), *For a later generation: The transformation of tradition in Israel, Early Judaism, and Early Christianity* (pp. 183–195). Harrisburg, PA: Trinity Press International.
Obadyāh b. Abraham b. Moses Maimonides. (1981). *The Treatise of the Pool / Al-Maqāla al-Ḥawḍiyya* (Paul Fenton, Trans.). Octagon Press.
Paul, M. (2017). *An Oasis in time: How a day of life can save your life.* Rodale Books.
Peli, P. H. (1991). *The Jewish Sabbath: A renewed encounter.* Schocken.
Pignarre, P., & Stengers, I. (2005). *La sorcellerie capitaliste: Pratiques de désenvoûtement.* Découverte.
Rahman, J., Mackenzie, D., & Falcon, T. (2017). Climate and creation: A conversation with Imam Jamal Rahman, Pastor Don Mackenzie, and Rabbi Ted Falcon. *Orion Magazine* (May/June 2017). Retrieved from https://orionmagazine.org/article/climate-and-creation.
Read, P. K. (2018). A new mourning: Remembrance day for lost species. *Undark*, 10 April 2018. Retrieved from https://undark.org/2018/04/10/new-mourning-species-extinction.
Remembrance Day for Lost Species; https://www.lostspeciesday.org.
Roth, J. (1982). 'Melakhah u'shevut': A theoretical framework [of Shabbat norms]. *Conservative Judaism, 35*(3), 4–34.
Rutz, C., Loretto, M. C., Bates, A. E., Davidson, S. C., Duarte, C. M., Jetz, W., Johnson, M., Kato, A., Kays, R., Mueller, T., Primack, R. B., Ropert-Coudert, Y., Tucker, M. A., Wikelski, M. & Cagnacci, F. (2020). COVID-19 lockdown allows researchers to quantify the effects of human activity on wildlife. *Nature, Ecology and Evolution, 4*, 1156–1159. Retrieved from https://www.nature.com/articles/s41559-020-1237-z.
Shulevitz, J. (2011). *The Sabbath World: Glimpses of a different order of time.* Random House.
Sosis, R. (2007). Psalms for safety: Magico-religious responses to threats of terror. *Current Anthropology, 48*(6), 903–911.

Tannenbaum, N. (2000). Protest, tree ordination, and the changing context of political ritual. *Ethnology, 39*(2), 109–127.
Technology Shabbats. (2022) Retrieved from https://www.letitripple.org/about/tiffany-shlain/technology-shabbats.
Winkler, G. (2003). *Magic of the ordinary: Recovering the Shamanic in Judaism*. Berkeley, CA: North Atlantic Books.
Wirzba, N. (2006). *Living the Sabbath: Discovering the rhythms of rest and delight*. Grand Rapids, Mich.: Brazos.

# 19

# ECOLOGY AND RELIGION WALK HAND IN HAND TO CONSERVE GREATER ADJUTANTS IN INDIA

*Anant Deshwal, Purnima Devi Barman and Arvind Mishra*

## Introduction

This is a story of cultural transition, for conservation is not just about saving species but often about changing the perception of human beings towards other organisms. It is also a story of the significance of nature in relation to maintaining a sustainable biosphere on our planet – for those who demand justification, the plethora of arguments for wildlife conservation range from the beauty of nature to its scientific value, and from its religious and cultural importance to its economic value. Many of these anthropocentric arguments hinge on the general acceptance of a species by humans. Megafauna and charismatic species such as elephants, tigers, and lions often attract enough attention through media and ecotourism to support their conservation. However, the question arises, what do you do if a particular species is loved and considered sacred in one place, but in another region, it is detested? The species fails to meet common beauty standards. How does one conserve a species that has no cultural or economic significance? This was the dilemma that Purnima Devi Barman and Arvind Mishra faced when they decided to conserve such a species. Purnima had set out to study the ecology of the Greater Adjutant Stork (*Leptoptilos dubius*) in Assam for her PhD. However, on observing the plight of this bird, she decided to devote her time to securing a long-term future for this species.

## Conservation and Cultural Status of the Greater Adjutant (*Anant Deshwal's Perspective*)

Purnima Devi Barman, Arvind Mishra, and I are from different parts of India but have one thing in common – conservation of wildlife. For each one of us, conservation is not a job description but a way of life. I quit a lucrative career

DOI: 10.4324/9781003246190-23

as a civil engineer to pursue my dream of conserving wildlife and nature. In this pursuit, I often chose to help conservation-based organisations but struggled with manpower. I often work on organisms or landscapes that are neglected, e.g. I worked on giving snakes in south-eastern India an image make-over to mitigate human-snake conflicts. While working on snakes and four-horned Antelope conservation, I also realised that in south-eastern India, the shrub forests were classified as wastelands. These shrublands provided crucial habitats for several bird, reptile, and mammal species and were neglected. Through my PhD, I worked towards the conserving and changing perceptions of shrub forests in India (Deshwal et al. 2019). It was this idea of working for organisms that are neglected and are in desperate need of a voice, that immediately made Purnima and me click when we meet during the leadership conference conducted by the Conservation Leadership Programme (CLP; it is a partnership of three biodiversity conservation organisations, BirdLife International, Wildlife Conservation Society, and Fauna and Flora International, which direct funding to early career researchers in developing countries) in spring 2015. We both are CLP alumni (had received funding from CLP). We have ever since shared our struggles and our success stories with each other. Thereby supporting each other's conservation efforts. In 2020, I found that Arvind Mishra had been spearheading conservation efforts for Greater Adjutants in Bihar. The three of us now plan to join forces to conserve this species across landscapes and cultural transitions.

Although once common throughout Northern India, Bangladesh, Nepal, and southern Vietnam (Elliot et al. 1992, Singha et al. 2002), the Greater Adjutant is now confined only to a few spots within Assam and Bihar. Estimated to be in the hundreds of thousands in the late 1800s (Barman et al. 2020), by the early 1990s, only 400 remained (Perennou et al. 1994). The Greater Adjutant was declared endangered in 1994: a status it holds to this day (BirdLife International 2022). Persecution through hunting, poisoning, and active destruction of nesting sites led to their rapid decline. A close cousin of the better-known Marabou Stork of Africa (*Leptoptilos crumenifer*), Greater Adjutants feed on rotting carcasses and swallow bones, which is reflected in Assam, by the local name *Hargila*, which means "Bone Swallower". In this chapter, we will show how the ecology and behaviour of the same bird can earn reputations in different communities that are poles apart. Also, how religion and science have come together to play a critical role in saving this endangered species will be a focus for us. In Bihar, the local name for the Greater Adjutant is "Garud" – a vehicle of Lord Vishnu, an important deity in the Hindu pantheon – indicating its sacred status held there, while in its nearby state of Assam, its very presence is considered a bad omen.

In Assam, Greater Adjutants build their nests in tall trees, often near the homes and gardens of villagers (Barman et al. 2020). Up to 15 nests can be built in a single tree (Barman et al. 2020), and the birds bring fish, snakes, rotting meat, and bones to feed their chicks (Barman et al. 2020). As expected, some of the food items along with faecal matter fall to the ground resulting in a foul smell and potentially health issues for the people in those homes (Barman et al. 2020).

Greater Adjutants often feed on garbage dumps and landfill sites, a fact that has not helped with their image. Faced with such issues, the bird was in desperate need of an image make-over.

## Changing Hearts and Minds (*Purnima Barman's Story*)

In 2007, I set out to create a holistic conservation programme by developing community support for the Greater Adjutant. I focused on three villages (Dadara, Singimari, and Pachariya) within 12 km of Guwahati, Assam. My initial survey revealed that villagers often treated the Greater Adjutant as a bad omen because it feeds on carcasses. Recognising the difficulty of changing perceptions within this male-dominated society, I approached women, elderly people, and students to be champions of Greater Adjutant conservation. This involved developing a sense of ownership for the project among villagers. I initiated the "Hargila Army" (http://www.pashoopakshee.com/hargila-army), a rural women's group that was dedicated to Greater Adjutant conservation.

One major challenge in starting the Hargila Army was that women were initially shy and reluctant to participate in conservation programmes. They felt that they had no voice as decision making was seen as a part of the male-dominated world. I directed my efforts to designing programmes specifically directed at women. Knowing that appealing to the religious sentiments of local people was an effective way to increase the acceptance of the Greater Adjutant, I made this a central pillar of my strategy to involve local people in a community-driven conservation programme. I also knew that in Bihar, unlike in Assam, the Greater Adjutant was treated like God. I used this knowledge successfully, to motivate women in Assam to participate in Greater Adjutant conservation. Employing the local religious practice of Panchamrit (baby shower) to celebrate the birth of each Greater Adjutant, it soon became famous in the communities as Hargila Panchamrit (baby shower for the Greater Adjutant). Hargila Panchamrit also involved praying to Lord Krishna, the reincarnation of Lord Vishnu, to bless the Greater Adjutant nestlings (Barman et al. 2020).

By effectively combining people's maternal instincts with the religious significance of the Greater Adjutant, I heralded a new chapter in its conservation. Gradually, Hargila Panchamrit went from being a local event practised by just a few women to a mainstream practice, an event, in which more than 10,000 women participate today. Women were motivated to write Naam songs (prayer songs) with Greater Adjutant conservation messages. Songs such as "Greater Adjutant you are safe in our village" were specially written for Hargila Panchamrit celebrations. Women prayed to God for a better life for the Greater Adjutant and called the Greater Adjutant to come and breed in their village. By 2015, Greater Adjutants were a part of the largest religious celebration in the area – "Jonmastami" (Celebration of the birth of Lord Krishna). A "Bhagwat Geeta" (Hindu holy book) procession was conducted where more than 10,000 people marched for peace for all species on Earth. Greater Adjutant campaigns were an

**FIGURE 19.1** Purnima Devi Barman leads a procession to create awareness for Hargila Conservation. Hargila papier-mâché creations are worn by women during the procession as a sign of respect for the bird and to spread awareness. Photo credits: Anupam Nath.

integral part of these processions. Through the Hargila Army, women actively appealed to their husbands, brothers, and fathers to prevent them from disturbing Greater Adjutant nests or cutting down trees in which the birds nested. Hargila Panchamrit now sees a new trend. Men are increasingly and actively becoming engaged in this celebration and taking ownership of the conservation of the Greater Adjutant (Figure 19.1).

## Normalising Perceptions

Hargila Army and I now often participate in local village fairs and festivals such as Bihu (Barman et al. 2020). Bihu festivals are a set of three important festivals in Assam: Rongali or Bohag Bihu observed in April, Kongali or Kati Bihu in October, and Bhogali or Magh Bihu in January. Rongali celebrates the spring festival, Bhogali celebrates the harvest festival, and Kongali reflects the season of limited supplies. In these festivals, people often host various events such as cooking, craft, and folk music competitions. Bihu festivals often invite people, especially women, to participate spontaneously in discussions regarding the Greater Adjutant. People were often amazed to learn that Greater Adjutants had chosen only their village to breed. This knowledge coupled with associated religious sentiments played a crucial role in changing the bird's image and encouraging the formation of a special bond with it. The bad omen sentiment was now being replaced by a sense of pride. Once an outcast, this bird was now welcomed by local villagers. They made sure that there were no disturbances during the breeding season. This increased acceptance and made it possible for academics and conservationists to assess the ecological and economic importance of the birds to the lives of the local farming community.

I went on to finish my PhD on Greater Adjutants and am now actively using the academic knowledge of the Greater Adjutant I have gained, to conserve this species in new ways. In addition to my previous approach, I am also employing rigorous scientific techniques to understand more about the ecology of the bird. I am now setting up nest towers and platforms to observe Greater Adjutants at the nest, and nets placed under their nests catch any nestling that may fall out. Local villagers of both sexes, and of all ages and walks of life are now actively participating as citizen scientists to study relevant ecology and conserve this species, while also volunteering to build the enormous nest towers and platforms and monitoring the nets for fallen nestlings. The local community has stepped forward to form an information network. If any Greater Adjutant is found dead, it is readily collected, and its cause of death assessed. This may include testing for poisoning. Through empathy and cultural understanding, I have managed to channel the maternal instincts and religious beliefs of the people into a powerful tool for conservation, which has led to local people being more accepting of scientific approaches and studies, for conserving a bird which once was despised.

Within Assam, the Greater Adjutant conservation programme has not only influenced this bird but has played an important role in women's empowerment.

**FIGURE 19.2** Purnima Devi Barman leads a cultural procession for stork conservation. Gamosa (Traditional Assamese Scarf) with Stork Motifs and Hargila papier-mâché creations are worn by people during the procession as a sign of respect for the bird and to spread awareness. Photo credit: Rathin Barman.

The Hargila Army has enabled women to earn their livelihood, thus giving them some financial freedom. Women were provided with weaving looms, yarn, and sewing machines (Barman et al. 2020), allowing them to create textiles such as Gamosa (Traditional Assamese Scarf) with Stork Motifs. These Gamosa reflect Assamese culture and are loved by local people (Barman et al. 2020), earning the women who produce them a livelihood, while also being admired by government officials, media personnel, and ecology experts alike (see Hargila Army products at https://www.pashoopakshee.com/hargila-army) (Figure 19.2).

## Cultural Contrasts (*Arvind Mishra's Story*)

The state of Bihar that lies near Assam had a completely different perspective on the Greater Adjutant Stork. Since its re-discovery in Bihar in 2006–2007 (Mishra and Mandal, 2009), I fought for the Greater Adjutant to be re-recognised in this context as the Garud – *the vehicle of Lord Vishnu.* Many people around the world had begun associating the Garud with an Eagle, and this had permeated the Bihari cultural context as well. I had to make several strong references to the lost traditions of Bihar which connected Greater Adjutants with the Garud, to convince local people otherwise. In fact, Greater Adjutants were loved by the people of Bihar for the cleaning services that they provided, so much so that they had been traditionally incorporated into religious celebrations such as Ram Navmi and Jonmastami. In Bihar, Greater Adjutants inhabit partially dry wetlands with abundant fish, riverbeds, swamps, paddy fields, and stagnant pools. Tall trees in or near paddy fields are often favoured places for Greater Adjutants to nest, given the proximity to foraging grounds. However, financially strapped farmers were forced to cut down the nesting trees in and around agricultural lands to pay for educating their children or the marriages of their daughters. Furthermore, fishermen are known to use Carbofuran (Trade name: Furadan), which is a carbamate pesticide and which has severely affected the Greater Adjutants through pesticide poisoning. Therefore, despite its revered status, loss of nesting sites coupled with pesticide poisoning had led to the local extirpation of Greater Adjutants to some extent in Bihar.

2006–2007 heralded a new era for Greater Adjutants in Bihar. I had spotted an active nest in the Bhagalpur region of Bihar, inspiring me to start working to ensure that the species made a comeback in Bihar. My immediate task was to get farmers to protect the Greater Adjutants, something that had previously been ignored. Knowing that to succeed I had to work with young and old people alike to conserve this species, I had established two groups: namely Garud Saviours and Garud Guardians. Garud Saviours was aimed at youth and Garud Guardians was aimed at older people. This generational approach was considered best, with the view to keep each group engaged, since I understood that engaging local communities meant more than just running an ecological campaign. I used a multi-pronged approach with a religious campaign drawing on local sentiment, as well as an awareness campaign incorporating scientific and local or traditional ecological knowledge (people's knowledge of the birds as cleaners/scavengers) and understanding.

My religious campaign replicated the success that was observed by Purnima Barman in Assam. Since in Bihar the Greater Adjutant already held a sacred position, I was able to appeal to the religious sentiments of farmers. In the Hindu religion, Garud and Nagas (snakes) are seen as mutual enemies. Thus, the Greater Adjutant is recognised as safeguarding farmers and their families from snakes. Such arguments played an important role in creating an atmosphere of open discussion and acceptance of other campaigns run by me. Farmers now started taking pride in the fact that the rarest bird, the Greater Adjutant, was nesting in their agricultural lands. Rather than cutting down nesting trees, there was a drive to plant trees preferred by the Greater Adjutants. Farmers became more open to my ecological campaign and started paying attention to the ecology of the bird. As in Assam, local communities became invested in conserving this species. An example in this regard is as follows: a large Peepal tree (*Ficus religiosa*) hosted 23 Greater Adjutant nests. Setting up of high-power transmission lines and road expansion threatened the tree and the nesting birds. Through the support of the local community and the District Magistrate, I was able to work out an alternative route for the proposed road. Thus, saving the tree. Since three Greater Adjutants had died by electrocution from the high-power transmission lines, I was able to convince the state government of Bihar (as represented by the Additional Chief Secretary and the Principal Secretary, Department of Environment, Forest and Climate Change) to make the transmission lines underground near the same nesting tree.

With the help of the state government, the world's first Greater Adjutant rehabilitation centre was established in Bhagalpur (Bihar) with the provision of a temporary breeding centre in the breeding zone through my continued advocacy. Based on my suggestions, the state government has also provisioned an expert veterinary doctor in this centre, specifically to treat injured or sick Greater Adjutants. Local communities have started acting as protective guardians to nesting Greater Adjutants by protecting them from nomadic tribes that hunt and eat this bird and its nestlings or eggs. It is important to mention that we have taken precautions not to offer direct financial incentives to villagers in the name of the conservation of the Greater Adjutant to avoid generating the feeling that "conservation means money". Nevertheless, annually, we felicitate villagers at programmes at the district headquarter, Bhagalpur, which are organised by the forest department.

## Conclusion

More than 15 years of dedicated service by pioneers in the field of religion-oriented conservation, together with the openness of local people to the religious significance of this endangered species, has led to an increase in the local Greater Adjutant populations. Greater Adjutants went from 400 to 1,000 individuals and the number of active nests increased from 27 to 350 in Assam. Our story demonstrates that conservation cannot simply be achieved by short-term project engagement over a few years. Rather, it is a process that is sustainable only through local community involvement, and also by engaging imaginatively with the culture and concerns of local people. We have also demonstrated that religion and rigorous academic ecology can work together in a synergy that not only benefitted birds, but also led to women's empowerment in Assam and laid the foundation for a sustainable conservation programme that holds the promise of continuing for many years to come through the changed perceptions of people towards the Greater Adjutant. Religion and science working together can prove to be a most effective partnership to achieve sustainable benefits for all.

## Literature Cited

Barman, P. D., Sharma, D. K., Cockrem, J. F., Malakar, M., Kakati, B., & Melvin, T. (2020). Saving the greater adjutant stork by changing perceptions and linking to Assamese traditions in India. *Ethnobiology Letters, 11*(2), 20–29.

BirdLife International. (2022). *Species factsheet: Leptoptilos dubius*. Retrieved from http://www.birdlife.org on 13/03/2022.

Deshwal, A. (2019). *Ecology and conservation of Shrubland Bird Communities in the Eastern Ghats of India*. Fayetteville: University of Arkansas.

Elliot, A., Del-Hoyo, J., & Christie, D. (1992). *Handbook of the Birds of the World. Vol 1: Ostrich to Ducks*. Barcelona: Lynx Edicions.

Mishra, A., & Mandal, J. N. (2009). Discovery of a breeding ground of the Greater Adjutant *Leptoptilos dubius* and their conservation in the floodplains of Bihar, India. *Journal of the Bombay Natural History Society, 106*(2), 190–197.

Perennou, C., Mundkur, T., Scott, D. A., Follestad, A., & Kvenild, L. (1994). *The Asian Waterfowl Census 1987–91: Distribution and Status of Asian Waterfowl. Kuala Lumpur: Asian Wetland Bureau Publication No. 86*. International Waterfowl and Wetlands Research Bureau Publication No. 24. Kuala Lumpur, Malaysia and Slimbridge, UK.

Singha, H., Rahmani, A., Coulter, M., & Javed, S. (2002). Nesting ecology of the greater Adjutant Stork in Assam, India. *Waterbirds, 25*(2), 214–220.

# PART V

# Last Glance

# 20

# ADDITIONAL SNAPSHOTS OF FAITH-BASED CONSERVATION

DOI: 10.4324/9781003246190-25

# 20a

# A TWIST IN CENTRAL ASIAN SNOW LEOPARD CONSERVATION

## Restoring Ancient Ceremony in Support of Modern Conservation Methods

*Apela Colorado, Beth Duncan and Darla Hillard*

Protector of sacred mountains, unifier and source of spiritual power and wisdom, the Snow Leopard (*Panthera uncia*) is the nexus of ancient traditions to the Indigenous people of Central Asia. The leopards are also an important indicator of ecological health of mountain landscapes and cultural integrity of their communities. Territory encroachment, depletion of prey animals, and communities unable to protect their livestock from Snow Leopards that were forced to scavenge for food sources in and around these communities had driven leopard numbers down to levels alarming to conservationists and Indigenous Cultural Practitioners (ICPs) alike. Traditional Western scientists and ICPs working in isolation were failing to curb this trajectory.

A fateful meeting in 2010 between the Snow Leopard Conservancy (SLC) and Worldwide Indigenous Science Network (WISN) sparked the groundbreaking idea of bringing Traditional Ecological Knowledge (TEK) and Western science together—on truly equal footing—to address these critical conservation issues. As stated elsewhere, the time has come for indigenous science to stand alongside Western science in a dynamic and reciprocal partnership. Remembering together requires models that are deeply counter-colonial.

In 2013, we convened the first Land of the Snow Leopard Network (LOSL) meeting, bringing together scientists, shamans, sacred site guardians, herders, hunters, respected religious (shamanism, Ismali Islam, and Tibetan Buddhism) and government leaders, and others from five regions in Tajikistan, Kyrgyzstan, Mongolia, and Russia. The result of this first meeting was a collective ICP statement that was presented to the UN Forum on Snow Leopard Conservation during which LOSL member and Kyrgyz ICP Zhaparkul Raimbekov was invited to attend—and pray—for the delegation.

Rooted in indigenous understanding of the sacredness, cultural and environmental significance of the Snow Leopard, LOSL (now totalling 100+ members),

DOI: 10.4324/9781003246190-26

**FIGURE 20.1** Altaian Shaman Slava Cheltuev leads a ceremony at a sacred site in the Buryat Republic of Russia. Slava was a guest of the Soyot People at their annual ceremony to honour the Snow Leopard as a protector and unifier of the Soyot community. Photo credit: Beth Duncan.

is reviving traditional knowledge, culture, spirituality, and identity, and applying it towards this and related ecosystem conservation. Its growing international recognition is creating pathways for participation in even more scientific planning for Snow Leopard conservation.

Bringing these two ways of knowing together is a leading-edge approach that has never before been applied directly to Snow Leopard conservation—and it's showing signs of success.

For example, LOSL representatives were invited to participate in the Global Snow Leopard and Ecosystem Protection (GSLEP) programme mid-term forum (2017), and in the Society for Conservation Biology Conference in Malaysia (2019). LOSL staged youth performances at the 2018 World Nomad Games in Kyrgyzstan and Snow Leopard Day Festivals originating in the Altai Republic have now expanded to other regions. Furthermore, Snow Leopard educational programmes are sprouting up throughout the network (Figure 20.1).

Beyond awareness building, the biggest measure of the network's success: Snow Leopard populations are showing signs of expansion (Dr. Rodney Jackson, SLC, personal communication, November 21, 2020). There have been multiple success stories since 2013 of human-wildlife coexistence and rural communities championing wildlife. Between 2015 and 2020, for example, there were 11 incidents in Tajikistan of Snow Leopards being released back into the wild by local communities instead of being killed after being caught raiding livestock

(Qurbon Alamshoev, LOSL Country Coordinator, personal communication, December 27, 2020). Further encouragement comes from the reports of Snow Leopard showing up in territories where they have not been spotted in decades, such as recent sightings in Talas, Kyrgyzstan (Akipress, 2021).

To learn more about or support LOSL, visit: landofsnowleopard.org.

## Reference

Akipress. (2021). *Footprints of snow leopards found in Talas region*. Retrieved from https://akipress.com/news:654860:Footprints_of_snow_leopards_found_in_Talas_region/

# 20b

# BIOCULTURAL IMPORTANCE OF SACRED GROVES IN KURDISTAN, IRAN

*Zahed Shakeri*

In developing countries, preserving biodiversity is a great challenge, particularly in areas with constant war, weak governments, severe poverty, and ongoing climate change. While formal conservation plans have often failed, local people in Kurdistan do voluntarily protect certain parts of their environment for their religious and cultural values: Sacred Natural Sites (SNS) are places where local ecological knowledge and environmental protection meet traditional myths, legends, and stories. In Kurdistan, SNS are rooted in ancient religions like Mithraism and Zoroastrianism. Despite the conversion of Kurdish people to Islam in the 16th and 17th centuries, almost every village has maintained its own sacred site until recently, e.g., a woodland, valley, mountain summit, or spring and its surroundings. While the local Kurdish people are well aware of SNS, scientific research on them is scant.

Most of Kurdistan's SNSs are found in forested areas and are used as burial grounds for locals. These sites are believed to be places of rest for the bodies and souls of the ancestors (Figure 20.2). As a result, these sites are protected as sacred groves. Generally, they are numerous, small (average size is 1 ha), flat, and easily accessible by road. Each village in the region has between one and three sacred groves. Some include a tomb of a saint referred to as a "Shakhs" (which means holy grave in Kurdish). Because people attach greater spiritual significance to these groves, they tend to be larger in size and better preserved than the other sacred groves. In some cases, Shakhs became pilgrimage sites with thousands of annual visitors.

Sacred groves often form near-natural islands that are surrounded by degraded oak wood pastures and traditional farmlands that developed through a long history of civilisation in the Zagros Mountains of Iran. Local communities have traditionally protected sacred groves by establishing strict rules and taboos that

DOI: 10.4324/9781003246190-27

**FIGURE 20.2** Tomb stones inside Nameshir's sacred grove in Nameshir village, Baneh, Kurdistan, Iran. Photo credit: Zahed Shakeri.

prohibit felling of trees, hunting, livestock grazing, collecting fodder, firewood, or any plant products, and avoid disturbances such as fire. These groves are important not only for spiritual values, which are foundational for social values, but also for biodiversity, inspiration, and cultural heritage values to local Kurdish people (Plieninger et al., 2020).

The Kurdish sacred groves make up less than 1% of the forest area of 1.5 million ha, but they shelter more than 50% of the flora in this region (Shakeri et al., 2021). Many of these species are endemic and have high conservation value, making these sacred groves an important resource to protect threatened plants, insects, and animals on a national and global level. The values, taboos, and practices of sacred groves are reflected in significantly different conservation status, plant diversity, vegetation composition, and environmental conditions as compared to more intensively used woodlands. Most notably, sacred groves hold higher taxonomic diversity and harbour many vulnerable and endangered plant species. The vegetation diversity and composition of sacred groves are an outcome of biotic factors such as litter depth, crown canopy cover, soil organic carbon, and nitrogen as well as active protection by local people (Shakeri et al., 2021).

Social taboos restricting natural resource use are commonly shared by people and thus a powerful pillar of conserving biodiversity in sacred groves. Despite the region's economic and social development, taboo practices have remained intact in Kurdistan. Key holders of these values and taboos are elderly people, women, and people with traditional lifestyles, who must be empowered and/or affirmed by conservation programmes to defend and revitalise their cultures, customs, values, and taboos. While sacred groves are small in extent, they are numerous, persistent, and rich in biodiversity and ecosystem services. Kurdish sacred groves are the only remains of old-growth forests in the border regions of Iran and Iraq and they are a perfect "people and nature" model of conservation.

## References

Plieninger, T., Quintas-Soriano, C., Torralba, M., Mohammadi-Samani, K., & Shakeri, Z. (2020). Social dynamics of values, taboos and perceived threats around sacred groves in Kurdistan, Iran. *People and Nature, 2*, 1237–1250.

Shakeri, Z., Mohammadi-Samani, K., Bergmeier, E., & Plieninger, T. (2021). Spiritual values shape taxonomic diversity, vegetation composition, and conservation status in woodlands of the Northern Zagros, Iran. *Ecology and Society, 26*, 30–38.

# 20c

# FAITH-BASED ENVIRONMENTAL ACTION BY PROTESTANT CHURCHES IN SINGAPORE

*Chua Ying Xuan*

Religions are becoming more influential in the global environmental movement, as evidenced by recent research on religious environmental activism in Asia (Sponsel, 2020). Despite extensive literature analysing the relationship of Christianity and environmental action globally, there is a research gap in the Singaporean context. Existing studies in Singapore covered how religion influences environmental attitudes, without an emphasis on any particular faith (Chan and Islam, 2015).

This study aims to fill a gap in the literature by examining environmental action by Protestant Christians in Singapore. Using data from semi-structured interviews through targeted and chain referral sampling, I investigated the processes and motivations by which Protestant churches initiate environmental activities. The Resource Mobilisation Theory (RMT) is used to explain how Protestants utilise material, human, spiritual, and organisational resources to mobilise environmental action. RMT looks at why groups mobilise around a problem or issue, and how a group galvanises to pursue their collective goals (Pinard, 2011). Sociologists of religion have applied RMT to religious movements, with examples being the Unification Movement (Bromley and Shupe, 1976) and the role of black churches in the Civil Rights movement (MacAdam, 1982). RMT is used to analyse how groups mobilise on environmental issues (Bomberg and Hague, 2018).

Among the four aforementioned resources from the RMT, spiritual resources are the most important resource, as it is the basis upon which human, material, and organisational resources are mobilised. Creation Care activities are performed with the understanding of biblical doctrine. The importance of spiritual resources is seen in how local churches relied on Creation Care experts for seminars because Creation Care initiators were hindered by a lack of theological

DOI: 10.4324/9781003246190-28

understanding for Creation Care. In this example, we can see that the reason behind relying on external support (organisational resources) is spiritual in nature—to provide members a biblical perspective of Creation Care. Human resources are also valued for their biblical knowledge in Creation Care, and their actions are motivated by their spiritual beliefs. In their process of leading and serving, there is a strong spiritual element of glorifying God through this ministry. Most churches agreed that material resources are secondary to Creation Care activities and believe that material resources will be secured once they are able to obtain the support from leadership (organisational resources).

Research findings also show that Singapore's Creation Care movement is nascent with sporadic environmental initiatives by Protestant churches and was first initiated by parachurches—Christian faith-based organisations that carry out their mission independent of church oversight. However, these parachurches are informal groups without any full-time ministry staff running Creation Care initiatives. Over the years, Creation Care has gained some traction among local churches, with Methodist churches and Presbyterian churches being more prominent in Creation Care. As part of mapping Creation Care activities among Protestant churches, I designed a matrix (see below) to compare how local churches run Creation Care activities differently, with the type of activities conducted and the level of involvement by church leadership as the basis of comparison.

By shedding light on necessary resources for Creation Care activities, this study aims to build capacity among Protestant churches and advance the Creation Care movement in Singapore (Figure 20.3).

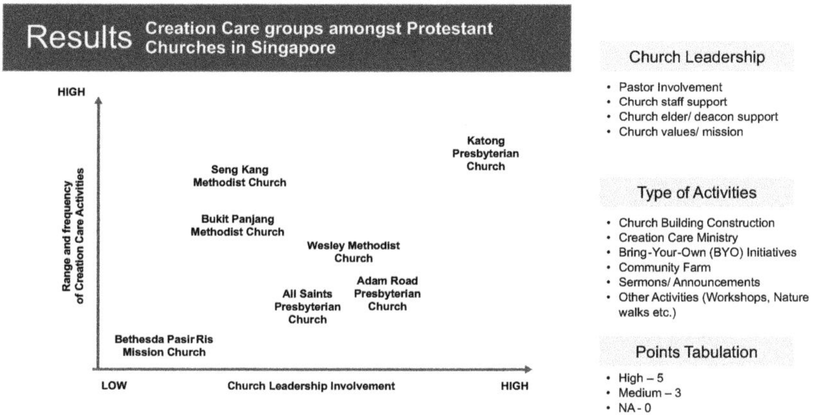


**FIGURE 20.3** Creation Care activities and leadership involvement among Protestant Churches in Singapore. The column on the right shows the criteria that measures involvement of leadership and type of Creation Care activities, where a score would be given based on the level of involvement from church leadership and regularity of activities. Credits: Chua Ying Xuan.

## References

Bomberg, E., & Hague, A. (2018). Faith-based climate action in Christian congregations: Mobilisation and spiritual resources. *Local Environment, 23*(5), 582–559.

Bromley, D., & Shupe, A., Jr. (1976). *The Moonies in America*. Beverly Hills, CA: Sage.

Chan, A., & Islam, M. D. (2015). State, religion, and environmentalism: Fostering social cohesion and environmental protection in Singapore. *Environmental Sociology, 1*(3), 177–189.

MacAdam, D. (1982). *Political process and the development of black insurgency*. Chicago, IL: University of Chicago Press.

Pinard, M. (2011). *Motivational dimensions in social movements and contentious collective action*. Montreal: McGill-Queen's University Press.

Sponsel, L. (2020). Introduction to "Religious Environmental Activism in Asia: Case Studies in Spiritual Ecology". Department of Anthropology. *Religions, 11*(7), 1–3.

# 20d

# GUARDIANS OF THE HEART OF THE WORLD

## Indigenous Cosmovision and Conservation of Colombia's Sierra Nevada de Santa Marta

*Ati Gunnawi Viviam Villafaña and Fiona Wilton*

For the Indigenous peoples living on the slopes of Colombia's Sierra Nevada de Santa Marta, sustaining the balance of the spiritual and ecological world is their sacred task. They call themselves the Elder Brothers, the guardians of the Earth, and that non-Indigenous societies are the Younger Brothers who are Se'muke (meaning an ingenuous or innocent state of mind) leading them to carry out exploitative practices, destroying the mountain's ecosystem and, by extension, the rest of the planet.

This is the highest isolated coastal mountain in the world—a massif of some 17,000 km2 that rises from the sandy shores of the Caribbean to an altitude of 5,755 m and snow-covered peaks. The full range of climatic zones for tropical America can be found here, with many endemic species in the mangrove swamps, tropical forests, and páramos (high grasslands). The massif is also a vital watershed for people and planet, with 35 rivers providing freshwater for the Indigenous communities as well as the non-Indigenous town and city dwellers and extensive agricultural lands that occupy the lower slopes.

The Sierra Nevada de Santa Marta is recognised as a vital part of an interconnected natural system. With good reason, the Kogi, Wiwa, Arhuaco, and Kankuamo, the four Indigenous groups of this region, believe that the Sierra is the beating heart of the world: what happens here happens everywhere, and when its rivers run dry, its ice caps melt, and its endemic species disappear, so do the rest of the world's (Figure 20.4.1).

> The fathers and mothers of everything on this planet are to be found in the Sierra. Knowledge of and respect for its rules allows it to act and to exist. There resides the life and spirit of all the elements that, when they come

DOI: 10.4324/9781003246190-29

**FIGURE 20.4.1** In this photo are three mamos (spiritual and political figures that unite the community) of the Indigenous peoples of the Sierra Nevada de Santa Marta, Colombia. From left to right: Mamo Kuncha (Arhuaco), Mamo Jacinto Zalabata (Kogui), Mamo Ramón Gil (Wiwa). Photo credit: Danilo Villafañe.

> together, extol the Universe. We see this manifested in the water of the sea and the rivers, the moon, the stars, the sun, the plants, the animals and us humans, the morning dew, the rainbows, the thunder, the air, blood and our ideas.
>
> *(Mamo Kuncha)*

Descendants of the Tayrona civilisation, the over 50,000 Indigenous population continue to inhabit the region in accordance with their traditions and life-ways that date back to pre-colonial days. Their spiritual and traditional leaders, known as Mamos or Mamas, maintain their deep commitment to restoring equilibrium to the Earth through daily meditations and ritual practices. This vigilance continues even as the Younger Brothers continue to encroach into the Sierra Nevada with logging, mineral extraction, commercial plantations, drug-crop cultivation, and expansive plans for damming rivers and building a coastal port.

The ancestral territory is delimited by 348 SNSs around the base of the mountain forming what is symbolically known as the *Linea Negra* ("black line"). Each site has a specific function and provides knowledge that the Mamas use to guide the use of the territory. Here, offerings are made and requests to ensure that all

**FIGURE 20.4.2** Map: Adapted by Ati Gunnawi Viviam Villafañe.
*Source*: National Natural Parks, Ministry for Environment and Sustainable Development, Colombia (2014).

human activities are compatible with the Law of Origin (Ley Sé/Sey). Here too, the sick are healed, conflicts resolved, and the Earth's vital cycle is celebrated.

A Presidential Decree 1500 was issued in 2018 recognising the interconnected sacred spaces of the *Linea Negra* as defining the ancestral territory of the Kogi, Wiwa, Arhuaco, and Kankuamo, and their Law of Origin, and with mechanisms for protection of the ancestral territory due to its spiritual, cultural, and environmental importance. It was a long-sought achievement, and yet the biodiversity, the water sources, and the sacredness of the Heart of the Earth remain under threat (Figure 20.4.2).

## References

ABColombia. (2019). *Emblematic Case: Sierra Nevada de Santa Marta* (24 July). Retrieved from https://www.abcolombia.org.uk/sierra-nevada-de-santa-marta/

Mallarach, J.-M. (Ed.). (2008). *Protected landscapes and cultural and spiritual values. Volume 2 in the series Values of Protected Landscapes and Seascapes, IUCN, GTZ and Obra Social de Caixa Catalunya.* Heidelberg: Kasparek Verlag. Retrieved from https://portals.iucn.org/library/sites/library/files/documents/2008-055.pdf

Sierra Nevada de Santa Marta. (May 4, 2009). *Sacred Land Film Project*. Retrieved from https://sacredland.org/sierra-nevada-de-santa-marta-colombia/

Wisdom from the Heart of the World: The voice of an Arhuaco youth environmental activist, Ati Viviam Villafaña. Living Nature event, organized by Invisible Dust and Flourishing Diversity, 28 October 2021. Retrieved from https://invisibledust.com/projects/living-nature#Viviam

# 20e

# TU BE'SHVAT—THE JEWISH FESTIVAL OF TREES

*Idit Pintel-Ginsberg*

Tu Be'Shvat, the 15th day of the month of Shevat, is first mentioned in a 3rd century CE Jewish canonical text, referring to it as "the New Year of the Trees" for taxation purposes. Occurring between mid-January to mid-February, it coincides in Israel with the blossoming of almond trees.

From the Middle Ages, pointing at the festiveness of "the New Year of the Trees" and the deep bond that already existed between Diaspora Jewry and the Land of Israel, a few customs surrounding it are mentioned: eating fruits of the Holy Land, excluding prayers of mourning, and abstaining from fasting.

Among Jewish communities whose descendants fled from the Spanish expulsion, the custom of eating fruits became elaborate and central to its observance. It is called the "Tu Be'Shvat Seder," for it carries a slight resemblance to the "Passover Seder" as it includes the drinking of four glasses of wine and the reading of given texts. The fruits listed include 21 different types of fruits, six of them native to Israel. This custom is described in "Hemdat Yamim"—an anonymous treatise based on Kabballistic Mysticism published in Izmir, Turkey in the 18th century.

The beginning of the 20th century marked a most significant turning point in Tu Be'Shvat's development. The secular Zionist movement (a national Jewish movement for the return of Jews to their homeland, Israel) turned it into the "Festival of Trees" while celebrating it with a new ritual: planting tree seedlings in Israel's barren lands. This ritual accurately fitted the Zionist narrative of returning the degraded Land of Israel to its supposedly former, flourishing glory, as the biblical "Land of Milk and Honey."

The first recorded tree planting on Tu Be'Shvat was in 1884 at the village of Yesod Hama'alah in northern Israel. The planters saw their acts as reflecting God's deeds in the Garden of Eden.

DOI: 10.4324/9781003246190-30

In 1908, initiated by the school system, it became the "planting day of the pupils of Israel." Schools participated in the mass effort of "changing the desert into a flourishing meadow."

**FIGURE 20.5** Children in Jerusalem, next to a blossoming almond tree, holding seedlings and hoes, Tu Be'Shvat 1912. Photo Credits: Yaakov Ben Dov (PHG/1017692) The Central Zionist Archives, Jerusalem Israel.

It is now Israel's official "Arbor Day," and as such is actively promoted by the Jewish National Fund (JNF)—an NGO acting as Israel's forest service. Each year, JNF organises nation-wide tree planting ceremonies, Tu Be'Shvat Seders, and other outdoor activities. JNF distributes educational and promotional materials concerning the value and function of trees in preserving Israel's fragile ecosystems. They also distribute, free of charge, over 100,000 seedlings annually to municipalities, schools, youth movements, and communities wishing to hold their own ceremonies.

In the 1950s and 1960s, tree planting ceremonies revolved around the establishment of new forests and commemorative groves and were based primarily upon three tree species. Today however, tree planting ceremonies centre on the restoration and rehabilitation of burnt forests and groves decimated by pests or drought, utilising a large variety of native and exotic tree species to create a more biodiverse forest (Figure 20.5).

Since January 2020, a more sophisticated tree planting campaign was launched: the public were invited to purchase and plant trees online. The invitation was made virtually through the JNF website, thus ensuring the public's participation in JNF afforestation efforts, through the work of its professional forestry staff.

As it is well known, ceremonies, rituals, and customs deepen and strengthen the beliefs and the values of society. It seems that the customs around Tu Be'Shvat do just that: the faith that planting trees by man is actually a reflection of God's deeds and is valued as a divine act, the desire to improve the condition of the land and its landscape is conveyed into a simple and direct human act achievable even for children. Thus, for over 120 years, Tu Be'Shvat ceremonies have constituted a tangible and participatory form of nature conservation.

# 20f

# CASE STUDY OF FOREST OF GRACE IN MONGOLIA

*Hang Ryeol Na*

Asian dust, also known as yellow dust, has been a public health issue in Korea since at least the 20th century. Coupled with smog, the particulate matter (PM) has negative effects on respiratory, cardiovascular, and neurological systems, causing diseases, including Parkinson's. These environmental health issues, among others, are behind the motivation for the Forest of Grace project.

The Forest of Grace is an afforestation project with the goal of proffering an ecological solution to the intensified desertification in Mongolia, where high-speed surface winds and dust storms kick up dense clouds of dry soil particles. From the deserts of Mongolia and northern China, yellow dust is carried eastwards by winds and passes over China, Korea, and Japan, especially in springtime. On a 30 ha area of barren land in Argalant, 82 km west of Ulaanbaatar, the capital of Mongolia, tree planting was initiated by the Korea Christian Environmental Movement Solidarity for Integrity of Creation (KCEMS) in 2009, in collaboration with Korean churches and a Mongolian non-governmental organisation named Green Silkroad.

Since the beginning of the project, approximately 28,000 trees of 11 different species have been planted (KCEMS, 2019). In addition to mitigating the desertification itself, benefits include windbreaks to slow the wind, which create favourable conditions for soils, crops, livestock, wildlife, and local people. Fruit trees were also chosen to provide economic and community benefits for local livelihoods in Mongolia (Figures 20.6.1 and 20.6.2).

A driving force for the Forest of Grace project has been a heartfelt desire to fulfil Christian Mission (Yim, 2011), in which South Korea is one of the most active countries in the world. Of any country, South Korea is surpassed only by the United States in the number of Christian missionaries it trains and sends abroad. These missionaries recognise an understanding of eco-theology as essential to

DOI: 10.4324/9781003246190-31

**FIGURE 20.6.1** Before the project.* Photo credit: Grace Grove 2009. https://www.facebook.com/gracegrove2009

**FIGURE 20.6.2** After the project. * Photo credit: Grace Grove 2009. https://www.facebook.com/gracegrove2009

their message of hope for the world. Indeed, the church's concern for the natural environment appears to reinforce their enthusiasm for global mission.

> I will put in the desert the cedar and the acacia, the myrtle and the olive. I will set junipers in the wasteland, the fir and the cypress together, so that people may see and know, may consider and understand, that the hand of the LORD has done this...

These verses from Isaiah 41:19–20 (NIV) are a typical example of what the KCEMS quotes for their activities.

Proponents of the Forest of Grace are aware that environmental health and Christian Mission have, in the past, been viewed as incompatible due to a binary perspective that perceived humans and nature as disparate domains. Under this perspective, human health was prioritised for attention over that of the environment because of a misunderstanding that physical, mental, and spiritual health of the individual is inseparable from the health of God's creation of which humans are a part. This understanding is reflected in the first principle of the 1992 Rio Declaration on Environment and Development which states, "Human beings are at the centre of concerns for sustainable development. They are entitled to a healthy and productive life in harmony with nature" (UN, 1992). This, more holistic, Christian theology, regards the Forest of Grace project as firmly rooted in Christian faith, such that anthropocentrism should be replaced by a God-ordained ecocentrism or deicentrism. This changed perception replaces an attitude of human conquest over nature with one of care and coexistence with humans living in harmony with the ecosystems of which they are a part, thus emphasising the spirit of stewardship.

## References


Korea Christian Environmental Movement Solidarity for integrity of Creation (KCEMS). (2019). *10th anniversary seminar for the Mongolian Forest of Grace Project.*

UN. (1992). *Rio Declaration on Environment and Development.* UN Doc. A/CONF.151/26, Principle 1.

Yim, H. M. (2011). The ecological mission of 'the Grove of Grace' for Mongolia. *Mission and Theology, 27*, 47–77. (Original in Korean:임희모 (2011) 몽골 '은총의 숲' 생태선교 : 북한 산림녹화 생태선교에 주는 함의, 선교와 신학, 27, 47–77)

# 20g

# JAINISM AND NATURE CONSERVATION

*Nita Shah and Anant Deshwal*[1]

Jainism has no concept of creator or hierarchy, simply, all life forms are considered equal. There is no discrimination based on sex or species. Both living and non-living beings have souls, and all are considered equal. Jain philosophy has three main facets: Ahimsa (non-violence), Anekantvad (viewing truth in its diverse aspects) and Aparigrah (non-possession). Souls are divided into two classes: Stahvara (immobile) and Trasa (mobile). Immobile souls consist of earth, water, fire, and vegetation and only have a sense of touch, while mobile souls can have two to five senses depending upon the organism (Jaini, 2000).

In Jain beliefs, Ahimsa is the central tenet for wildlife conservation. Jainism teaches ethics of Ahimsa that can be practised by following five precautions: preservation of speech (Vagupati), preservation of mind (Manogupti), care in walking (Irya), care in picking and laying things (Adana-niksepana) and taking care in eating and drinking (Alokitapana-bhojana) (Dunda, 2002). Jainism recognises that violence can be physical, verbal, and emotional and so its Ahimsa teachings relate to all forms of human action, professing empathy for all living beings: plant, animal, or human.

Aparigrah (non-possession or simplicity) is a mental attitude of non-attachment to possessions, curtailment of one's wants while not depriving others of their legitimate possessions. It promotes following a minimalist or ascetic lifestyle, which has strongly influenced the present authors. Nita Shah feels privileged to have been born a Jain as her beliefs have allowed her to follow the path of her dreams while being protectionist towards nature by following a minimalist lifestyle and being non-ritualistic. Anant Deshwal, however, was not born a Jain but was captivated by the teachings of Mahavira (the last Tithankara: Tithankaras show the path from "the phenomenal world" to attain "liberation/moksha").

DOI: 10.4324/9781003246190-32

Nita's life has been a testimony to Jain belief system. Her actions, be it little deeds or major conservation projects, have been aimed at helping other life forms. She has devoted herself to protecting the habitat in the Rann of Kutch in India, and prevented this saline desert from being denotified, leading to the conservation of several characteristic taxa of the habitat, such as the Indian Wild Ass (*Equus hemionus khur*), flamingos, the spiny-tailed lizard, the wintering Houbara Bustard and two species of migratory cranes. Her intense top-down and bottom-up advocacy strategies have been crucial in evolving policies that saved the three Gyps vultures in the Indian Sub-continent, creating conservation ripples globally. She has also been instrumental in conserving wild equids, the endemic Kashmiri Stag (*Cervus hanglu hanglu*, a subspecies of Central Asian Red Deer), and the Tibetan Antelope (Chiru) (*Pantholops hodgsonii*). Once on a peak winter evening on the Northern Tibetan Plateau in east Xinjiang Province, Nita spotted a Tibetan Wolf (a prime female) (*Canis lupus chanco)* caught in an inhumanely designed leg-hold trap (laid by local herders). Despite the inherent difficulties of the situation, she was successful in rescuing and setting her free. All these acts were, in a sense, 'instinctive'. The embodiment of Jain beliefs provides guidance to limit one's possessions, for compassion, and for acceptance that there exist multiple paths to reconciliation, which bring synergy and harmony between the inner (mental) and outer (physical) worlds.

Following the teachings of Mahavira, Anant quit a lucrative career as a civil engineer to pursue his dream of conserving wildlife and nature. In this pursuit, he often chose to help conservation-based organisations but struggled with manpower. He has followed a minimalist lifestyle and has devoted his life to wildlife conservation. Before starting his PhD, his only possessions were his motorcycle and a backpack of clothes. He often chose to help conservation-based organisations but struggled with manpower. Both Nita and him, while growing up, have independently provided shelter and care to dying, old, or injured animals. Jainism and its philosophy have influenced people over generations, for conserving wildlife even before there was a dire need for it.

Jainism and its roots in conservation have existed in India for over two millennia. Accounts of Jains entering the political arena to advocate for animal protection are aplenty. The most notable account was that of the Śvetāmbara monk Hīravijaya, who convinced the Mughal emperor Akbar to order a prohibition on the killing of animals around the sacred places and on festive days of the Jains (Jaini, 1998: 283–284). There are numerous reputable charitable animal hospitals (pāñjarāpol) run by Jains in accordance with Jain ideology. The embodiment of Jain beliefs provides guidance to limit one's possessions, for compassion and for acceptance that there exist multiple paths to reconciliation, which bring synergy and harmony between the inner (mental) and outer (physical) worlds. It is invigorating to note that despite Jainism being an ancient religion, it offers an open platform for discussion, learning, and awareness, even in the 21st century.

## Note

1 We would like to mention that both authors have taken the lead on this piece at various junctures - this was needed due to the situation caused by the Covid pandemic.

## References

Dunda, P. (2002) *The Jains (2nd Edition)*. London and New York: Routledge.
Jaini, P. S. (1998) *The Jaina Path of Purification*. New Delhi: Motilal Banarsidass.
Jaini, P. S. (2000). *Collected papers on Jaina studies*. Delhi: Motilal Banarsidass.

# 20h

# FAITH AND SCIENCE DIRECTING A ROCHA'S MARINE CONSERVATION WORK IN KENYA

*Robert D. Sluka*

Tropical marine ecosystems link with terrestrial forests through watersheds and are impacted by changes in land use. Watamu Marine National Park (WMNP) is linked with Mida Creek mangrove forests and the Arabuko-Sokoke Forest conserving marine habitats on Kenya's coast. Much of the research and tourism focus is on the iconic coral reefs which harbour diverse marine life. Little has been documented about the biodiversity of non-charismatic species and habitats. Motivated by biblical commands and love of God (Sluka, 2012), A Rocha Kenya began in 2011 to study the species and habitat biodiversity of WMNP. Even though the ocean occupies 71% of the planet, most of the literature and thinking around a Christian response to caring for the planet has focused on terrestrial concerns. The A Rocha team wanted to understand better how Christian thought applied to the ocean and WMNP was an experimental laboratory not just for science, but for our thinking on how to implement holistic Christian principles in a particular place (Figure 20.7).

A Rocha Kenya and International staff along with many volunteers/interns systematically studied taxonomic groups within marine habitats from 2011 to 2016. This led to a detailed understanding of those species and habitats which were important for conservation (Cowburn et al., 2018). We identified a coral species, *Anomastraea irregularis*, which is only found in this area of the world, but which appears to have no economic value to humans. These corals live in nearshore rockpools that had been overlooked for conservation. Local young men, known as Beach Boys, carve out a living by trying to convince tourists into booking low quality tours of these rockpools giving dubious information and anecdotes. The A Rocha Kenya team tried to model a holism, integrating Christian love for God and neighbour, including our non-human neighbours, by developing a conservation and training programme that started in 2016 helping

DOI: 10.4324/9781003246190-33

**FIGURE 20.7** A Rocha Kenya researchers investigate the intertidal zone of Watamu Marine National Park. Photo credit: Sarah Sluka.

these Beach Boys to better understand the habitat and desist from environmentally destructive behaviours.

The A Rocha team learned much in attempting this type of holistic project and we certainly would not claim success yet.[1] But this has given us an example of how to try and implement a marine conservation project which takes science and theology seriously. Students were trained in the process, and A Rocha staff wrote extensively on the scientific and theological lessons we were learning (Sluka, 2016). Ultimately, this gave shape to a wider Marine Conservation Programme[2] within A Rocha and the Christian motivation to love God and neighbour has augmented efforts globally attempting to let both science and theology drive our conservation efforts and their implementation.

## Notes

1 In 2018, A Rocha began a partnership with St Andrews University, Scotland which seeks to extend love of neighbour more generally to the local community and continues in 2021 to investigate livelihood development and participatory management of WMNP.

2 www.arocha.org/marine

## References

Cowburn, B., Musembi, P. M., Sindorf, V., Kohlmeier, D., Raker, C., Nussbaumer, A., … Rogers, A. D. (2018). The habitats and biodiversity of Watamu Marine National Park: Evaluating our knowledge of one of East Africa's oldest Marine protected areas. *Atoll Research Bulletin, 618*, 1–45.

Sluka, R. D. (2012). *Hope for the Ocean: Marine biodiversity, poverty alleviation and blessing the Nations* (pp. 1–28). Cambridge: Grove Books Limited.

Sluka, R. D. (2016). The hidden things of God in the Ocean. *Anglican EcoCare Journal of EcoTheology, 2*, 41–50.

# 20i

# FAITH IN CONSERVATION IN PAPUA NEW GUINEA

*Tweedy Malagian*

Biblical theology and Indigenous (tribal) land ethics have more in common with each other than the extractive, utilitarian land ethic of the cash-based world and the development paradigm. Papua New Guinea (PNG) offers a unique opportunity to showcase this as the country is predominantly Christian (96%), while according to the constitution and the ancestors, the land is held in customary tenure collectively by clans and tribes and is unalienable.

There are many sacred sites and landscapes known as "masalai ples" all over PNG that are taboo for hunting and cutting down trees. People believe that spirits inhabit these sacred places. Early missionaries had destroyed a lot of traditional rituals including spiritual ornaments and "masalai ples" during the missionary era but as Papua New Guineans began to receive good education, many see their duty towards God the Creator as involving the preservation of His flora and fauna and all living things that would normally be destroyed by man's greed. And these sacred places help in a big way to protect them to recover, recuperate, rejuvenate, and reproduce. They still believe, cherish, and hold onto their traditional customary land tenure system which is their heritage, and are exploring what Christian Faith has to say about the closely linked problems of their land, sea, and people and the effects of climate change and sustainable living in PNG.

However, corruption in high places is rife and there are corrupt foreign loggers and miners who do not have any respect at all for the sacred places. Hence, there are a number of local environmental groups which are standing up to loggers and miners and are also inspired by faith. Gildipasi Environmental Association in Madang is currently fighting against an Asian logging company which intruded into their protected and sacred areas and has logged hundreds and thousands of virgin trees (Figure 20.8).

DOI: 10.4324/9781003246190-34

FIGURE 20.8 Members of the Gildipasi Association demanding a stop to the Asian logging company on their land. Photo credit: Mathew Lawun (Spokesman for Gildipasi).

There is still a very high respect for "*masalai ples*" (sacred places) all over PNG and the need to educate the people about their importance is more urgent now than ever before.

In general, PNG faith traditions demand respect towards our land, mountains, trees, rivers and sea, and its creatures. When a new garden is going to be made, it is common for a gardener to talk to the trees before cutting them down: "I am sorry I am going to cut you down so that I can make my new garden. Thank you for your branches which I will use for my house, firewood and the trunk for my new canoe." They talk to the soil as they dig, asking the soil to bless their crops to yield a bigger harvest in the months to come.

"We are belly-button Christians! Our umbilical cords are still connected to the earth, our mother." So claimed Yat in a speech, an environmentalist from Caritas Papua New Guinea, who is providing advice and support to the Gildipasi Environmental Association in the North Coast of Madang. "But when our tribal people leave the rainforest and go into the city, we say that they have lost their bellybuttons; they no longer have a living connection to the earth."

"Our land was never for sale!" said Yat. "The earth is our mother. You never sell your mother." Yat, a committed, life-long, seminary-trained Catholic, continued, "One hundred years ago, my ancestors gave the church permission to use the land. That was all. Our land was never sold."

Bernard Narakobi wrote in his book "The Melanesian Way":

> Our vision sees the human person in his totality with the spirit world as well as the animal and the plant world. This human person is not the absolute master of the universe but an important component in the interdependent world of the person with the animal, the plant and the spiritual. However, he came to be, the Melanesian is!

## Reference

Narakobi, B. (1983). *The Melanesian Way*. Boroko and Suva: Institute of Papua New Guinea Studies.

# 20j

# VULTURES, GRIEF, AND INTERSPECIES RITUALS

*Will Tuladhar-Douglas*

If one were to survey the few remaining zones of interaction between humans and vultures, it would be hard to grasp how old and intimate this relationship has been. Vultures have helped humans confront death quite possibly for as long as we have been human.

Six species of vulture are found in South Asia: *Gyps. fulvus*, *G. tenuirostris*, *G. bengalensis*, *G. indicus*, *Neophron percnopteris*, and *Sarcogyps caalvus;* on the Himalayan plateau, one finds *Gyps himalayensis* (which ranges down into Nepal and Bhutan) and *Gypaaetus barbatus*. All of the South Asian species declined sharply after 1990. The die-off was clearly caused by a veterinary medicine (Diclofenac) in carrion, and it appears that *G. himalayensis* was also affected at least in the southern part of its range.

I am pursuing comparative research into three sites in Asia where humans have recently fed vultures:

1. The Towers of Silence (*dhakma*), where Parsis purified their corpses with the help of vultures. These fell into disuse as a result of the near extinction of vultures in South Asia from 1990 onwards due to diclofenac poisoning.
2. The bird offering (*bya gtor*) sites in Tibet, where Tibetans who had died clean deaths were offered to vultures. These are presently contested in a number of ways between Tibetans and Han Chinese authorities and tourists.
3. "Vulture restaurants" established by conservation biologists after 1990 in a desperate attempt to save the remaining populations of vultures in South Asia.

The first two of these sites need to be contextualised historically; contrary to some accounts, there is no evidence that the Tibetans adopted this practice from

DOI: 10.4324/9781003246190-35

earlier Zoroastrians (the Iranian community that is the source of the Parsis in India). Rather, there is good evidence for the antiquity of exposing dead bodies to scavengers in early South Asian literature—horrific descriptions of charnel grounds and battlefields are a feature of both Sanskrit and Pali literatures—and we have good archaeological evidence for ritual offering of dead bodies and even dancing as or with vultures from sites such as Catal Huyuk and indeed all the way back to late Neanderthal sites in Europe.

In theoretical terms, however, we can gain critical insights into our own presuppositions by studying these three sites together. There are surprising commonalities. In particular, the complex interactions between human and vulture sociabilities are mediated through spatial and social boundaries. There are specific places surrounded by strong social and physical barriers, but there are also specific human subgroups who carry a double burden: they process the grief of the community where someone has died, and they are also shunned and regarded as unclean or polluted. Among the Parsis, these are the *nussesalars*; among the Tibetans, these are the corpse-cutters; and, surprisingly, among humans in Nepal where I conducted fieldwork, these are highly educated upper-caste scientists who have to negotiate the impurity of handling the carcasses of buffalo and other bovines. In all three cases, the mourning and grief that accompanies death is managed at liminal sites, by liminal specialists. While the conservation biologists who manage the vulture restaurants in Nepal are not managing individual human deaths, they are willing to challenge the profound social stigma associated with bovine carcasses and vultures because of the grief they feel at the horrific die-off of vultures across South Asia (and now Africa as well). A final and especially interesting commonality is that all three groups of vulture experts must be aware of the internal social organisation of the vulture communities that come to feed, though obviously in different ways—for example, for the Tibetans, the vultures are sky-flying spirits and powerful teachers, led by an especially wise vulture. Each community of vulture experts has evolved gestures and rituals that allow them to call the vultures as needed, and negotiate with these remarkably skilled and hygienic cleaners.

Hence, our assumptions about what constitutes a sacred site, for the purposes of conservation biology, blur badly in this case. The heartfelt grief of the conservation biologists, and their acute awareness of the possibility of extinction, is just as effective a sentiment as the more traditionally religious or spiritual concerns of the Parsis and Tibetans. All three communities have evolved an architecture and a set of signals shared between vultures and humans; and, importantly, in all three cases, the place where they interact is a protected area created through other effective means that is rich in biodiversity.

## Recommended Reading

Bush, R. (2013). *Vultures of Tibet*. Retrieved from https://vimeo.com/141606509.

Cuthbert, R. J., et al. (2016). Continuing mortality of vultures in India associated with illegal veterinary use of diclofenac and a potential threat from nimesulide. *Oryx, 50*(1), 104–112.

Mistry, C. (2014). *Chronicle of a Corpse Bearer*. New Delhi: Rupa.
Ogada, D., et al. (2016). Another Continental Vulture Crisis: Africa's Vultures collapsing toward extinction. *Conservation Letters, 9*(2), 89–97.
Russell, N. (2019). Spirit Birds at Neolithic Çatalhöyük. *Environmental Archaeology, 24*(4), 377–386.
Van Dooren, T. (2011). *Vulture*. London: Reaktion Books.

# EPILOGUE

*Andrew Gosler, Radhika Borde, Alison Ormsby, Stephen Awoyemi*

To spend time in nature is to be restored, body and soul. This simple wisdom, embodied in the opening words of Psalm 23 of the Abrahamic faiths,[1] is the rationale for this book and underpins the values and understanding of many of the authors included within it. Its apparent simplicity, however, belies a bewildering tangle of what, for some, are contentious issues. What, for example, do we mean by 'nature' and what is the 'soul'? Both have profoundly different meanings for each of us as our understanding of the world is experienced, shaped and interpreted through the prism of personalised culture and worldview, formal and informal education. If the relationship between humanity and the biosphere were healthy and sustainable, the concept of conservation should be inherent. Yet, the very complexity of our relationships with the concepts of 'nature' and 'soul' underwrites the extinction crisis that is the Anthropocene, and is one of the key topics explored in this book.

## Challenges of the Anthropocene

*Crisis* is a Greek word meaning judgement or discernment. Whether we take it to imply a divine judgement of humanity, or a practical judgement by humanity of the necessary steps forward, there is no doubt that we find ourselves at a turning point. This is because we have not merely entered the earth's sixth mass global extinction,[2] but we are its cause (Kolbert, 2014). The sixth extinction is unique. It is unique not only in its comprehensive nature, affecting almost every species on the planet, ecosystems, climate, atmospheric and marine composition. Neither does its uniqueness rest only on its causation by the diverse activities of a single animal species. Whilst, to the best of our knowledge, both are true, the sixth extinction is unique in that through its own science (global and local;

DOI: 10.4324/9781003246190-36

Sillitoe, 2006), this causative species understands what it is doing. Hence in the anthropogenic sixth extinction, the judgement is not merely a scientific one of the *best* course of action, but a moral one of the *right* course of action. In this crisis, our worldview comes into sharp focus as literally *critical*. Our worldview will deliver answers to ultimate and moral questions such as: To whom does the planet ultimately belong? And, what right do we have to destroy phenomena of such beauty and complexity of form and function as living organisms and ecosystems, which have taken millions of years to create? That religious responses to these questions lead to life-sustaining solutions should tell us something about the nature of truth expressed through the age-old wisdom of religious understanding. We should not be surprised, therefore, to find sacred natural sites across the globe to be the sanctuaries where biodiversity not only persists, but thrives, as has been reported time and again within this book and elsewhere (Verschuuren *et al.*, 2010; Berkes, 2017; Liljeblad & Verschuuren, 2019).

The Anthropocene crisis therefore reflects a fundamental question in the human psyche: What is the place of humans in the world/biosphere/creation? It is a question that reveals the environmental crisis to be a manifestation of a deeper human spiritual crisis, for whilst we are clearly of this world, we seem curiously 'other'. Science and scriptures tell us we are part of, and rooted in, the greater community of life. Yet whilst evolved out of that greater community, the nature of humans, physical (e.g., fully bipedal), mental (e.g., the extent of, or capacity for reasoned thought) and moral (see Katz, 2000), casts us like no other organism. There is, however, a risk in a human focus on our own distinctiveness that can lead to a sense of human exceptionalism. This can lead to a belief that only humans are conscious or that humanity can persist independently of the biosphere of which it is but a part. The first of these misjudgements is challenged daily through an engagement with other species, including pets, and the second by the environmental crisis of the Anthropocene.

## Terminology and Awareness

Expressing this deeper spiritual crisis raises concerns about current thinking and policy in conservation. It is this: If the restorative power of nature for our well-being should be at the heart of the human conservation motive, why has the language and policy of a globalised professional conservation community become so dominated by quasi-economic concepts such as ecosystem services, which reflect neither the true relationship between humans and the biosphere, nor the true motivation of most conservationists to conserve? We suggest that the language and perceptions of faith-based conservationists speaking through this volume and elsewhere better reflect those true relationships.

A globalised humanity (or what may be called the 'Younger Brothers' – see Villafaña and Wilton in this volume) today goes about its business largely oblivious of the non-human world. Numerous studies from around the world demonstrate that nature is less salient to young people today than it was to their grandparents

(Spoon, 2014; Gosler & Tilling, 2021). Furthermore, their ignorance of nature makes them more threatening to the well-being of other species. But this is not true of all humans. The traditional and local ecological knowledge held by elders in Indigenous communities (the 'Elder Brothers' – see Villafaña and Wilton as above) demonstrates extensive experience of a diverse range of plant and animal species, and the names coined for them speak of their relationships with people. Also, this disconnect has not always existed. For example, Desfayes (1998) lists more than 100,000 local names for wild birds (c. 600 species) that were collected from across Europe, mostly prior to the 20th century. However, because knowledge is retained on a need-to-know basis, it must be passed between generations if it is to persist and be viable. Amongst the Elder Brothers, knowledge of nature is bound up with religious or spiritual knowledge and awareness (Verschuuren & Brown, 2019; Liljeblad & Verschuuren, 2019; Berkes, 2017). The declining salience of nature amongst young people has many causes, including urbanisation and an increasing attention to the internet (summarised in Gosler and Tilling, 2021). It may also parallel declining religious attendance (Franck & Iannaccone, 2014), and a declining interest, of young people in particular, in religious matters.[3] The reasons for this decline are also complex, and correlation does not imply causation, but coupled also with a rise in mental health issues amongst young people (Sellers *et al.*, 2019), the question arises of whether these trends express the ancient wisdom with which we started this chapter. Namely, if a connection with nature has the power to heal (Maybe, 2008; Harkness, 2020), may we not be harmed, physically and mentally, by a disconnection from the life-sustaining grounding given through a connection with the non-human world?

The wisdom of the Elder Brothers suggests that the human connection with nature is a profoundly spiritual one. This suggests that to deny our spiritual nature, as has become fashionable in many Western secular and academic circles, may deny something of fundamental significance for our understanding of life, and effectiveness as conservationists. We might reflect on the significance of sacred natural sites as centres of biodiversity, and recognise the need to radically change our practical, detached approach to nature conservation. Across the world, nature reserves hold great significance as sanctuaries for endangered species. Yet, many are threatened by visitors who, whilst attracted to such places, do not understand the conservation importance of the site or species conserved (Honey & Frenkiel, 2021). Paradoxically, that, in turn, can result in sites of conservation interest suffering greater damage than had they remained ungazetted. If sacred natural sites have high biodiversity, might we not support conservation better by recognising sites of high biodiversity as sites to be revered? Sacred natural sites deserve respect and imply a 'proper' way to behave. These are the very same ways of being that are required urgently by many of our nature reserves if they are to continue to offer sanctuary for rare species in fragile habitats.

In ecology, every species has its niche. Intimately associated with its evolved anatomy, physiology, diet and behaviour, every species is found to have a 'place' and to influence the web of life. Furthermore, as we delve more deeply into the nature of ecological relationships, we find them to be most readily characterised as engagements of mutual dependency, be they fungi and plants (Sheldrake, 2020), plants and insects (Hughes, 2022), plants and vertebrates (Fleming & Kress, 2011; Horn *et al.*, 2011), even predator and prey (Terbough & Estes, 2010) and ultimately Gaia (Lovelock, 2016). This stands in stark contrast to the assumptions of earlier generations of naturalists who perceived only competition (Darwin, 1859) as the driving force in nature. But if the mutual dependencies that are being described can be taken to be an ecological universal, what of humans? Through the growing awareness of ecological mutuality and human embeddedness in nature, earlier conceptualisations of ecosystem services, which were rendered ecologically unsustainable by their explicit anthropocentricity, are being revised so as to recognise human services to nature as well as ecosystem services to humanity (Comberti *et al.*, 2015). Ecologically, as an omnivore, people might be seen to fill several ecological roles, including both seed disperser and driver of ecological cascades. However, if the place, role and significance of humanity in the world have to do with the 'nature' of humans, we must acknowledge all aspects of the human condition. Perhaps the most significant aspect of human nature is that life is lived in the mind, where a concern for matters spiritual is ever-present. It is essential, therefore, if we are truly to discern the place, the 'purpose', of humanity, that we give due regard to the significance of spiritual and religious matters in human affairs.

## Practices and Relationships

The religious life is concerned as much with ways of being and doing as it is with ways of seeing and understanding. Involving a respect for the self as well as for communities of being, these considerations often involve a sense of the 'right' ways to be or do, in order to be regarded as a 'real' human being. Evolved through an ongoing dialogue with the rest of nature, it is then not surprising that the experience of time spent observing nature resonates with many themes of religious experience and practice. Themes such as silence and stillness, awe, surprise, delight and joy, an intense awareness of the senses and of the self as well as a respect for the other and need for self-discipline permeate both experienced domains (i.e., the experience of nature and the experience of religion). Acts of mindfulness, meditation and contemplation combine to produce a sense of the sacred space in which a focus on the nature of one's relationship with the other, perceiving the sanctity of life, is heightened. Central also to human engagement with non-human organisms is a sense that whether they ignore us, seek protection from us or flee from us; what they tell us about ourselves is invariably honest. Through such relationships, we are confronted with truth, a fact which resonates with the absolute need for self-awareness and honesty in the presence of the divine.

And what of the rituals, festivals and holy days of religion, reflected, for example, in the chapter by Sheth in this book? These may mark the passage of the seasons, springtime and harvest, the gifts of the earth, of friendship and gift of one to another, as well as our own significant life events from birth to death. Religious observance is also typically rich with symbolic meaning (which requires learning), standing for the relationship between humans, other creatures and the existent realities of the numinous domain of potential, guidance and morality. Thus, the communal events, worship and festivities of religion are themselves a celebration of life. Viewed in this way, through the diverse traditions and practices of religion, a role for humanity comes into focus. We might perceive that humanity is that part of the web of life that has emerged to celebrate life itself. The love of life and the drive to celebrate it express an understanding also of the need to nurture and sustain life in all its expressions. Here then might be discerned the natural roots of the human motivation to conserve. Here also can be found the most profound understandings of the meaning of loss, whether personal bereavement or the extinction of life or its diversity. Through all this is the centrality in human affairs of love and relationship within families and communities and potentially beyond the human community to a broader community of life. Since, as has been observed, we shall not save what we do not love (Gosler *et al.*, 2013), in the Anthropocene, the repeating themes of hope, love and loss have a particular poignancy.

## Reverence for Species and Habitats

Throughout the world's varied religious and spiritual traditions, a great diversity of organisms finds religious and cultural significance. See, for example, the chapter by Colorado *et al.* on snow leopards in this book, and one might think of the diverse roles of birds as messengers of God or the gods (Deshwal in this book; Tidemann & Gosler, 2010; Sax, 2021). But one group presents repeatedly and with overwhelming symbolic poignancy: trees. From the *Tree of Life* and the *Tree of the Knowledge of Good and Evil* in the Biblical account of the Garden of Eden (Holy Bible: Book of Genesis), to the Festival of Trees (Pintel-Ginsberg in this volume), and their significance for sacred places, sacred groves and forests throughout the world, trees have overwhelming importance. The ecological significance of trees in all their diversity is unparalleled, not only since without them there would be no forests, but because by virtue of their physiology, phenology, interactions with other organisms and influence on both soils and atmospheric gas and water content, they conduct the whole rhythm and symphony of the forest. The greatest terrestrial biocultural diversity on the planet is found in woodland or forest (MEA, 2005), and forest is the only ecosystem that contributes to all human needs as ecosystem services (MEA, 2005; Ninan, 2009). Nevertheless, their persistence is one of the most pressing issues of the Anthropocene. Reflecting the overwhelming importance of forest, this volume contains chapters about their spiritual significance globally (Barrow chapter), in Japan (Massey and colleagues' chapter), India (Ormsby and Krishnan's chapter) and Ethiopia (chapter by Massey and colleagues). In the Jewish tradition, the symmetrical form of the written account of creation in Genesis 1 recalls a symbolism of days represented physically in the symmetrical form of the seven branched lampstand or *Menorah*. Reminiscent of an evolutionary tree, the Menorah symbolises the Tree of Life. Its Biblical description, Exodus (25:31–40), cites the significance of the almond tree (whose Hebrew name is itself rich with symbolic meaning) and the olive. If the religious perception of the human role in nature affirms its conservation, through the story and symbolism of the *Tree of Hope* (Coreth *et al.* this volume), we must recognise that the shared concern for life on earth has the power to diffuse and resolve cultural and religious tensions. Perhaps recognising the multifarious means by which trees express the principle of mutual dependency (fungi, invertebrates, vertebrates, etc.), the trees stand as testimony to the fundamental peacebuilding roles of conservation and faith. As recognised in the Holy Qur'an of Islam that along with all other creatures, every tree is worshipping Allah'(Surah Al-hajj – 18). Through the eyes of diverse faiths, the deforestation of the world, destroying millions of years' worth of creative evolution for financial gain, is the greatest of sins and truly it is fuelling a crisis.

## Final Thoughts

We would end with a reflection on a significant message of this volume and on its title. A repeated theme of this book has been a sense of the stewardship role of humans within ecosystems, captured powerfully in the words of a prayer from the Atharva Veda.[4] But it is more than this. Within the context of a religious perception of the sacrality of life, the stewardship role becomes a sacred duty. Conservation, the prevention of extinction and of creating the best conditions for diverse communities of life, can be perceived as part of that sacred duty of human stewardship. Whilst the role of scientifically trained conservation professionals must remain significant in helping in the biosphere's healing, it has become clear that conservation, this sacred duty, has been undertaken for thousands of years by local communities in the context of their faiths. The challenge going forward is to recognise the value of both local and global perspectives to work together with humility in a spirit of mutual respect.

An early working title for this book was *Faith in Conservation*. We delighted in both its simplicity of style and its obvious ambiguity, both of whose meanings we find empowering. We were convinced by our publisher, however, to change the title on the grounds of differences in the understanding of faith versus religion. Nevertheless, we find value in the title, for what is faith but the very ground for hope? It is the motivating power of religion and spirituality to give proponents hope for the future in the darkest of circumstances. The underpinnings of hope found daily through faith themselves form the ground for an ever-deepening faith through our diverse traditions and practices, and so also, for the motivation to conserve.

## Notes

1 The opening three verses of Psalm 23 capture a sense of the peace that is to be found in nature. Dating from the time of King David, about 3000 BP, it is one of the best-known prayers in Judaism, Christianity and Islam. In English translation (NRSV): *The Lord is my shepherd, I shall not want. He makes me lie down in green pastures; he leads me beside still waters; he restores my soul. He leads me in right paths for his name's sake.*

2 As discerned from the fossil record of the last 540 million years, after Raup, D. & Sepkoski, J. (1982). Mass extinctions in the marine fossil record. *Science* 215, 1501–1503, and Rohde, R.A. & Muller R.A. (2005). Cycles in fossil diversity. *Nature* 434, 208-210.

3 A google scholar search on 'young people less interested in religion' found 1,540,000 results – January 17, 2022.

4 A prayer from the *Atharva Veda* (Book 12, Hymn 1, Mantra 35) of the Hindu faith (dating from roughly 3000 BP) reflects the human concern to conserve and safeguard the earth's potential for replenishment. In English translation: *Let what I dig from thee, O earth, rapidly spring and grow again. O Purifier, let me not pierce through thy vitals or they heart.*

## References


Berkes, F. (2017). *Sacred ecology*. New York: Routledge.

Comberti, C., Thornton, T. F., Wyllie de Echeverria, V., & Patterson, T. (2015). Ecosystem services or services to ecosystems? Valuing cultivation and reciprocal relationships between humans and ecosystems. *Global Environmental Change, 34*, 247–262.

Darwin, C. (1859). *The origin of Species* (1st edn.). London: John Murray.

Desfayes, M. (1998). *A thesaurus of bird names: Etymology of European Lexis through paradigms* (Vols. 1 & 2). Switzerland: Musée cantonal d'histoire naturelle.

Fleming, T. H., & Kress, W. J. (2011). A brief history of fruits and frugivores. *Acta Oecologica, 37*, 521–530.

Franck, R., & Iannaccone, L. R. (2014). Religious decline in the 20th century West: Testing alternative explanations. *Public Choice, 159*, 385–414. https://doi.org/10.1007/s11127-013-0103-9

Gosler, A. G., Bhagwat, S., Harrop, S., Bonta, M., & Tidemann, S. (2013). Chapter 6: Leadership and listening: inspiration for conservation mission and advocacy. In D. Macdonald & K. J. Willis (Eds.), *Key topics in conservation biology 2* (pp. 92–109). Oxford: John Wiley & Sons, Ltd.

Gosler, A. G., & Tilling, S. M. (2021). Knowledge of nature and the nature of knowledge: Student natural history knowledge and the significance of birds. *People & Nature*, 1–16. https://doi.org/10.1002/pan3.10265

Harkness, J. (2020). *Bird therapy*. London: Unbound Books.

Honey, M., & Frenkiel, K. (2021). *Overtourism: Lessons for a better future*. Washington, DC: Island Press.

Horn, M. H., Correa, S. B., Parolin, P., Pollux, B. J. A., Anderson, J. T., Lucas, C., … Goulding, M. (2011). Seed dispersal by fishes in tropical and temperate fresh waters: The growing evidence. *Acta Oecologica, 37*, 561–577.

Hughes, B. (2022). *Pollination biology and ecology*. New York: Syrawood Publishing House.

Katz, L. (2000). *Evolutionary origins of morality: Cross disciplinary perspectives*. Thorverton: Imprint Academic.

Kolbert, E. (2014). *The sixth extinction: An unnatural history*. London: Bloomsbury.

Liljeblad, J., & Verschuuren, B. (Eds.). (2019). *Indigenous perspectives on sacred natural sites: Culture, governance and conservation*. Abingdon: Earthscan.

Lovelock, J. (2016). *Gaia: A new look at life on Earth*. Oxford: OUP.
Maybe, R. (2008). *Nature cure*. New York: Vintage Books.
MEA. Millennium Ecosystem Assessment (Program). (2005). *Ecosystems and human well-being*. Washington, DC: Island Press.
Ninan, K. N. (2009). *Conserving and valuing ecosystem services and biodiversity economic, institutional and social challenges*. London: Routledge.
NRSV. (1989). *The Holy Bible: Containing the old and new testaments: New revised standard version*. Zondervan: Grand Rapids.
Sax, B. (2021). *Avian illuminations: A cultural history of Birds*. London: Reaktion Books.
Sellers, R., Warne, N., Pickles, A., Maughan, B., Thapar, A., & Collishaw, S. (2019). Cross-cohort change in adolescent outcomes for children with mental health problems. *Journal of Child Psychology and Psychiatry, 60*(7), 813–821.
Sheldrake, M. (2020). *Entangled life: How fungi make our worlds, change our minds and shape our futures*. London: The Bodley Head.
Sillitoe, P. (Ed.). (2006). *Local science vs global science: Approaches to indigenous knowledge in International development*. Oxford: Berghann Books.
Spoon, J. (2014). Quantitative, qualitative, and collaborative methods: Approaching indigenous ecological knowledge heterogeneity. *Ecology and Society, 19*, 33. https://doi.org/10.5751/ES-06549-190333
Terbough, J., & Estes, J. A. (Eds.). (2010). *Trophic cascades: Predators, prey and the changing dynamics of nature*. Washington, DC: Island Press.
Tidemann, S., & Gosler A. (Eds.). (2010). *Ethno-Ornithology – Birds, indigenous peoples, culture and society*. London: Publ. Earthscan Ltd. *(Hardback)*.
Verschuuren, B., & Brown, S. (Eds.). (2019). *Cultural and spiritual significance of nature in protected areas: Governance, management and policy*. Abingdon: Earthscan.
Verschuuren, B., Wild, R. G., McNeely, J. A., & Oviedo, G. (Eds.). (2010). *Sacred natural sites: Conserving nature and culture*. Abingdon: Earthscan.

# ACKNOWLEDGEMENTS

We would like to start by thanking all of the faith-inspired conservation practitioners whose work has made this book possible. Without them, quite simply, we, as editors and authors, would not have gotten the opportunity to share the stories contained in this book. Next, we thank each other, for walking together on what has been a long path that we traversed through COVID-19, accidents, and various geopolitical catastrophes – thank you for staying inspired despite all of this. This has been a collaborative project, especially so between authors who have provided feedback to each other on their chapters – we express gratitude to all of the book's authors for working so well together as peers. We would especially like to thank Stephen Awoyemi for conceiving of this publication project and bringing the editorial team together. We would also like to thank Marcos Amend for providing one of his many stunning photographs of nature for the book's cover, and Prasanjit Dasgupta for helping us work with the chapter photos. Finally, we would like to thank Katie Stokes, Hannah Ferguson, and John Baddeley from Routledge, for helping us bring this publication project to fruition.

# INDEX

Note: **Bold** page numbers refer to tables; *italic* page numbers refer to figures and page numbers followed by "n" denote endnotes.

For Product Safety Concerns and Information please contact our EU
representative GPSR@taylorandfrancis.com
Taylor & Francis Verlag GmbH, Kaufingerstraße 24, 80331 München, Germany

www.ingramcontent.com/pod-product-compliance
Ingram Content Group UK Ltd.
Pitfield, Milton Keynes, MK11 3LW, UK
UKHW022030061126
477531UK00003B/132
* 9 7 8 1 0 3 2 1 5 9 0 0 3 *